DIGITAL COLOR IMAGE PROCESSING

DIGITAL COLOR IMAGE PROCESSING

Andreas Koschan
Mongi Abidi

WILEY-INTERSCIENCE
A JOHN WILEY & SONS, INC., PUBLICATION

Published by John Wiley & Sons, Inc., Hoboken, New Jersey.
Published simultaneously in Canada.

Library of Congress Cataloging-in-Publication Data:

Koschan, Andreas, 1956–
Digital color image processing / by Andreas Koschan and Mongi A. Abidi.
p. cm.
ISBN 978-0-470-14708-5 (cloth)
1. Image processing—Digital techniques. 2. Color. I. Abidi, Mongi A. II. Title.
TA1637.K678 2008
621.36'7—dc22 2007027369

Printed in the United States of America.

10 9 8 7 6 5 4 3 2 1

to my daughter Andrea (Andreas Koschan)
in memory of my father Ali (Mongi Abidi)

TABLE OF CONTENTS

PREFACE

Color information is gaining an ever-greater importance in digital image processing. Nevertheless, the leap to be mastered by the transition from scalar to vector-valued image functions is not yet generally addressed in most textbooks on digital image processing. The main goal of this book is to clarify the significance of vector-valued color image processing and to introduce the reader to new technologies. The present state of the art in several areas of digital color image processing is presented in regard to a systematic division into monochromatic-based and newer vector-valued techniques. The potentials and the requirements in vector-valued color image processing are shown.

This text is organized in regard to advanced techniques for three-dimensional scene analysis in color images. It is structured into four parts. The first four chapters illustrate the fundamentals and requirements for color image processing. In the next four chapters, techniques for preprocessing color images are discussed. In subsequent chapters, the areas of three-dimensional scene analysis using color information and of color-based tracking with PTZ cameras are viewed. In the final two chapters, the new area of multispectral imaging and a case study on applications of color image processing are presented. For selected areas of digital color image processing such as edge detection, color segmentation, interreflection analysis, and stereo analysis, techniques are discussed in detail in order to clarify the respective complexity of the algorithms.

Chapter 12 on multispectral imaging addresses an emerging area in the field of image processing that is not yet covered in detail in textbooks. It is further augmented by a subsection on face recognition using multispectral imaging. The three case studies presented in the final three chapters summarize the results and experience gained by the authors in luggage inspection, video surveillance, and biometrics in research projects that have been funded by the National Safe Sky Alliance, the National Science Foundation, and the U.S. Department of Energy over multiple years. Several algorithms have been tested and evaluated under real conditions in a local airport.

This text is written at a level that can be easily understood by first and second year graduate students in Electrical and Computer Engineering or Computer Science as well as by researchers with basic knowledge in image processing who want to extend their understanding in the area of color image processing. The book instructs the reader beyond the standard of image processing and is a complement to existing textbooks in its field. Furthermore, the three application chapters on assisting screeners in luggage inspection in airports, video surveillance of high security facilities, and multispectral face recognition for authentication address recent problems of high importance to current safety and security issues. These chapters significantly augment the book's content.

This material is based on lectures and courses that have been taught by the authors at (1) the University of Tennessee, Department of Electrical and Computer Engineering, Knoxville, Tennessee and (2) the Technical University of Berlin, Department of Computer Science, Berlin, Germany between 1991 and 2007. Currently, Andreas Koschan is a Research Associate Professor, and Mongi Abidi is a Professor and Associate Department Head. Both are with the Department of Electrical and Computer Engineering, University of Tennessee. The techniques and algorithms have been tested by Masters students and Ph.D. students in Berlin, Germany and Knoxville, Tennessee and the figures illustrate the obtained results.

Andreas Koschan
Mongi Abidi

Knoxville, April 2008

ACKNOWLEDGMENT

The authors are indebted to a number of colleagues in academic circles as well as in government and industry who have contributed in various important ways to the preparation of this book. In particular, we wish to extend our appreciation to Besma Abidi, Gunter Bellaire, Karl-Heinz Franke, Ralph Gonzalez, Walter Green, Andrei Gribok, Reinhard Klette, Heinz Lemke, David Page, Joonki Paik, Dietrich Paulus, Volker Rehrmann, Werner Ritter, Volker Rodehorst, Kartsten Schluens, and Horst Voelz.

The many investigations and results presented in this book could not have been achieved without the readiness of many students to grasp our ideas and suggestions. We would particularly like to name Vivek Agarwal, Alexander Bachem, Faysal Boughorbel, Hong Chang, Klaus Curio, Peter Hannemann, Harishwaran Hariharan, Tobias Harms, Ralf Huetter, Sangkyu Kang, Kannan Kase, Ender Oezguer, Rafal Salustowicz, Wolfram Schimke, Kathrin Spiller, Dirk Stoermer, Sreenivas Sukumar, Kay Talmi, Axel Vogler, Yi Yao, Mingzhong Yi, and Yue Zheng. We thank all of them cordially for their commitment.

We thank Becky Powell, who helped immensely with the translation of the research and teaching material, which was previously available only in German, into English. Moreover, we thank Justin Acuff for his efforts with the formatting of the book and the update of some of the figures. Last, but not least, special thanks goes to George Telecki, Rachel Witmer, and Melissa Yanuzzi at Wiley. Their assistance and patience during the production of this book are truly appreciated.

AK
MA

1 INTRODUCTION

In our daily life, our vision and actions are influenced by an abundance of geometry and color information. When crossing a street, we identify a technical apparatus by its geometry as a traffic light. However, only by analyzing color information do we subsequently decide whether we are to continue, if the light is green, or stop, if the light is red. A camera-assisted driving information system should be able to evaluate similar information and either pass the information on to the driver of a vehicle or directly influence the behavior of the vehicle. The latter is of importance, for example, for the guidance of an autonomous vehicle on a public road. Something similar to this applies to traffic signs, which can be classified as prohibitive, regulatory, or informative signs based on color and geometry.

The assessment of color information also plays an important role in our individual object identification. We usually do not search in a bookcase for a book known to us solely by its title. We try to remember the color of the cover (e.g., blue) and then search among all of the books with a blue cover for the one with the correct title. The same applies to recognizing an automobile in a parking lot. In general, we do not search for model X of company Y, but rather we look for a red car, for example. Only when we see a red vehicle do we decide, according to its geometry, whether that vehicle is the one for which we are looking. The search strategy is driven by a hierarchical combination of color and form. Such hierarchical strategies are also implemented in automatic object recognition systems.

While in the past color image processing was limited essentially to satellite imagery, it has gained importance in recent years on account of new possibilities. This is due, among other things, to the high information level that color images contain in relation to gray-level images. This information allows color image processing to succeed in areas where "classical gray-level image processing" currently dominates. The decision confidence level for various techniques can be greatly improved by the additional classification markers color can provide. The applied procedures are thereby made simpler, more robust, or even applicable in the first place.

The fundamental difference between color images and gray-level images is that in a color space, a color vector (which generally consists of three components) is assigned to a pixel of a color image, while a scalar gray value is assigned to a pixel of a gray-level image. Thus, in color image processing, vector-valued image functions are treated instead of the scalar image functions used in gray-level image processing. Color image processing techniques can be subdivided on the basis of their principal procedures into two classes:

1. *Monochromatic-based techniques* first treat information from the individual color channels or color vector components separately and then combine the individual results.
2. *Vector-valued techniques* treat the color information as color vectors in a vector space provided with a vector norm.

The techniques from the first class can also be designated as *rental schemes* [Zhe et al. 93], since they frequently borrow methods from gray-level image processing and implement them separately on each color component. Thereby the dependencies between the individual color components (or vector components) are usually ignored. The monochromatic-based techniques make it clear that the transition from scalar to vector-valued functions, which can be mastered with color image analysis, is not yet generally known.

Color attributes such as hue or saturation are also used in monochromatic-based techniques. However, the analysis or processing of color information occurs separately for each component, for example, only the hue component or only the saturation component is treated (as in a gray-level image). In contrast, vector-valued techniques treat the color information in its entirety and not separately for each vector component.

While monochromatic-based techniques were predominantly regarded in the early days of color image processing, in recent times vector-valued techniques are being more frequently discussed. The difference between the two techniques serves as a systematization of the procedure in order to point out the respective conditions of developments from monochromatic-based techniques to vector-valued techniques. Better or more robust results are often attained with monochromatic-based techniques for color image processing than with techniques for gray-level processing. The monochromatic-based techniques, however, do not define a new way of image processing but rather demonstrate only transference of known techniques to color images. In contrast, the analysis and processing of vector-valued image information establishes a new step in image processing that simultaneously presents a challenge and a new possibility for analyzing image information. One difficulty with vector-valued techniques has been that the signal-theoretical basics for vector-valued color signals have not yet been presented.

In the past, the application of techniques for color image processing was restricted by additional factors. One factor was limited data memory and the "slow" processors: a three-channel color image of 1024×1024 pixels occupies, for example, 3 MB. For a geometric stereo analysis technique at least two images (6 MB) are needed, and for a photometric stereo analysis technique generally three images (9 MB) are necessary. These must be treated at a processing speed appropriate for the requirements of the application. Using more modern computers, the limitations on memory space and processing speed are not totally eliminated; however, the importance of this problem continues to decrease. Thus, the processor requirements for implementing digital color image processing today are satisfied.

Another factor that limited the applicability of color image processing in the past was color camera technology. In recent years, the availability of robust and low-cost color CCD cameras has made the acquisition of high-quality color images feasible under many varying acquisition conditions. However, in spite of enormous advances in camera technology there is a lack, as already mentioned, of extensive signal-theory investigations of vector-valued color signals. Here an urgent need for basic research exists.

In areas such as photogrammetry and remote sensing, images with more than three "color" channels are frequently analyzed. Newer areas of application analyze color images that represent three-channel spectral transmissions of visible light. Knowledge of the processing occurring in the human eye and brain of the signals that come from the three sensitive (with regard to different wavelengths) receptors in the retina can be used for the development and evaluation of techniques for color image processing.

The three different receptor types in the human retina are also the reason that commercial CCD-color cameras likewise implement measurements in three different wavelength areas of visible light. These cameras deliver a three-channel signal and the three channels are represented separately on a monitor or screen for the observer. Furthermore, the color attributes hue and saturation are defined only within the spectral area of visible light. In this book, techniques for the analysis of three channel color images are presented whose spectral transmissions lie within the visible area of light.

As an example, correspondence analysis in stereo images shows that red pixels do not correspond with blue pixels, even when their intensity values are similar. The segmentation of color images based on classification of color values is generally substantially more differentiated than segmentation based exclusively on intensity values.

The evaluation of color information in the image creates additional new possibilities for solving problems in computer vision. Many image processing techniques still assume that only matte (Lambertian) surfaces in the scene are analyzed. This assumption does not hold for real scenes with several reflecting (non-Lambertian) surfaces. However, this limitation can be overcome under certain conditions by highlight elimination in color images. Furthermore, physically determined phenomena, such as shadows or interreflections, can be analyzed more easily in color images than in gray-level images. For this, predominantly vector-valued image processing techniques are used that employ reflection models derived from physical optics for modeling image functions. These techniques are denoted as *physics-based vision techniques*. The invariant extraction of color information in relation to varying lighting conditions and description of image characteristics represents another problem in computer vision. Here promising vector-valued techniques for so-called color constancy can make an important contribution.

1.1 GOAL AND CONTENT OF THIS BOOK

Color information is gaining an ever-greater meaning in digital image processing. Nevertheless, the leap to be mastered by the transition from scalar to vector-valued image functions is not yet generally known. One goal of this book is to clarify the significance of vector-valued color image processing. The present state of the art in several areas of digital color image processing is represented in regard to a systematic division into monochromatic-based and newer vector-valued techniques. The more recent potentials and the requirements in vector-valued color image processing are shown. Here references will be made to the fundamentals lacking in many areas of digital color image processing.

While a terminology for gray-level image processing has been established for the most part, corresponding terms do not yet exist for vector-valued color images. Fundamental ideas in color image processing are specified within the context of this work. Monochromatic-based techniques still dominate in many practical applications of digital color image processing, such as in medicine, agriculture, and forestry, as well as industrial manufacturing. A few examples of monochromatic-based and vector-valued techniques of color image analysis in practical usage are presented in Section 1.3.

This book is organized in regard to advanced techniques for three-dimensional scene analysis in color images. In the first four chapters, the fundamentals and requirements for color image processing are illustrated. In the next four chapters, techniques for preprocessing color images are discussed. In subsequent chapters, the area of three-dimensional scene analysis using color information is viewed. In the final three chapters, case studies on application of color image processing are presented. For some selected areas of digital color image processing, such as edge detection, color segmentation, interreflection analysis, and stereo analysis, techniques are discussed in detail in order to clarify the respective complexities of the solution for the problem.

Knowledge of the human visual system is frequently utilized for designing procedures in digital image processing (see, e.g., [Mar82], [Ove92], and [Wat88]). This also applies for digital color image processing. In Chapter 2, an introduction to human color vision is presented whereby color blindness of a section of the population and the phenomenon of color constancy are given special attention. For the representation and treatment of color images, a suitable form of representation for the data must be selected. Different color spaces used in color image processing are presented in Chapter 3. Chapter 4 contains the technical requirements for color image processing (color camera, color filter, standard illuminants, color charts, etc.) as well as techniques of photometric and colorimetric calibration that are necessary for the further treatment of color images.

Techniques for noise suppression and contrast enhancement in color images are the subject of Chapter 5. An important task in preprocessing color images is the extraction of edges in the image. Various procedures for color edge detection are discussed in Chapter 6. A comparison of the results of one monochromatic-based and two vector-valued color edge operators are also given. An overview of different techniques for color image segmentation is presented in Chapter 7. There, a robust technique for the segmentation of color images based on the watershed transformation is presented.

An interesting challenge and at the same time a new possibility of color image processing is the analysis of physical phenomena, such as the analysis of highlights and interreflections. In Chapter 8, an overview of the techniques for highlight analysis and a new method for minimizing interreflections in real color images is presented. In addition, different procedures for achieving color constancy are discussed.

A detailed description of the use of color information for static stereo analysis is given in Chapter 9. There, investigations for edge-based as well as area-based color stereo techniques can be found. Also shown is how stereo matching results can be significantly improved by projecting color-coded light patterns onto the object. The inclusion of color information into dynamic and photometric stereo analysis is the subject of Chapter 10.

Chapter 11 addresses case studies of color use in an automated video tracking and location system that is under development at the University of Tennessee's Imaging, Robotics and Intelligent Systems (IRIS) Laboratory in Knoxville, Tennessee. Chapter 12 discusses the acquisition and analysis of multispectral images. Their use in face recognition is outlined as an example of multispectral image processing. The application of color coding in x-ray imaging is the subject of Chapter 13.

1.2 TERMINOLOGY IN COLOR IMAGE PROCESSING

There is agreement concerning the terminology used in the processing of gray-level images [HarSha91]. In contrast, a corresponding transference onto vector-valued color images does not yet exist. For example, it has not yet been established what a color edge is, what the derivative of a color image is, or what should be understood as the contrast of a color image. In color image processing, the terms are used very differently and also somewhat imprecisely. In the following section, terminology used in color image processing is established.

1.2.1 What Is a Digital Color Image?

The central terminology of color image processing is that of the digital color image. A *digital image* is defined for image pixels that are assumed in the real plane or could be elements of a discrete set of points. A gray-level image E assumes an *image value* $E(\mathbf{p}) = E(x, y)$ in an *image pixel* $\mathbf{p} = (x, y)$ as a uniquely determined function value, approximately a numerical *gray value* u, which characterizes a determined gray tone. For this, $E(x, y) = u$ is written formally. (Note that for the sake of simplification, the double parentheses is omitted in the coordinate equation $E(\mathbf{p}) = E((x, y))$ for $\mathbf{p} = (x, y)$.) The triple $(x, y, E(x, y)) = (x, y, u)$ is indicated as *pixel* (from *picture element*), where x and y are the coordinates in the image plane. The points in the image plane are converted by the image acquisition equipment into integer-valued, device-dependent coordinates of the row and column position.

Discrete image pixels and discrete image values distinguish a digital image. The index domains $1 \leq x \leq M$ and $1 \leq y \leq N$ are presupposed. The values M and N mark the image resolution. The value $A = M \cdot N$ marks the image size. For the possible image values $E(x, y)$ of a digital gray-level image E, $G_{\max} + 1$ gray values, $G_{\max} \geq 1$, are assumed. The representation of (continuously distributed) image values and gray tones into a limited number of gray values is called *quantization*. For the $G_{\max} + 1$ gray values, a connected interval of non-negative integers is assumed. For an integer gray value u holds

$$0 \leq u \leq G_{\max} .$$

The standard value for gray-level images is $G_{\max} = 255$.

A *color image* corresponds intuitively to the perceived representation of our colored environment (i.e., to one's individual visual sensory perception). Computationally, a color image is treated as a vector function (generally with three components). The range of the image function is a vector space, provided with a norm that is also called a *color space*. For a (three-channel) *digital color image* $\mathbf{C}$, three vector components u_1, u_2, u_3 are given for one image pixel (x, y):

$$\mathbf{C}(x, y) = \left(u_1(x, y), u_2(x, y), u_3(x, y)\right)^T = \left(u_1, u_2, u_3\right)^T . \tag{1.1}$$

The colors represented by concrete value combinations of the vector components u_1, u_2, u_3 are relative entities. Each of the vectors $(u_1, u_2, u_3)^T$ with the generally integer components $0 \leq u_1, u_2, u_3 \leq G_{max}$ characterizes a color in the basic color space. Examples of color spaces are the *RGB* color space, which is used for representing a color image on a monitor (additive color mixture), or the *CMY*(*K*) color space, which is used for printing a color image (subtractive color mixture).

A color image is denoted as *true-color image* if the vector components of the digitalized color image represent spectral transmissions of visible light. The generation of a true-color image results as a rule by using a color CCD camera, which commercially has a quantization of eight bits per color channel and/or vector component (see Section 4.1).

A *false-color image* corresponds essentially to a true-color image, however, with the difference that areas of wavelengths outside the visible light are also allocated to the vector components of the color image. An example of that is an infrared image whose information content does not come from visible light. For its representation and visualization, the information of the infrared spectrum is transformed into the area of visible light.

The term *pseudocolor image* is used if selected image pixels are recoded or colored, that is, for these image pixels, the associated image value (gray value or color vector) is replaced by a given color vector. The original image can be a gray-level image in which the significant areas should be recoded into color (e.g., areas in a digital x-ray image to be used for aiding the radiologist in a diagnosis). The selection of the color vectors is often arbitrary and serves solely for better visualization of different image domains.

Another example of a pseudocolor image is a true-color image in which color vectors were recoded. This can be used for the special emphasis (coloring) of certain image areas or for reducing the number of differing color vectors in the image. The last case is implemented for reducing color quantization (e.g., to 256 colors). While in early years many workstations could represent only 256 colors, most workstations today offer a true-color representation with a quantization of eight bit per color component (i.e., altogether 24 bits per image pixel or ca. 16 million colors). Reducing the number of differing color vectors in the image can also be used for reducing the amount of image data to be stored. An image in 8-bit mode needs less storage space than an image in 24-bit true-color mode. Less data needs to be transferred for representing an image in the Internet saved with 8-bit color quantization.

A color quantization is realized in general by using *indexed colors*. After, for example, 256 color vectors are selected for an image (based on a quantization algorithm), these are placed on a *colormap* or *palette*. For each image pixel the associated index number is listed. On the basis of this number the indexed color is selected for representing the color image on a monitor. In the graphic data formats GIF (Graphics Interchange Format) and TIFF (Tagged Image File Format), the associated colormap is contained along with the indexed color image. In general, a colormap of this type contains *RGB* entries suited to the nonlinear monitor that are meant for the direct representation of a color image (without additional correction) on the monitor. By using indexed colors for true-color images, the color information of the image is reduced and in the process the quality of the color image is also impaired. Such color images are just barely suitable for further treatment with image analysis techniques.

In the image examples discussed so far, color vectors with three components or three color channels were always observed so that we could talk of three-channel images. This technique can also be expanded to n (color-) channels. It concerns, then, a so-called *multichannel* or *multiband image*,

$$\mathbf{C}(x,y) = \left(u_1(x,y), u_2(x,y), \ldots, u_n(x,y)\right)^T = \left(u_1, u_2, \ldots, u_n\right)^T , \tag{1.2}$$

whose special case for $n = 1$, for example, can be a gray-level image or intensity image and for $n = 3$ can be a three-channel true-color image.

Another special case is the *multispectral image*, in which data is acquired of a given scene in a number of more than three different spectral bands. Some (or all) of the spectral bands may lie outside the visible light (e.g., in LANDSAT images with the spectral areas 500 - 600 nm (blue-green), 600 - 700 nm (yellow-red), 700 - 800 nm (red-infrared), and 800 - 1100 nm (infrared)). The image values in a LANDSAT image are represented by vectors with four components. Other examples of multichannel images are radar images in which the individual channels represent the received signals for differing wavelengths and polarizations. Recent research activities also include the acquisition, representation, and processing of multispectral color images with more than three channels of information for the visible light spectrum. Images with, for example, six color bands can be visualized with very high fidelity when special hardware is used. Digital images with more than a hundred spectral bands are called *hyperspectral images*. However, there exists no common agreement on the minimum number of spectral bands in a hyperspectral image. The acquisition and analysis of multispectral images will be presented in more detail in Chapter 12.

1.2.2 Derivative of a Color Image

For a color component or a gray-level image $E(x,y)$ the *gradient* or the *gradient vector* is given by

$$grad(E) = \left(\frac{\partial E}{\partial x}, \frac{\partial E}{\partial y}\right)^T = \left(E_x, E_y\right)^T . \tag{1.3}$$

Here, the indexes x and y are introduced as abbreviations that indicate the respective partial derivative of the function, that is, it holds

$$E_x = \frac{\partial E}{\partial x} \quad \text{and} \quad E_y = \frac{\partial E}{\partial y}.$$

The absolute value of the gradient,

$$\|grad(E)\| = \sqrt{\left(\frac{\partial E}{\partial x}\right)^2 + \left(\frac{\partial E}{\partial y}\right)^2}\,, \tag{1.4}$$

is a measurement for the "height change" of the gray-level image function. It takes on the extreme value of zero for a constant gray-level plateau (in the ideal case $E(x,y) = const$).

A three-channel color image can be described by a function $\mathbf{C}: Z^2 \to Z^3$. This definition can be easily expanded to *n*-channel color images. However, color images with three vector components will be examined in this book. The differential of function **C** is given in matrix form by the *functional matrix* or *Jacobian matrix* **J**, which contains the first partial derivatives for each vector component. For a color vector in a color space with $\mathbf{C}(x,y) = (u_1, u_2, u_3)^T$ the derivative is described at a location (*x*,*y*) by the equation $\Delta\mathbf{C} = \mathbf{J}\Delta(x,y)$. It holds

$$\mathbf{J} = \begin{pmatrix} \frac{\partial u_1}{\partial x} & \frac{\partial u_1}{\partial y} \\ \frac{\partial u_2}{\partial x} & \frac{\partial u_2}{\partial y} \\ \frac{\partial u_3}{\partial x} & \frac{\partial u_3}{\partial y} \end{pmatrix} = \begin{pmatrix} grad(u_1) \\ grad(u_2) \\ grad(u_3) \end{pmatrix} = \begin{pmatrix} u_{1x} & u_{1y} \\ u_{2x} & u_{2y} \\ u_{3x} & u_{3y} \end{pmatrix} = (\mathbf{C}_x, \mathbf{C}_y). \tag{1.5}$$

Both vectors are indicated with $\mathbf{C}_x$ and $\mathbf{C}_y$:

$$\mathbf{C}_x - (u_{1x}, u_{2x}, u_{3x})^T \text{ and } \mathbf{C}_y = (u_{1y}, u_{2y}, u_{3y})^T .$$

1.2.3 Color Edges

While in gray-level images a discontinuity in the gray-level function is indicated as an edge, the term *color edge* has not been clearly defined for color images. Several different definitions have been proposed for color edges. A very old definition [Rob76] states that an edge exists precisely in the color image if the intensity image contains an edge. This definition ignores, however, possible discontinuities in the hue or saturation values. If, for example, two equally light objects of various colors are arranged in juxtaposition in a color image, then the edges determining the object geometry cannot be determined with this technique. Since color images contain more information than gray-level images, more edge information is expected from color edge detection in general. However, this definition delivers no new information in relation to gray-value edge detection.

A second definition for a color edge states that an edge exists in the color image if at least one of the color components contains an edge. In this monochromatic-based definition, no new edge detection procedures are necessary. This presents the problem of accuracy of the localization of edges in the individual color channels. If the edges in the color channels are detected as being shifted by one pixel, then the merging of the results produces very wide edges. It cannot be easily determined which edge position in the image is the correct one.

A third monochromatic-based definition for color edges [Pra91] is based on the calculation of the sum of absolute values of the gradients for the three color components. A color edge exists if the sum of the absolute values of the gradients exceeds a threshold value. The results of the color edge detection by the two previously named definitions depend heavily on the basic color spaces. An image pixel that, for example, is identified in one color space as an edge point must not eventually be identified in another color space as an edge point (and vice versa).

All previously named definitions ignore the relationship between the vector components. Since a color image represents a vector-valued function, a discontinuity of chromatic information can and should also be defined in a vector-valued way. A fourth definition for a color edge can result by using the derivative, described in the previous section, of a (as a rule in digital color image processing three-channel) color image. For a color pixel or color vector $\mathbf{C}(x,y)=(u_1,u_2,u_3)^T$, the variation of the image function at position (x,y) is described by the equation $\Delta\mathbf{C}=\mathbf{J}\Delta(x,y)$. The direction along which the largest change or discontinuity in the chromatic image function is detected is represented in the image by the eigenvector $\mathbf{J}^T\mathbf{J}$ corresponding to the largest eigenvalue. If the size of the change exceeds a certain value, then this is a sign for the existence of a color edge pixel.

A color edge pixel can also be defined applying vector ordering statistics or vector-valued probability distribution functions. The various techniques for the extraction of edges in color edges are the subject of Chapter 6.

1.2.4 Color Constancy

The colors of the surfaces of an object represent important features that could be used for identifying the object. However, a change in lighting characteristics can also change the several features of the light reflected from the object surfaces to the sensor. *Color constancy* is the capability of an invariant color classification of surfaces from color images with regard to illumination changes.

The human visual system is nearly color constant for a large area of surfaces and lighting conditions. As an example, a red tomato appears red in the early morning, at midday, and in the evening. The perceived color is therefore not the direct result of the spectral distribution of the received light, which was the assumption for many years (see [Zek93] for a detailed representation). A brief introduction to this subject is presented later in Section 2.4.

Color constancy is likewise desirable for a camera-based vision system when its use should occur under noncontrollable lighting conditions. Achieving color constancy in digital color image processing is, however, a problem that is difficult to solve since the color signal measured with a camera depends not only on the spectral distribution of the illumination and the light reflected on the surface, but also on the object geometry. These characteristics of the scene are, as a rule, unknown. In digital image processing, various techniques are identified for the numerically technical realization of color constancy. Color constancy techniques (in digital color image processing) can be classified into three classes with regard to the results that they intend to obtain:

1. The spectral distribution of the reflected light is to be estimated for each visible surface in the scene.
2. A color image of the acquired scene is to generate in the way it would appear under known lighting conditions.
3. Features are to be detected for the colored object surfaces in the image that are independent from lighting conditions (invariant to illumination changes).

The examination of all three techniques or procedures for achieving color constancy is the subject of Section 8.3.

1.2.5 Contrast of a Color Image

The term *contrast* is used ambiguously in the literature. In the following, several examples (without claiming completeness) are introduced.

1. Contrast describes the relation between the brightness values in an image or section of an image. As measurement for the size of the contrast, for example, the Michelson Contrast $(I_{max} - I_{min} / I_{max} + I_{min})$ is used [Gil94], whereby the largest-appearing brightness value is indicated by I_{max} and the smallest-appearing brightness value is denoted by I_{min}. This is described as *relative brightness contrast*.
2. The perceptual phenomenon of brightness perception of a surface in dependence on the lightness of the background is likewise indicated as contrast. For the illustration of this phenomenon, a gray surface surrounded by a white surface and a gray surface of the same lightness surrounded by a black surface is used. The gray-on-white background is perceived as somewhat darker than the gray-on-black background. This phenomenon is called *simultaneous brightness contrast* [Gil94]. An example is given in Fig. 1.1.
3. In a color image with low brightness contrast, details can be distinguished from the background on the basis of differing color saturation. The relation between the saturation values in a color image can be described as *relative saturation contrast*.

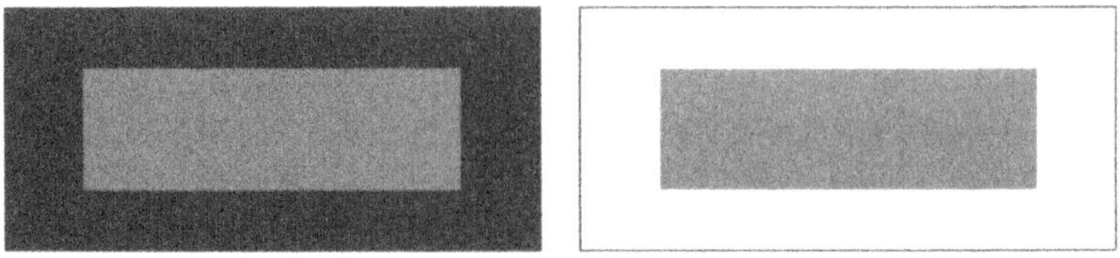

Figure 1.1. *Example of simultaneous (brightness) contrast: The left-hand grey rectangle appears lighter than the right-hand one.*

4. The detection of a colored surface depends likewise on the color of the surface surrounding it. A gray surface surrounded by a red ring appears, e.g., bluish-green [Zek93]. For the description of *induced color*, influenced by the color of the surrounding surface, the opponent color model is frequently implemented [Kue97]. This type of contrast is also denoted as *simultaneous color contrast*. Davidoff [Dav91] describes the effect of color contrast as the change of color constancy in a systematic manner.
5. Another type of contrast is the *successive (color) contrast*. This occurs when a colored area is observed over a long period of time and a neutral area is subsequently fixed. An *afterimage* of the previously observed area appears either in the opponent colors (negative afterimage) or approximately in the previously observed colors (positive afterimage) [Kue97]. Afterimages appear also with closed eyes.

Apart from the contrast definitions named here, the question is posed for digital color image processing as to what should be affected by the computer-aided change of contrast of a color image. The goal of enhancing the contrast in an image is generally to improve the visibility of image details. Only in rare cases is the goal of the technique the systematic influence of color constancy.

In many technical-based books, the contrast of a color image is regarded solely as brightness contrast in the sense of definition 1 (see, e.g., [Poy96]). Most display devices have implemented this definition for contrast control. On a color monitor (or television) the (nonlinear) area between the darkest and lightest pixel is adjusted with the "contrast control." With the "lightness control," a positive or negative offset for the lightness to be represented is established according to the adjustment. Also in the image-editing software program Adobe Photoshop TM the function of contrast change refers to the lightness values of the image.

Digital color image processing offers the opportunity of changing the relative brightness contrast as well as the possibility of including perception-based observations if the need arises. In addition, color attributes such as saturation and intensity can also be set in relation to each other in the vector-valued color signals. A fact to be remembered is that the term *contrast of a color image* should not be used without the use of an adjective (e.g., *relative* or *simultaneous*) or an appropriate definition of the term.

1.2.6 Noise in Color Images

Until now, not much has been published on the subject of noise in color images. It is generally assumed [BarSan97] that the individual components of the vector-valued color signal are degraded separately from each other by noise and that not all components are equally affected. This can be described, for example, by various additive overlays of the signals in the individual color components by malfunctions or Gaussian noise. Here the model

$$\mathbf{y} = \mathbf{x} + \mathbf{n}$$

is used as a basis, whereby $\mathbf{x}$ denotes the undisturbed image vector at a position (i,j) in the color image. The corresponding vector with noise is indicated by $\mathbf{y}$ and $\mathbf{n}$ is an additive noise vector at position (i,j) in the image.

It cannot be concluded from the assumption of the existence of differing overlays in the individual color components that monochromatic-based techniques for separate noise suppression in the individual color components provide the best results. Vector-valued techniques allow, in general, a better treatment of noise in color images (see, e.g., [PitTsa91], [Pla et al. 97], and [Zhe et al. 93]). Vector-valued techniques are dealt with later in Section 5.3.

1.2.7 Luminance, Illuminance, and Brightness

The terms *luminance*, *lightness*, and *brightness* are often confused in color image processing. To clarify the terminology we borrow three definitions from Adelson [Ade00]:

1. *Luminance* (usually L in formulas) is the amount of visible light that comes to the eye from a surface. In other words, it is the amount of visible light leaving a point on a surface in a given direction due to reflection, transmission, and/or emission. *Photometric brightness* is an old and deprecated term for luminance. The standard unit of luminance is *candela per square meter* (cd/m^2), which is also called *nit* in the United States, from Latin *nitere* = "to shine" (1 nit = 1 cd/m^2).
2. *Illuminance* (usually E in formulas) is the amount of light incident on a surface. It is the total amount of visible light illuminating (incident upon) a point on a surface from all directions above the surface. Therefore illuminance is equivalent to *irradiance* weighted with the response curve of the human eye. The standard unit for illuminance is *lux* (lx), which is *lumens per square meter* (lm/m^2).
3. *Reflectance* is the proportion of incident light that is reflected from a surface. Reflectance, also called *albedo*, varies from 0 to 1, or equivalently, from 0% to 100%, where 0% is ideal black and 100% is ideal white. In practice, typical black paint is about 5% and typical white paint about 85%. (For the

sake of simplification, we consider only ideal matte surfaces, for which a single reflectance value offers a complete description.).

Luminance, illuminance, and reflectance are physical quantities that can be measured by physical devices. There are also two subjective variables that must be discussed:

1. *Lightness* is the perceived reflectance of a surface. It represents the visual system's attempt to extract reflectance based on the luminances in the scene.
2. *Brightness* is the perceived intensity of light coming from the image itself, rather than any property of the portrayed scene. Brightness is sometimes defined as perceived luminance.

1.3 COLOR IMAGE ANALYSIS IN PRACTICAL USE

In many practical applications the analysis of gray-level images is not sufficient for solving the problems. Only by evaluating color information in the images can the problem be solved or be resolved considerably more easily than in gray-level images. Even now the monochromatic-based techniques predominate in practical applications. Only in recent times have vector-valued techniques been discussed. In the following, examples are presented in which the necessity of analysis of color images arises directly from the demands of the applications. None of the posed tasks could be solved with the techniques from gray-level image processing. In order to clarify the differences and common features, categorization is introduced for the techniques. The following nomenclature indicates:

M: Monochromatic-based techniques, and
V: Vector-valued techniques.

Furthermore, it will be differentiated in this section as to whether

α: The techniques deliver better results by evaluating color information than by evaluating gray-level information, or
β: The techniques are possible only by the evaluation of color information.

For example, a $V\beta$-technique is a vector-valued technique that is possible only by evaluating color information. One difficulty in assigning a technique to one of the classes listed above is that no one class of techniques will be followed continually in every case. For example, the vector-valued color signal can be evaluated in one processing step while in another processing step only gray-level information is analyzed. For systematization only the part of the procedure that refers to the evaluation of color information is used as a basis. The technique in this example is denoted as a V-technique.

Another difference between the techniques can result from the use of true-color or pseudocolor images. If not mentioned otherwise, the use of true-color images is always assumed in the following. The information on the basic color space for the representation of color values in the image is without further specification. The discussion of color spaces is, as previously mentioned, the subject of Chapter 3. The following examples should illustrate the diverse possibilities of using color image processing.

There are a roughly equal number vector-valued and monochromatic-based techniques in these examples. However, this does not reflect the actual level of development. In fact, nearly all the vector-valued techniques of color image analysis in practical usage known to the authors are presented here, while only a few examples of monochromatic-based techniques used in practice are named. The reason for this is that, according to our estimation, the vector-valued techniques are the more interesting of the two. As previously mentioned, better results are frequently obtained with monochromatic-based techniques than with techniques of gray-level image analysis, but the techniques used are as a rule identical or similar to the known techniques from gray-level image analysis. On the other hand, the vector-valued approaches of color image analysis present a new procedural class that obtains special consideration in this work.

1.3.1 Color Image Processing in Medical Applications

In many medical applications, x-rays, which traditionally exist as gray-level images, must be evaluated for a diagnosis. By transferring the gray values into pseudocolors the visualization of small nuances can be improved considerably, especially in x-rays with 12-bit quantization. The application of color coding used in x-ray imaging is the subject of Chapter 13.

Some research studies exist on the use of color image processing in the classification of skin tumors. An accurate evaluation of a pigment sample and a hue typical of a melanocyte is necessary for the classification. In [Ros et al. 95], the automatic classification of skin tumors is discussed without practical realization. Two $M\alpha$-procedures can be found in [Sto et al. 96] and [Umb et al. 93]. In both techniques, principal component analysis is first implemented in order to obtain less correlated values. In [Sto et al. 96], a best channel for a gray-value segmentation is subsequently selected. For the color classification the centers of gravity of the intensity values within each segmented region are compared in every color channel. In [Umb et al. 93] a quantization (in four colors) for the segmentation of a skin cancer image is implemented applying principal component analysis. An $M\beta$-technique is proposed in [Xu et al. 99]. There, a gray-level image is created for skin cancer image segmentation. The gray-level image is obtained after mapping colors into intensities in such a way that the intensity at a pixel is proportional to the CIELAB color distance of the pixel to the average color of the background. Another $M\beta$-technique is presented in [Gan et al. 01], where several components of the *RGB*, the CIELAB, and the *HSI* color space are used for melanoma recognition.

Peptic ulcers (*Ulcera ventriculi*) represent a frequent and serious illness in humans. Approximately 1 - 5 % of stomach ulcers are malignant. Here, early detection is necessary for a successful cure. By evaluating the contour of ulcers in color endoscope images, a doctor can be aided considerably in his or her diagnosis of an ulcer (malignant or benign). In [Pau et al. 93], a vector-valued color variant of the Sobel operator is suggested for determining the contour. In order to calculate the difference between the color vectors in the *RGB* space, a distance measurement similar to the Euclidian distance is used. The individual vector components are, however, without more exact motivation, weighed differently. This technique constitutes a $V\alpha$–procedure.

An $M\beta$-procedure for a quantitative description of the severity of an inflammation of the larynx (laryngitis) is presented in [Sch et al. 95]. The severity of the illness is assessed by the doctor subjectively on the basis of redness of the mucous membrane of the larynx in a laryngoscopic color image. The finding can be evaluated using color information in the CIELUV color space. In [Sch et al. 95], the classification of the redness is implemented solely by an observation of the U component of the CIELUV color space.

1.3.2 Color Image Processing in Food Science and Agriculture

The visual appearance of food is a deciding factor in assessing its quality. An important part of quality control in the food industry is, therefore, based on visual inspection. This is traditionally carried out by the human eye. Apart from the absence of reliable quantitative assessment criteria, visual assessment by human beings is time consuming and cost intensive. Until now, the tools needed for implementing automatic quality control using color criteria were lacking. The introduction of color image analysis has decisively changed this. By using analysis in the production process, it can be automatically determined, for example, whether baked goods have the desired size and color appearance [Loc et al. 96].

Another application of color image processing in food control is automatic counting of the number of pepperoni slices and olives on a pepperoni pizza [Ste95]. At first sight, this application does not seem sensible. But if one considers that each customer who buys a pepperoni pizza containing only one slice of pepperoni will probably never buy another pizza from this company again, the economic damages caused by this type of situation become obvious. In [Ste95], a $V\beta$–procedure is presented for segmentation (e.g., of pepperoni slices and olives) in the image with the help of color vector comparisons in the *RGB* space. Another $V\beta$-technique for automatic pizza quality evaluation applies segmentation in the *HSI* space [SunBro03].

At the University of Genoa, an agriculture robot with a color stereo camera system is tested [Bue et al. 94]. Its purpose is to monitor tomato cultivation in a hothouse. Tomatoes ripe for the harvest should be selected with the help of segmentation of color images. Simultaneously, a possible fungus attack should be detected and automatically treated with pesticides. For segmentation in the *HSI* space, a *Mβ*-procedure is suggested that fixes the regions by separate threshold value formation in the *H* and *S* components of the image. Subsequent stereo matching of the segmented regions (for determining the distance between grasping arm and tomato) results without considering color information.

1.3.3 Color Image Processing in Industrial Manufacturing and Nondestructive Materials Testing

To avoid any possibility of confusion and to enable a clear identification, colored markings are used in the electronics industry and pharmaceutical industry. For example, electrical resistors [Asa et al. 86] or ampoules filled with medicine [Bre93] can be automatically determined and selected by an analysis of their color code. In [Asa et al. 86], evaluation of the color code occurs with the help of a monochromatic-based subdivision of the hue and saturation components in the *HSI* color space. In [Bre93], no information on the selection process is given. The information from the signal processors used for increasing the processing speed suggests, however, a monochromatic-based technique.

Furthermore, for the identification of medicine a pharmaceutical code is employed that is composed of a variable number of thick and thin rings applied to ampoules. The use of color image processing is important in this case for legibility since many colors (e.g., yellow) do not have sufficient relative lightness contrast in the gray-value representation. In each case, a defectively marked ampoule must be automatically detected and removed. The use of color image processing can ensure this [Bre93].

1.3.4 Additional Applications of Color Image Processing

A cost-efficient inspection and monitoring of the air quality is another example of a use for color image processing. The active examination of lichens (e.g., *Parmelia sulcata* and *Hypogymnia physodes*) produces a valuable indicator for this [BonCoy91]. Direct conclusions about air quality can be drawn from irregularities in growth, form, or coloring of the lichens. In general (see [BonCoy91]), a series of tests over a period of seven days is conducted, whereby the abovenamed criteria (growth, form, and coloring) are recorded daily. Digital color image processing serves as an effective aid for the automatization of these mass screenings.

Bonsiepen and Coy [BonCoy91] combine the individual components of the color vectors in the *RGB* color space into a scalar feature and segment the scalar feature image produced by this as a gray-level image. More exact segmentation results can be expected here by using a vector-valued technique.

Another possible application is the digitization of maps. These are generally read by flatbed scanners. Chromatic distortions result through errors in the mechanical adjustment and chromatic aberration of the scanner's lens system, by which brown or blue lines in the maps are no longer represented as a blue or a brown, but rather by a class of blue and brown tones. Automatic classification of colored lines requires that the chromatic distortions first be removed. A *Vβ*-technique for this is based on determining eigenvectors in the *RGB* color space [KhoZin96].

1.3.5 Digital Video and Image Databases

Just as the CD has replaced the long-playing record in recent years, the videotape is now being replaced by the DVD ("digital versatile disc" or "digital video disc"). This results in another new range of applications results for color image processing. The main activities in this area still relate at the present to an efficient coding and decoding of color images. This extensive subject area is not covered further here. Interested readers are referred to the following publications on this subject: [ArpTru94], [CarCae97], [Che et al. 94], [MemVen96], [Mit et al. 96], [Ove et al. 95], [Sag et al. 95], [Sch95], [VauWil95], [Wu96], [ZacLiu93], and [ZhaPo95]. A detailed representation of techniques for digital image coding is presented in [RaoHwa96]. Activities in this area are also influencing the development and design of techniques for videophones, teleconferences, and digital cinema.

Additional research deals with the retrieval of image sequences or individual images in image databases (*image retrieval*). For example, at the Massachusetts Institute of Technology, image content oriented search techniques are being researched (see [Pen et al. 96] and [Pic95]). Additional research in the area of color image retrieval deals with search techniques based on histograms of features in the *HSI* color space [RicSto96], with the selection of a "best" color space (*RGB, HSV, YUV*, or Munsell [WanKuo96]; see Chapter 3 for the definition of color space), or various definitions of the *RGB* color space [Lu96] for representing color images using fuzzy techniques in connection with color histograms [StrDim96], distinction of color images in image databases [FauNg96], [GevSme96], [GonSak95], and special techniques for color indexing [SawHaf94], [SmiCha96]. The techniques of color indexing employed here or of color histogram evaluation are similar to those that are also used in color object recognition.

1.4 FURTHER READING

An introduction to various color spaces and the transformations between the spaces are given in [Pra91]. Very worth reading is the (968-page) standard book on color by Wyszecki and Stiles [WysSti82]. The treatment of color information in the human visual system is presented in detail by Zeki [Zek93]. An extensive presentation of techniques for digital image coding (JPEG, MPEG, fractal coding, etc.) can be found in [RaoHwa96]. Mathematical foundations for vector analysis are contained, for example, in [Mat96] and [Sha97].

An interesting overview of the fundamentals of physics-based color image processing has been published by Healey, Shafer, and Wolff [Hea et al. 92]. This is a compilation of 28 selected publications from a number of authors. A technical introduction to the area of digital video is presented by Poynton in [Poy96]. Also recommended is an overview by Poynton of various technical questions regarding color, which can be found on the Internet at http://www.poynton.com/Poynton-color.html. This site also contains links to other color related sites.

1.5 REFERENCES

[Ade00] E.H. Adelson. Perception and lightness illusions. In: M. Gazzaniga (ed.), *The New Cognitive Neurosciences.* MIT Press, Cambridge, Massachusetts, 2000, pp. 339-351.

[ArpTru94] R.B. Arps, T.K. Truong. Comparison of international standards for lossless still image compression. *Proc. of the IEEE* **82** (1994), pp. 889-899.

[Asa et al. 86] T. Asano, G. Kenwood, J. Mochizuki, S. Hata. Color image recognition using chrominance signals. *Proc. 8th Int. Conference on Pattern Recognition*, Paris, France, 1986, pp. 804-807.

[BarSan97] A.J. Bardos, S.J. Sangwine. Recursive vector filtering of colour images. *Proc. 4th Int. Workshop on Systems, Signals and Image Processing*, M. Domanski, R. Stasinski (eds.), Poznan, Poland, 1997, pp. 187-190.

[Bar et al. 95] M. Barni, V. Cappellini, A. Mecocci. A vision system for automatic inspection of meat quality. *Proc. 8th Int. Conference on Image Analysis and Processing*, San Remo, Italy, 1995, pp. 748-753.

[BonCoy91] L. Bonsiepen, W. Coy. Stable segmentation using color information. *Proc. 4th Int. Conference on Computer Analysis of Images and Patterns*, R. Klette (ed.), Dresden, Germany, 1991, pp. 77-84.

[Bre93] B. Breuckmann. Applikationsberichte Grauwert- und Farbbildverarbeitung. In: B. Breuckmann (Hrsg.), Bildverarbeitung und optische Meßtechnik in der industriellen Praxis. Franzis-Verlag Munich, Germany, 1993, pp. 176-199 (in German).

[Bue et al. 94] F. Buemi, M. Magrassi, A. Mannucci, M. Massa, G. Sandini. The vision system for the agrobot project. *Proc. 5th ASAE Int. Conference on Computers in Agriculture*, Orlando, Florida, 1994, pp. 93-98.

[CarCae97] D. Carevic, T. Caelli. Region-based coding of color images using Karhunen-Loeve transform. *Graphical Models and Image Understanding* **59** (1997), pp. 27-38.

[Che et al. 94] *Y.-S. Chen, H.*-T. Yen, W.-H. Hsu. Compression of color image via the technique of surface fitting. *Computer Vision, Graphics, and Image Processing: Graphical Models and Image Processing* **56** (1994), pp. 272-279.

[Dav91] J. Davidoff. *Cognition through Color.* MIT Press, Cambridge, Massachusetts, 1991.

[FauNg96] D.S. Faulus, R.T. Ng. EXQUISI: An expressive query interface for similar images. *Proc. SPIE* **2670**, San Jose, California, 1996, pp. 215-226.

[Gan et al. 01] H. Ganster, A. Pinz, R. Röhrer, E. Wildling, M. Binder, H. Kittler. Automated melanoma recognition. *IEEE Transaction on Medical Imaging* **20** (3) 2001, pp. 233-239.

[GevSme96] T. Gevers, A.W.M. Smeulders. Color-metric pattern-card matching for viewpoint invariant image retrieval. *Proc. 13th Int. Conference on Pattern Recognition* **3**, Vienna, Austria, 1996, pp. 3-7.

[Gil94] A. Gilchrist. Introduction: Absolute versus relative theories of lightness perception. In: A. Gilchrist (ed.): *Lightness, Brightness, and Transparency*. Lawrence Erlbaum, Hillsdale, New Jersey, 1994, pp. 1-34.

[GonSak95] Y. Gong, M. Sakauchi. Detection of regions matching specified chromatic features. *Computer Vision and Image Understanding* **61** (1995), pp. 263-269.

[HarSha91] R.M. Haralick, L.G. Shapiro. Glossary of computer vision terms. *Pattern Recognition* **24** (1991), pp. 69-93.

[Hea et al. 92] G. Healey, S.A. Shafer, L.B. Wolff (eds.). *Physics-Based Vision Principles and Practice Color*. Jones and Bartlett, Boston, 1992.

[KhoZin 96] A. Khotanzad, E. Zink. Color paper map segmentation using eigenvector line-fitting. *Proc. IEEE Southwest Symposium on Image Analysis and Interpretation*, San Antonio, Texas, 1996, pp. 190-194.

[Kue97] R.G. Kuehni. Color: An Introduction to Practice and Principles. Wiley, New York, 1997.

[Loc et al. 96] P. Locht, P. Mikkelsen. K. Thomsen, Advanced color analysis for the food industry: It´s here now. *Advanced Imaging*, November 1996, pp. 12-16.

[Lu96] G. Lu. On image retrieval based on colour. *Proc. SPIE* **2670**, San Jose, California, 1996, pp. 310-320.

[Mar82] D. Marr. Vision - A Computational Investigation into the Human Representation and Processing of Visual Information. W.H. Freeman, San Francisco, 1982.

[Mat96] P.C. Matthews. *Vector Calculus*. Springer, Berlin, 1996.

[MemVen96] N.D. Memon, A. Venkateswaran. On ordering color maps for lossless predictive coding. *IEEE Transactions on Image Processing* **5** (1996), pp. 1522-1527.

[Mit et al. 96] S. Mitra, R. Long, S. Pemmaraju, R. Muyshondt, G. Thoma. Color image coding using wavelet pyramid coders. *Proc. IEEE Southwest Symposium on Image Analysis and Interpretation*, April 1996, San Antonio, Texas, pp. 129-134.

[Ove et al. 95] L.A. Overturf, M.L. Comer, E.J. Delp. Color image coding using morphological pyramid decomposition. *IEEE Transactions on Image Processing* **4** (1995), pp. 177-185.

[Ove92] I. Overington. *Computer Vision - A Unified, Biologically-Inspired Approach*. Elsevier, Amsterdam, Netherlands, 1992.

[Pau et al. 93] D.W.R. Paulus, H. Niemann, C. Lenz, L. Demling, C. Ell. Fraktale Dimension der Kontur endoskopisch ermittelter Farbbilder von Geschwüren des Magens. *Proc. 15th DAGM-Symposium Mustererkennung*, S.J. Pöppl, H. Handels (eds.), Lübeck, Germany, 1993, pp. 484-491 (in German).

[Pen et al. 96] A. Pentland, R.W. Picard, S. Sclaroff. Photobook: Content-based manipulation of image databases. *Int. J. of Computer Vision* **18** (1996), pp. 233-254.

[Pic95] R.W. Picard. A society of models for video and image libraries. *Technical Report No. 360*, Media Laboratory Perceptual Computing, MIT, 1995.

[PitTsa91] I. Pitas, P. Tsalides. Multivariate ordering in color image filtering. *IEEE Transactions on Circuits and Systems for Video Technology* **1** (1991), pp. 247-259.

[Pla et al. 97] K.N. Plataniotis, D. Androutsos, S. Vinayagamoorthy, A.N. Venetsanopoulos. Color image processing using adaptive multichannel filters. *IEEE Transactions on Image Processing* **6** (1997), pp. 933-949.

[Poy96] C.A. Poynton. *A Technical Introduction to Digital Video*. Wiley, New York, 1996.

[Pra91] W.K. Pratt. *Digital Image Processing*, 2nd ed., Wiley, New York, 1991, pp. 548-553.

[RaoHwa96] K.R. Rao, J.J. Hwang. *Techniques and Standards for Image, Video and Audio Coding*. Prentice Hall, New Jersey, 1996.

[RicSto96] R. Rickman, J. Stoneham. Content-based image retrieval using colour tuple histograms. *Proc. SPIE* **2670**, San Jose, California, 1996, pp. 2-7.

[Rob76] G.S. Robinson. Color edge detection. *Proc. SPIE Symposium on Advances in Image Transmission Techniques* **87**, 1976, pp. 126-133.

[Ros et al. 95] T. Ross, H. Handels, J. Kreusch, H. Busche, H.H. Wolf, S.J. Pöppl. Automatic classification of skin tumors with high resolution surface profiles. *Proc. 6th Int. Conference on Computer Analysis of Images and Patterns*, Prague, Czech Republic, 1995, pp. 368-375.

[Sag et al. 95] J.A. Saghri, A.G. Tescher, J.T. Reagan. Practical transform coding of multispectral imagery. *IEEE Signal Processing* **12** (1995), pp. 32-43.

[SawHaf94] H.S. Sawhney, J.L. Hafner. Efficient color histogram indexing. *Proc. 1st Int. Conference on Image Processing*, Austin, Texas, November 1994.

[Sch95] P. Scheunders. Genetic optimal quantization of gray-level and color images. *Proc. 2nd Asian Conference on Computer Vision* **2**, Singapore, 1995, pp. 94-98.

[Sch et al. 95] I. Scholl, J. Schwarz, T. Lehmann, R. Mösges, R. Repges. Luv-basierte Bestimmung der Rötung in digitalen Videolaryngoskopie-Bildern. *Proc. 1st Workshop Farbbildverarbeitung*, V. Rehrmann (Hrsg.), Koblenz, Fachberichte Informatik 15/95, Universität Koblenz-Landau, 1995, pp. 68-73.

[Sha97] R.W. Sharpe. *Differential Geometry*. Springer, Berlin, 1997.

[SmiCha96] J.R. Smith, S.-F. Chang. Tools and techniques for color image retrieval. *Proc. SPIE* **2670**, San Jose, California, 1996, pp. 310-320.

[Ste95] B. Steckemetz. Quality control of ready-made food. *Proc. 17th DAGM-Symposium Mustererkennung*, G. Sagerer, S. Posch, F. Kummert (eds.), Bielefeld, Germany, 1995, pp. 153-159.

[Sto et al. 96] W. Stolz, R. Schiffner, L. Pillet, T. Vogt. H. Harms, T. Schindewolf, M. Lanthaler, W. Abmayr. Improvement of monitoring of melanocytic skin lesions with the use of a computerized acquisition and surveillance unit with a skin surface microscopic television camera. *J. Am. Acad. Dermatology* **2** (1996), pp. 202-207.

[StrDim96] M. Stricker, A. Dimai. Color indexing with weak spatial constraints. *Proc. SPIE* **2670**, San Jose, California, 1996, pp. 29-40.

[SunBro03] D.-W. Sun, T. Brosnan. Pizza quality evaluation using computer vision--Part 2: Pizza topping analysis. *J. of Food Engineering* **57** (2003), pp. 91-95.

[Umb et al. 93] S.E. Umbaugh, R.H. Moss, W.V. Stoecker, G.A. Hance. Automatic color segmentation algorithms with application to skin tumor feature identification. *IEEE Engineering in Medicine and Biology* **12** (1993), pp. 75-82.

[VauWil95] V.D. Vaughn, T.S. Wilkinson. System considerations for multispectral image compression designs. *IEEE Signal Processing* **12** (1995), pp. 19-31.

[WanKuo96] X. Wan, C.-C. J. Kuo. Color distribution analysis and quantization for image retrieval. *Proc. SPIE* **2670**, San Jose, California, 1996, pp. 8-16.

[Wat88] R. Watt. Visual Processing - Computational, Psychophysical and Cognitive Research. Lawrence Erlbaum, Hove, 1988.

[WysSti82] R.G. Wyszecki, W.S. Stiles. *Color Science: Concepts and Methods, Quantitative Data and Formulae*, 2nd ed., Wiley, New York, 1982.

[Wu96] X. Wu. YIQ vector quantization in a new color palette architecture. *IEEE Transactions on Image Processing* **5** (1996), pp. 321-329.

[Xu et al. 99] L. Xu, M. Jackowski, A. Goshtasby, C. Yu, D. Roseman, S. Bines, A. Dhawan, A. Huntley. Segmentation of skin cancer images. *Image and Vision Computing* **17** (1999), pp. 65-74.

[ZacLiu93] A. Zaccarin, B. Liu. A novel approach for coding color quantized images. *IEEE Transactions on Image Processing* **2** (1993), pp. 442-453.

[Zek93] S. Zeki. *A Vision of the Brain.* Blackwell Scientific, Oxford, England, 1993.

[ZhaPo95] Y. Zhang, L.-M. Po. Fractal coding in multi-dimensional color space using weighted vector distortion measure. *Proc. 2nd Asian Conference on Computer Vision* **1**, Singapore, 1995, pp. 450-453.

[Zhe et al. 93] J. Zheng, K.P. Valavanis, J.M. Gauch. Noise removal from color images. *J. Intelligent and Robotic Systems* **7** (1993), pp. 257-285.

2 EYE AND COLOR

Knowledge gained from human color perception is frequently included in the evaluation or processing of digital color images. Apart from techniques that use solely mathematical color spaces and color metrics, there exist also a large number of techniques that are based on physiological and psychological insights into the processing of information in the human visual system. First, the attempt is made to transfer this knowledge from human color perception into a computer-supported model; second, this knowledge serves as a motivation for a number of proposed algorithms. The knowledge of differences in eye sensitivity with regard to differing wavelengths is of importance to color image enhancement, color image coding, and color image display. Furthermore, the opponent color space is implemented in some techniques for color image segmentation and in geometrical color stereo analysis. Knowledge of the cortical coding of color information is necessary for understanding some techniques for solving the color constancy problem.

The use of technical terms borrowed from perception psychology in digital color image processing is thus a direct result of this adaptation. Since the description and the understanding of "perception-based" techniques are not possible without this terminology, a short introduction to theories of human *chromatopsy* (color vision; Greek *chroma* = color, Greek *opsis* = vision) is provided in the following. A more detailed representation of what is known so far about human visual color perception can be found in [Gil94] or [Zek93].

Further information on the effect of colors on the human psyche is still not well known. The poet and natural philosopher Johann Wolfgang von Goethe (1749 - 1832) assumed in his color theory a connection between the human psyche and the colors surrounding us (humans). Conclusions can then be drawn from the psychic state of a human about possible causes of illness. Although Goethe's physical explanations have been refuted by the physics standards of the present, even today his color theory influences many interesting and controversial discussions. Here it should be indicated that color perception includes indeed more than just color vision. However, this is not the subject of this chapter.

Furthermore, it should be mentioned that seeing a color and naming a color represent two separate processes. We may agree on naming an object *orange* although we see it differently. Color naming is also based on cultural experience and it is applied differently in different languages (see, e.g., [Lin et al. 01], [Osb02], and [Sch01]). Nevertheless, a description of the area of color naming is excluded in this chapter.

2.1 PHYSIOLOGY OF COLOR VISION

The eye is the sensory organ of vision. It reacts to stimulations by electromagnetic radiation with a wavelength between 380 and 780 nm (nanometer, 1 nm = 10^{-9} m) and with a frequency between $4.3 \cdot 10^{14}$ and $7.5 \cdot 10^{14}$ Hz (Hertz). The relation between wavelength λ and frequency ν can be directly given since the product $\lambda \cdot \nu = c$ is constant and the speed of light c is specified by $c = 2.9979246 \cdot 10^8 \, m/\sec$. An illustration of this relation can be found in Fig. 2.1.

Human sensitivity also occurs when stimulated by wavelengths of neighboring areas. For example, infrared light is felt as warm and ultraviolet light leads to a reddening or browning of the skin. Nevertheless, these wavelengths cannot be detected with the eye. Thus, only the wavelengths within the spectrum of visible light are of importance for (human) color vision. In the following, if it is not expressly mentioned otherwise, the term *light* is used to mean *visible light*. The structure of the human eye is now outlined. A schematic cross-section of the right eye of a human is presented in Fig. 2.2.

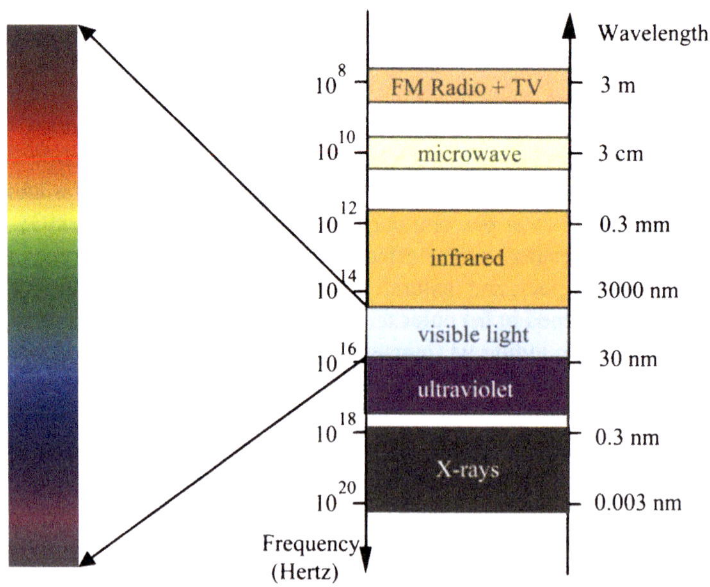

Figure 2.1. *Excerpt from the electromagnetic spectrum.*

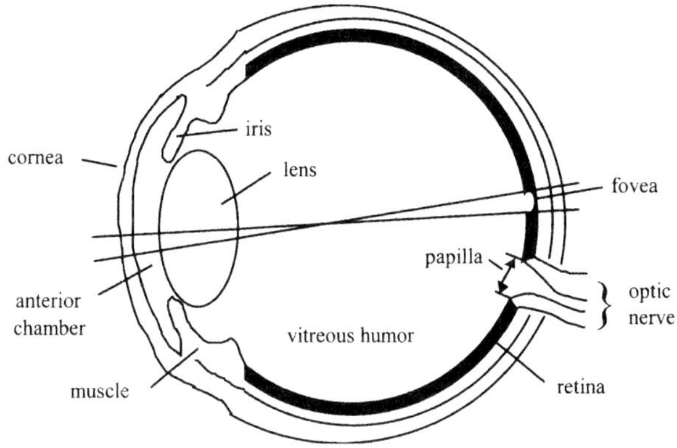

Figure 2.2. *Simplified representation of the human eye.*

The retina consists of photoreceptors, glia cells, pigment cells, and four different classes of nerve cells. The photoreceptors can be subdivided morphologically into two classes: rods (about 120 million), and cones (about 6 million). In the fovea centralis, which is the area of the eye's sharpest vision, the retina contains only cones. In color vision, the cones absorb the light striking the retina (the entering visible light). This information is assessed in three overlapping spectral areas and subsequently passed on in the form of electrical impulses over four different layers of nerve cells (the horizontal cells, bipolar cells, amacrine cells, and ganglia cells) to the visual paths. Color information is passed from there over the corpus geniculatum laterale to the primary visual cortex (V1, *visual area* 1) and from there further to higher cortical regions (specifically V4 for color vision; see [Zek93]). Finally, the interlinking result of the evaluation of the color information results in color sensation in the brain.

Color information is represented and coded in at least three different forms on the way from the eye to the higher brain regions. Color information in this chapter is indicated, according to its whereabouts in the visual path, as *receptoral*, *postreceptoral*, and *cortical* color information. Several theories on the known representational forms of color information are described in the following sections. Note that the different color theories presented here are not competing with each other. They are true for different areas along the visual path.

2.2 RECEPTORAL COLOR INFORMATION

The colors that we perceive in our environment are divided into two classes: *chromatic* and *achromatic*. The gray levels that go from white to black are denoted as achromatic colors. The chromatic colors, which we perceive on the surfaces of objects, can be characterized by three components: *hue*, *saturation*, and *luminance* (or *brightness*). These three color components are introduced here since they are necessary for the description of color vision. A detailed representation can be found in Chapter 3.

Hue describes the type of chromaticity a color has and is indicated generally with words such as *red*, *yellow*, and *blue*. Hues can also be represented in a closed series from red to orange, yellow, green, blue, violet, purple, then to red in a color circle. Chromaticity describes the dissimilarity of a color to an achromatic color of equal luminance (i.e., to an equal light, gray tone). Saturation describes the purity of a color, or the measure of the degree to which a pure color is diluted by white light. As saturation decreases, colors appear more faded. Luminance indicates the strength of light sensitivity as it is connected to each color sensitivity. The greater the strength of the lighting, the lighter the color appears.

The first step of representing color information occurs in the retina. The differences between day and night visions in humans must first be distinguished. The vision process under daylight lighting conditions is denoted as *photopic vision,* whereby the cones function in this process as receptors. They are stimulated by the daylight. Vision occurring under nighttime lighting conditions is called *scotopic vision*. In this case it is the rods that are stimulated. In the time of transition (dawn), neither of the two receptor classes dominates. This condition is denoted as *mesopic vision*.

Visual acuity and color vision are very well marked in photopic vision, and the location of the greatest visual acuity lies in the center of the fovea centralis. In contrast, only achromatic colors can be perceived in scotopic vision. Functional color blindness exists during night vision due to the low sensitivity of the cones. As mentioned before, only cones are located in the area of the fovea centralis and not rods. That is why it is very difficult to focus during scotopic vision, for example, to read a book under moonlight condition. The location of the greatest visual acuity and the greatest sensitivity of the retina lies on the edge of the fovea centralis and not in its center.

Thomas Young and Hermann von Helmholtz proposed the hypothesis that color vision is based on three different cone types that are especially sensitive toward long-, middle-, and short-wave light, respectively. This hypothesis is also called the *three-color theory* or *trichromatic theory* since the cone types sensitive to long-, middle-, and short-wave light are also designated as red, green, and blue cones. The latter, frequently used designation can lead to confusion since the absorption of long-wave light is not identical to the sight of the color red. Each of the three cone types works as an independent receiver system of photopic vision. The signals are included together in a neuronal light-dark system and neuronal color system. In 1965, there came experimental confirmation of a long-expected result. There are three types of color-sensitive cones with differing pigments in the retina of the human eye, corresponding roughly to red-, green-, and blue-sensitive detectors. This is generally regarded as a proof of the trichromatic theory.

Trichromates and Dichromates

All hues of the color circle can be represented either by certain spectral colors in the spectrum of visible sunlight or by an additive color mixture of two spectral colors. *Additive color mixture* results when light of differing wavelengths falls on an identical place of the retina. In contrast, *subtractive color mixture* describes how the light-absorbing properties of materials mix to make colors in reflected light. The latter is the case, for example, when watercolors are mixed together or when several color filters of differing spectral transmissions are inserted one after the other into a beam of light. Fig. 2.3 illustrates the two different color mixtures.

Each color $\mathbf{C}_X$ that can be produced by primary light sources can be generated for the color normal by the additive color mixture of three suitable colors $\mathbf{C}_1$, $\mathbf{C}_2$, and $\mathbf{C}_3$. Here a definite sensory equation applies, which can be represented in vector notation by

$$\alpha \mathbf{C}_1 + \beta \mathbf{C}_2 + \gamma \mathbf{C}_3 \cong \delta \mathbf{C}_X . \quad (2.1)$$

In this equation the symbol $\cong$ means *visual equivalent.* Two color samples are designated *metameric* if they differ spectrally but they yield the same or similar color sensation under at least one set of viewing conditions (i.e., they look the same). Metamerism implies that two objects that appear to have exactly the same color may have very different colors under differing lighting conditions. The wavelengths of the *primary colors* $\mathbf{C}_1$, $\mathbf{C}_2$, and $\mathbf{C}_3$ are standardized internationally. They are the spectral colors with the wavelengths 700 nm (red), 546 nm (green), and 435 nm (blue). A detailed description of the individual color models and color distances can be found in Chapter 3.

As mentioned previously, the hues of luminous colors are unambiguously defined by maximally three constants for the color normal according to Eq. (2.1). For the largest part of the population, the constants α, β, and γ in Eq. (2.1) are practically equal (*normal trichromates)* for the generation of a hue. Deviating constants (*anomal trichromates*) apply for a small percentage of the population. Roughly 2% of the population are *dichromates* who are born with only two classes of cone receptors. For them all colors can be described by an equation with two constants:

$$\alpha \mathbf{C}_1 + \beta \mathbf{C}_2 \cong \delta \mathbf{C}_X . \quad (2.2)$$

The perceived color values in dichromates are substantially less differentiated than in trichromates [StoSha00]. The dichromatic effects of color vision as well as anormal trichromacy are genetically determined [GrüGrü85]. The most commonly occurring color blindness (more precisely, color-deficient vision) is the red-green blindness. This appears if the cones are lacking either the red or the green photoreceptor. In very rare cases, color blindness is caused by lack of the blue photoreceptor. Note that most investigations with dichromates took place

Figure 2.3. *(a) Additive and (b) subtractive color mixture.*

in Europe and North America. Thus, 2% of the population is to be read as 2% of the Caucasian population. Some experiments in the United States have shown that dichromacy occurs less frequently in other ethnic groups.

More common than the complete lack of a photoreceptor is the appearance of a significant difference in response behavior of color receptors opposite the color normal. In one case, the greatest sensitivity of the (red) long-wave photopigment lies very close to that of the (green) middle wave (protanomaly); in another case, the greatest sensitivity of the (green) middle-wave photopigment is shifted in the direction of the (red) long wave (deuteranomaly). The interlinking result is a reduction of the individual ability to distinguish small color differences, especially those with less brightness (see [Mur86]). Fewer than 0.005% of humans are totally color blind (complete *monochromates*). They can only perceive gray levels since they are genetically determined to have only one photoreceptor system.

An additional limitation of color perception appears with aging. This does not concern a reduction of the ability to distinguish small color differences, as with color-deficient vision, but rather is a reduction of the range of the perceived color spectrum. In other words, this means that the ability to perceive red hues deteriorates with increasing age. This influence of color perception is not caused by a change of the response behavior of photoreceptors, but rather it is a result, among other things, of the gradual yellowing of the lens with age. Therefore, the bandwidth range of the wavelengths passing through the lens decreases along with its permeability. In Fig. 2.4, the permeability of the lens for ages 20, 45, and 63 years is presented.

The sensitivity of the photoreceptors is influenced, however, by the flow of Na^+, K^+, and Ca^{2+} ions. Humans with high blood sugar content (e.g., diabetics) have a lower color perception than the general population. This reduction of the sensitivity of photoreceptors appears in diabetics above all in short-wave light.

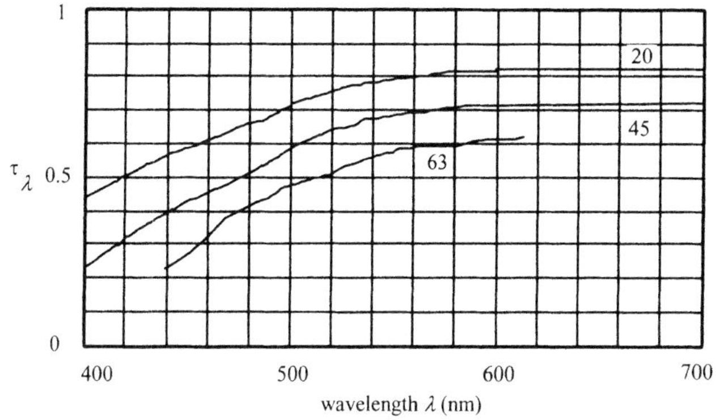

Figure 2.4. *Spectral transmission degree τ_λ of the human lens in different ages (according to [LeG75]).*

A detailed representation of this relation goes beyond the scope of this section. The reader interested in this subject can refer to [Kur et al. 94] or [Kin et al. 72].

2.3 POSTRECEPTORAL COLOR INFORMATION

Many phenomena of color vision, such as color-deficient vision caused by the loss of a cone type or the sensation of a "mixed color" produced from different spectral areas, can be well explained with the help of the trichromatic theory. However, in contrast, other phenomena of color vision point to the fact that the representation of color information generated by the three cone types are not directly passed to the brain, but rather is subject beforehand to yet another "processing." The trichromatic theory cannot explain, for example, why the four colors red, yellow, green, and blue are perceived as especially pure colors, although according to this theory only three colors are involved. No perceived color appears red or green simultaneously, or similarly, yellow and blue. Furthermore, the color black appears equal to white, although black distinguishes itself by the absence of the signal.

The physiologist Ewald Hering attempted to explain these observations. In his *opponent color theory* or *four-color theory* of color vision, antagonistic neuronal processes, with the opponent colors red-green and yellow-blue and a likewise antagonistically organized light-dark system, are named as the cause for color vision. The notation "red-green system" is common and therefore is used here, although the designation "long-middle-wave system" would be more suitable. The term "red-green system" can lead to misunderstandings since it implicitly contains the assumption that the reaction of the cell itself leads to the perception of a surface as red or green. However, this is not the case.

Hering's opponent color theory and Young and Helmholtz's three-color theory were long viewed as being competing theories. From today's standpoint, both theories and models represent color information in different processing steps of the visual path. After the coding by the three cone types, an additional coding of the color information occurs in accordance with Hering's opponent color theory. Ganglia cells present in the retina and corpus geniculatum laterale are made responsible for this additional transformation of the color signal. The functional organization of ganglia cells is illustrated in the following.

2.3.1 Neurophysiology of Retinal Ganglia Cells

Neuronal elements that are connected with *one* nerve cell are designated as a receptive unit. The receptive unit of a ganglia cell surrounds a greater area in the retina. Each ganglia cell of the eye is in turn assigned to a small area from which a stimulation and/or a moderation can be triggered with suitable light stimuli. This area is called the *receptive field* (RF). Within the retina the receptive fields of neighboring ganglia cells overlap considerably.

In principle, two large neuron classes can be distinguished: the *ON-center neurons* and the *OFF-center neurons*. In the mammalian retina ON- and OFF-center neurons have receptive fields that can be subdivided functionally into RF center and RF periphery. The exposure of the RF center causes stimulation in the on-center neurons (i.e., the impulse frequency of the action potential rises). On the other hand, the exposure of the RF periphery causes a moderation (i.e., the impulse frequency of the action potential becomes smaller [GrüGrü85]). If the RF center as well as the RF periphery are exposed simultaneously, then as a rule the reaction from the RF center dominates. If the light stimulus is eliminated, then in the RF center a moderation is released and in the RF periphery a stimulation (see Fig. 2.5).

The receptive fields of the OFF-center neurons illustrate an almost mirror-image of the receptive fields of the ON-center neurons: Exposure of the RF center causes a moderation; elimination of the light stimulus in the RF center causes an activation. Conversely, the exposure of the RF periphery causes activation and the elimination of the light stimulus in the RF periphery causes a moderation.

2.3.2 Reaction of Retinal Ganglia Cells to Colored Light Stimuli

The retinal ganglia cells can be further differentiated with the help of the functional organization of receptive fields. This is based on animal experimentation, for example with rhesus monkeys, whose color vision, according to behavioral and biological measurements, is presumably similar to the color vision of humans. Nerve cells with partially color-specific reactions can be found in the visual pathway at the first central processing unit between the retina and

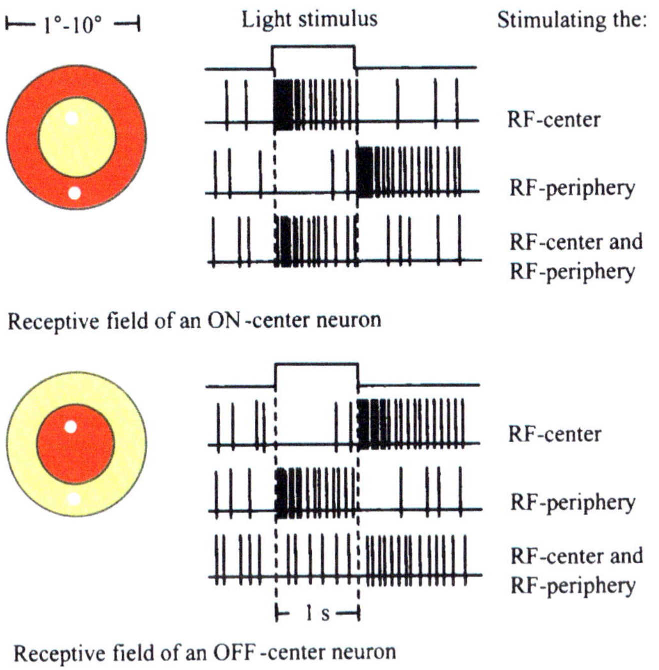

Figure 2.5. *Functional organization of receptive fields of ganglia cells in the mammalian retina. For analysis of the receptive fields white points of light are projected either into the RF center or into the RF periphery (after [GrüGrü85]).*

---

cerebral cortex, the corpus geniculatum laterale. These nerve cells can be subdivided into three classes:

1. The *ganglia cells* of the *light-dark system* (ON-center neurons or OFF-center neurons) show qualitatively the same reaction for the entire area of visible light (ca. 380 - 780 nm). They react independently from the wavelength of monochromatic light and in this cell class no difference exists between the spectral sensitivities of the RF periphery and RF center.
2. The ganglia cells of the *red-green system* show a moderation in the middle-wave spectral area during exposure of their RF center to monochromatic light stimuli and an activation in the long-wave spectral area. In contrast, the exposure of the RF periphery with-middle wave light stimuli causes a stimulation and with long-wave a moderation.
3. The ganglia cells of the *yellow-blue system* show a moderation during the exposure of their RF center to short-wave light stimuli and an activation with long-plus-middle-wave. The exposure of the RF periphery to short wave-light stimuli causes a stimulation and with long-plus-middle-wave light stimuli a moderation.

Important signal processing for color vision takes place in the neurons of the retina and in the corpus geniculatum laterale. The signals from the *three different cone types* are so interconnected with each other that, in addition to the light-dark system, two color-specific opponent neuron systems originate in the ganglia cell layer. Together these systems form a *four-color system* with the opponent color pairs yellow-blue and red-green.

2.4 CORTICAL COLOR INFORMATION

The processing of color information in the human brain is not fully described by either the three color theory or the opponent color theory. From a neurophysiological point of view, signal processing in the cortex is likewise important for color vision. In the fourth visual area (V4) of the cortex, cells were discovered (in monkeys) that react very selectively toward small areas of the visible wave spectrum. Here a very fine bandpass characteristic was discovered, that is, the cells are very sensitive to a wavelength area of on average 10 - 50 nm [Zek80].

Today, studies of human brain activity are possible by applying the method of positron emission tomography (PET). Thus, in color vision increased activity can be observed in the visual areas V1 and V4 of the cortex. However, how the processing of color information occurs in the visual cortex is still largely unknown. A detailed representation of some investigations and the results can be found in [Zek93].

2.5 COLOR CONSTANT PERCEPTION AND RETINEX THEORY

One of the most interesting phenomena of visual color perception is the ability of the human visual system to perceive colors nearly independent of changing lighting conditions. This *color constancy* (see Section 1.2.4) exists for a set of surfaces and lighting conditions. Thus a red object is seen as a red object under daylight conditions regardless of whether it is early morning or noon. The borders of the color constancy ability of the human visual system are surely known to anyone who has ever bought a shirt under neon lighting conditions and then later at home was surprised by the color effect of the clothing under daylight conditions. From psychological experiments it is known that visual color constancy is influenced by depth information in space [Ber94] and scene complexity [Kra et al. 02]. In this connection, color perception depends on depth perception rather than vice versa. The two surfaces of a monochromatic, right-angled folded piece of cardboard appear more similar in color than the same surfaces of a piece of cardboard unfolded and placed flat. A detailed representation of this subject and several psychological experiments can be found in [Ber94].

The first algorithm for describing the color constancy ability of the human visual system was proposed by Land and McCann [LanMcC71]. Their theory of color vision, which they called *retinex theory* by combining the two words *retina* and *cortex*, refers to a planar world of "Mondrian" images. These images, constructed in the style of the Dutch artist Piet Mondrian, consist of different, partly overlapping, monochrome rectangles (see Fig. 2.6). In various series of tests these images were illuminated with light of differing spectral composition and the color constancy of the observer was studied.

The retinex theory assumes that for each component of the color signal (in the *RGB* space) a separate monochromatic processing occurs. The combined result should depend solely on the characteristics of the surface reflectance and not on the lighting conditions. According to the retinex theory, the reflectance characteristics of different surfaces are each compared separately in the short-, middle-, and long-wave spectral areas. The result of the comparisons in a color channel is a *lightness record.* In the retinex theory, Land assumes that the brain compares the lightness records that are determined separately for a scene in each color channel. These occur independently of the spectral composition of the lighting (and thus independently of the relative intensity of the light). The construction of the colors of the surfaces in the brain is, according to the retinex theory, the result of "comparisons of comparisons" (see [Zek93]). The retinex theory distinguishes itself therefore fundamentally from other theories of color perception since it contains only comparisons and no mixtures or additions.

The retinex theory has been studied in great detail by Brainard and Wandell [BraWan86]. They implemented several variants of the retinex algorithm (see [Hor74] and [Lan86]) and applied these on different Mondrian images. In this connection, they changed the arrangement of the colored surfaces in the images

Figure 2.6. *Example of an image in the style of the Dutch painter Piet Mondrian (1872 - 1944).*

examined. While the colors of the individual surfaces perceived by the human observer were always the same for different arrangements in each investigation, the results determined by all variants of the retinex algorithm depended on the arrangement of the surfaces. However, a color constancy algorithm should provide the same values for each arrangement. Furthermore, the shaping of the retinex theory on planar, Mondrian images is a limitation that is of little interest for digital color image processing.

The retinex theory does not provide a full description of the human color constancy ability, nor is the use of calculation schemes sufficient to attain color constancy in digital color images. Because of this, a large number of publications propose techniques for the approximation of color constancy in digital color images under various boundary conditions. These are explained later in Section 8.3. Note that many different descriptions of retinex methods of lightness record computation exist.

A detailed and interesting representation of the processing of color information in the human visual system is given by Zeki [Zek93]. [Kai96] is recommended for an interactive introduction to the area of human color perception; it is the author's intention to continually expand and update the web book. An interesting overview of cognitive processes that are influenced by color information can be found in [Dav91]. A presentation of various perceptual processes and their causes is given in [Gil94].

2.6 REFERENCES

[Ber94] S.S. Bergström. Color constancy: Arguments for a vector model for the perception of illumination, color, and depth. In: A.L. Gilchrist (ed.), *Lightness, Brightness, and Transparency*. Lawrence Erlbaum, Hillsdale, New Jersey, 1994, pp. 257-286.

[BraWan86] D.H. Brainard, B.A. Wandell. Analysis of the retinex theory of color vision. *J. Optical Society America* **3** (1986), pp. 1651-1661.

[Dav91] J. Davidoff. *Cognition through Color*. MIT Press, Cambridge, Massachusetts, 1991.

[Gil94] A.L. Gilchrist (ed.). *Lightness, Brightness, and Transparency*. Lawrence Erlbaum, Hillsdale, New Jersey, 1994.

[GrüGrü85] O.-J. Grüsser, U. Grüsser-Cornehls. Physiologie des Sehens. In: R.F. Schmidt (ed.), *Grundriß der Sinnesphysiologie*. 5th ed., Springer, Berlin, 1985, pp. 174-241.

[Hor74] B.K.P. Horn. Determining lightness from an image. *Computer Graphics and Image Processing* **3** (1974), pp. 277-299.

[Kai96] P. Kaiser. *The Joy of Visual Perception: A WebBook*, at http://www.yorku.ca/eye/.

[Kin et al. 72] P.R. Kinnear, P.A. Aspinall, R. Lakowski. The diabetic eye and colour vision. *Trans. of the Ophthalmological Society U.K.* **92** (1972), pp. 69-78.

[Kra et al. 02] J.M. Kraft, S.I. Maloney, D.H. Brainard. Surface-illuminant ambiguity and color constancy: Effects of scene complexity and depth cues. *Perception* **31** (2002), pp. 247-263.

[Kur et al. 94] A. Kurtenbach, U. Wagner, A. Neu, U. Schiefer, M.B. Ranke, E. Zrenner. Brightness matching and colour discrimination in young diabetics without retinopathy. *Vision Research* **34** (1994), pp. 115-122.

[LanMcC71] E.H. Land. J.J. McCann, Lightness and retinex theory. *J. Optical Society of America* **61** (1971), pp. 1-11.

[Lan86] E.H. Land. Recent advances in retinex theory. *Vision Research* **26** (1986), pp. 7-21.

[LeG75] Y. Le Grand. Measurement of the visual stimulus. In: E.C. Carterette, M.P. Friedman (eds.), *Handbook of Perception* V: *Seeing*. Academic Press, New York, 1975, pp. 25-55.

[Lin et al. 01] H. Lin, M.R. Luo, L.W. MacDonald, A.W.S. Tarrant. A cross-cultural colour-naming study: Part II - Using a constrained method. *Color Research & Application* **26** (2001), pp. 193-208.

[Mar82] D. Marr. Vision - A Computational Investigation into the Human Representation and Processing of Visual Information. Freeman, New York, 1982.

[Mur86] G.M. Murch. Human factors of color displays. In: F.R.A. Hopgood, R.J. Hubbold, D.A. Duce (eds.), *Advances in Computer Graphics II*. Springer, Berlin, 1986, pp. 1-27.

[Osb02] R. Osborne. Telesio's dictionary of Latin color terms. *Color Research & Application* **27** (2002), pp. 140-146.

[Sch01] J.A. Schirillo. Tutorial on the importance of color in language and culture. *Color Research & Application* **26** (2001), pp. 179-192.

[StoSha00] A. Stockman, L.T. Sharpe. The spectral sensitivities of the middle- and long-wavelength-sensitive cones derived from measurements in observers of known genotype. *Vision Research* **40** (2000), pp. 1711-1737.

[Zek80] S. Zeki. The representation of colours in the cerebral cortex. *Nature* **284** (1980), pp. 412-418.

[Zek93] S. Zeki. *A Vision of the Brain*. Blackwell Scientific, Oxford, England, 1993.

3 COLOR SPACES AND COLOR DISTANCES

Color is a perceived phenomenon and not a physical dimension like length or temperature, although the electromagnetic radiation of the visible wavelength spectrum is measurable as a physical quantity. The observer can perceive two differing color sensations wholly as equal or as metameric (see Section 2.2). Color identification through data of a spectrum is not useful for labeling colors that, as also in image processing, are physiologically measured and evaluated for the most part with a very small number of sensors. A suitable form of representation must be found for storing, displaying, and processing color images. This representation must be well suited to the mathematical demands of a color image processing algorithm, to the technical conditions of a camera, printer, or monitor, and to human color perception as well. These various demands cannot be met equally well simultaneously. For this reason, differing representations are used in color image processing according to the processing goal.

Color spaces indicate *color coordinate systems* in which the image values of a color image are represented. The difference between two image values in a color space is called *color distance*. The numbers that describe the different color distances in the respective color model are as a rule not identical to the color differences perceived by humans. In the following, the standard color system *XYZ*, established by the International Lighting Commission CIE (*Commission Internationale de l'Eclairage*), will be described. This system represents the international reference system of color measurement.

3.1 STANDARD COLOR SYSTEM

The model of *additive color mixture* is used when light of differing wavelengths reaches an identical place on the retina or image sensor. The differing color stimuli are combined into one color through projective overlapping. Due to Grassmann's First Law of additive color mixture, any color stimulus can be uniquely related to a particular set of three primary color stimuli, as long as each primary stimulus is independent. By using vector notation, the unit length vectors of the primary colors can be viewed as a basis of a (not necessarily orthonormal) vector space.

Grassmann's First Law of color mixture can be written (applying vector addition) as

$$\mathbf{M} = R \cdot \mathbf{R} + G \cdot \mathbf{G} + B \cdot \mathbf{B}.$$

The quantities of primary colors **R**, **G**, **B** in the mixed color **M** are indicated by *R*, *G*, and *B*. However, not all colors can be produced with a single set of primary colors. This represents no injury to Grassmann's First Law of color mixture since the quantities of primary colors can be assumed to be negative in the color blending, for example:

$$\mathbf{M} + R \cdot \mathbf{R} = G \cdot \mathbf{G} + B \cdot \mathbf{B}.$$

The abovementioned principle of color blending is also valid for a set **M** of n colors. The following applies:

$$\mathbf{M} = \{\mathbf{M}_i \mid i = 1, \ldots, n\} \text{ with } \mathbf{M}_i = R_i \cdot \mathbf{R} + G_i \cdot \mathbf{G} + B_i \cdot \mathbf{B} \text{ and}$$

$$\mathbf{M}_{1 \ldots n} = \mathbf{R} \cdot \sum_{i=1}^{n} R_i + \mathbf{G} \cdot \sum_{i=1}^{n} G_i + \mathbf{B} \cdot \sum_{i=1}^{n} B_i .$$

By an analysis of the visible wavelength interval in n narrow intervals, each interval approximately forms the color stimulus of a spectral color. Note that the waveband lies in general between 2 and 10 nm. A scanning that is too fine would result in a light density that is too low and not further measurable. If now the $\mathbf{M}_i$ from both of the abovenamed equations are color values of the spectral color intervals, any color stimulus can be described by additive mixture of a subset of $\boldsymbol{M}$ that corresponds to the spectral composition. If the waveband is shifted toward zero, then a continuous color stimulus $M(\lambda)$ in the form

$$M(\lambda) = R(\lambda) \cdot \mathbf{R} + G(\lambda) \cdot \mathbf{G} + B(\lambda) \cdot \mathbf{B}$$

is produced. For standardization, the International Lighting Commission (CIE) set the employed monochromatic primary values as well as the color matching functions in 1931 as the definition of the (hypothetical) colormetric *2° CIE standard observer*. Here 2° indicates the size of the visual field, which, until the introduction of the 10° CIE standard observer in 1964, should guarantee color perception without stimulation of the rods. The wavelengths and the relative spectral powers of the primary values are listed in Table 3.1. The color stimuli standardized with $S(\lambda)$ produce the *spectral tristimulus values* $\bar{r}(\lambda)$, $\bar{g}(\lambda)$ and $\bar{b}(\lambda)$. The following applies:

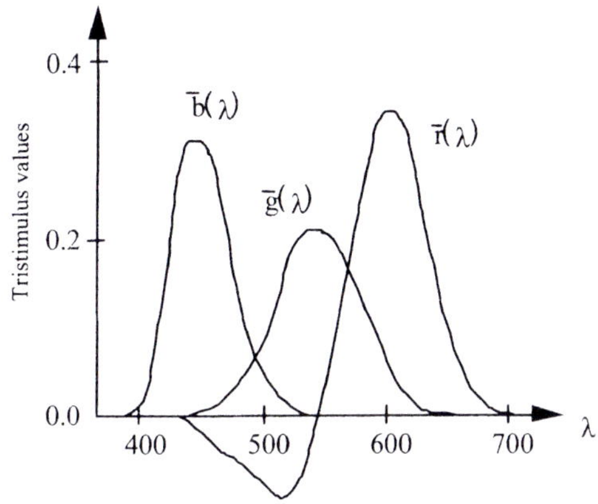

Figure 3.1. *The curves are representations of the CIE color matching functions* $\bar{r}$, $\bar{g}$, *and* $\bar{b}$ *for the 2° CIE standard observer.*

Primary	λ (in nm)	S
R	700.0	72.09
G	546.1	1.379
B	435.8	1.000

Table 3.1. *Wavelengths and the corresponding relative spectral power S of the CIE 1931 primaries.*

$$\overline{m}(\lambda) = \bar{r}(\lambda) \cdot \mathbf{R} + \bar{g}(\lambda) \cdot \mathbf{G} + \bar{b}(\lambda) \cdot \mathbf{B} .$$

The spectral values are dismantled in Fig. 3.1 as *color matching functions* over the visible wavelength area, whereby the units of the ordinates are arbitrarily fixed. See [TruKul96] concerning the problem of scanning the spectral value curves for certain uses in color image processing.

3.1.1 CIE Color Matching Functions

In order to obtain a color mixture that covers the entire visible wavelength area, the CIE defined virtual primary values X, Y, and Z (i.e., those that do not correspond to any physical spectral distribution). The conversion of the real spectral value curves into the positive, virtual color matching functions $\bar{x}(\lambda)$, $\bar{y}(\lambda)$ and $\bar{z}(\lambda)$ for the 2° CIE standard observer can be explained by the linear transformation

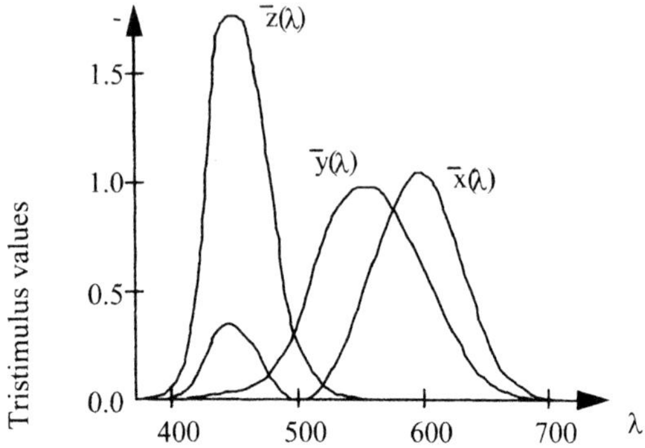

Figure 3.2. *The CIE color matching functions $\bar{x}, \bar{y}$, and $\bar{z}$ for the 2° CIE standard observer.*

$$\begin{pmatrix} \bar{x}(\lambda) \\ \bar{y}(\lambda) \\ \bar{z}(\lambda) \end{pmatrix} = \begin{pmatrix} 0.49000 & 0.31000 & 0.20000 \\ 0.17697 & 0.81240 & 0.01063 \\ 0.00000 & 0.01000 & 0.99000 \end{pmatrix} \cdot \begin{pmatrix} \bar{r}(\lambda) \\ \bar{g}(\lambda) \\ \bar{b}(\lambda) \end{pmatrix}.$$

The color matching functions that are assigned to the new spectral values can be taken from Fig. 3.2.

3.1.2 Standard Color Values

A radiation that causes a color stimulus in the eye is indicated as color stimulus function $\varphi(\lambda)$. For luminous objects, the color stimulus function is identical to the spectral power $S(\lambda)$. In comparison, in body colors, the color stimulus function is composed multiplicatively of the spectral power $S(\lambda)$ and the spectral reflection factor $R(\lambda)$. By observation of fluorescent samples, the fluorescence function $S_F(\lambda)$ is added for the color stimulus function of the body color. The following applies:

$$\varphi(\lambda) = \begin{cases} S(\lambda) & \text{for luminous objects} \\ S(\lambda) \cdot R(\lambda) & \text{for body colors} \\ S(\lambda) \cdot R(\lambda) + S_F(\lambda) & \text{for fluorescent samples} \end{cases}.$$

With that, the *standard color values* are determined by the equations

$$X = k \cdot \int \varphi(\lambda) \cdot \bar{x}(\lambda)\, d\lambda,$$
$$Y = k \cdot \int \varphi(\lambda) \cdot \bar{y}(\lambda)\, d\lambda \text{ , and}$$
$$Z = k \cdot \int \varphi(\lambda) \cdot \bar{z}(\lambda)\, d\lambda$$

with

$$k = \frac{100}{\int S(\lambda) \cdot \bar{y}(\lambda)\, d\lambda}.$$

The normalization factor k is used for body colors in order to achieve the value 100 for the standard color value $Y_{white\,body}$ for a pure pale, white body under illumination with any type of light. In practice the integral is changed into a sum on the basis of the finite measurements. For nonluminous objects, it should be taken into account that the color measurement depends on the measurement geometry. Therefore, a triple of the positive standard color values X, Y, and Z can be assigned to each spectral signal.

3.1.3 Chromaticity Diagrams

It is advantageous for a two-dimensional graphic representation to merely present an intersecting plane of the three-dimensional color space in a *chromaticity diagram*. Since all color values, whose position vectors are linearly dependent, differentiate themselves solely by brightness, simply the brightness information is lost. The color information that remains through elimination of brightness is indicated as *chromaticity*. In other words, chromaticity is defined by hue and saturation (without brightness). The commonly related intersection plane for the construction of the chromaticity diagram in the RGB cube is the unit plane $R + G + B = 1$, whereby an equilateral triangle results (see Fig. 3.3).

This triangle is also known as the *Maxwell color triangle*. M is the intersection between the observed color position vector and the unit plane. Using the Newtonian gravity formulation the following relationships result for a color M,

$$r_M = \frac{R}{R+G+B}, \quad g_M = \frac{G}{R+G+B}, \text{ and } b_M = \frac{B}{R+G+B}.$$

where $r_M + g_M + b_M = 1$. However, the Cartesian representation with r as abscissa and g as ordinate, by which blue lies in the origin, is more commonly used. The above relationships do not change by this. As an example of such a representation, the chromaticity diagram for the 2° CIE standard observer is shown in Fig. 3.4, in which two (of three) *2° chromaticity coordinates*, namely x and y, are represented. The position of the spectral colors from 400 - 700 nm is listed as *spectral color transmission*.

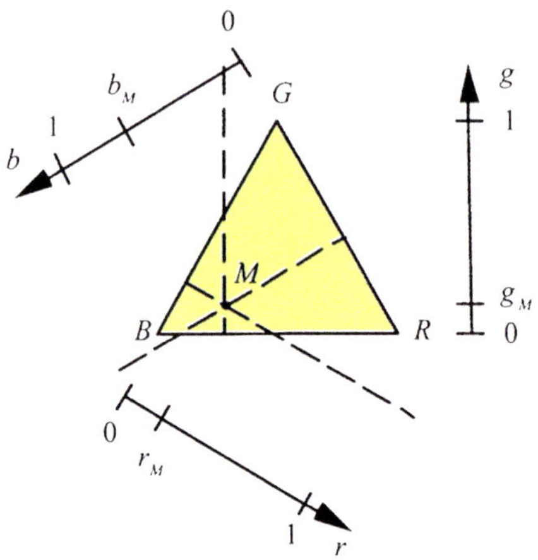

Figure 3.3. *Maxwell color triangle represented as a chromaticity diagram.*

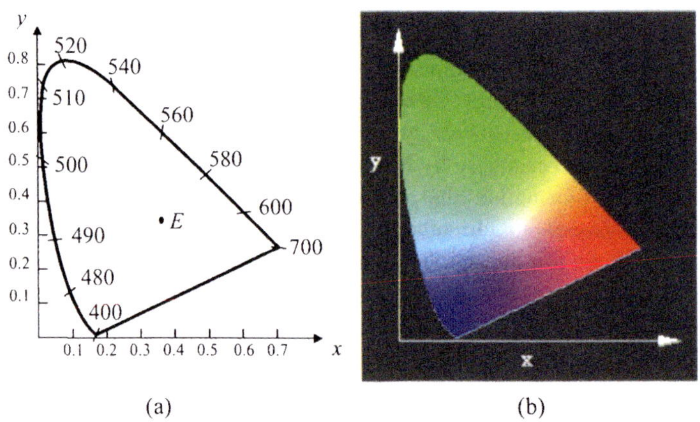

Figure 3.4. *The CIE chromaticity diagram of the 2° CIE standard observer (a) in numerical representation and (b) as color-coded illustration.*

The connection of the endpoints is called the *purple boundary*. All real chromaticities lie in the middle. The point that represents the equienergy spectrum $(x = y = z = 1/3)$, is indicated by E. The CIE chromaticity diagram can also be used for the definition of a *color gamut*, which shows the results of the addition of colors. By the addition of any two colors (e.g., I and J in Fig. 3.5) each color can be generated on its connecting line, in which the relative portions of both colors are varied. Through the use of a third color, K, a gamut with all the colors within the triangle IJK can be produced by mixing with I and J, in which again the relative portions are varied. On the basis of the form of the chromaticity diagram, it is obvious that not all colors can be produced through the mixture of the three colors red, green, and blue. No triangle can be constructed whose corners lie within the visible area and that covers the entire area.

The chromaticity diagram is also applied to compare color gamuts of different color monitors and different color printers [Fol et al. 94]. A comparison of different techniques for the reproduction of a device's color gamut in a color gamut of another device is explained in [MonFai97]. Since the color gamut of a color printer is relatively small in relation to the color gamut of a color monitor, not all colors can be printed on paper as seen on the monitor. A reduced gamut should be used for the monitor if the color images depicted on the screen are to be printed as true to the original as possible. Several image processing programs, like, for example, Adobe PhotoshopTM, support the user with a selection of a color gamut for the printing process.

3.1.4 MacAdam Ellipses

Since only brightness information is lost by the two-dimensional representation of colors, the chromaticity can be extracted from a chromaticity diagram by hue and saturation. A chromaticity can be described by specifying the wavelength of this spectral color and the ratio of the mixture. This is possible because the mixtures of two chromaticities of differing weighting lie on a line that connects these chromaticities. Also, according to Helmholtz, a color can be described as an additive mixture from a spectral color and an achromate (E) (with the exception of

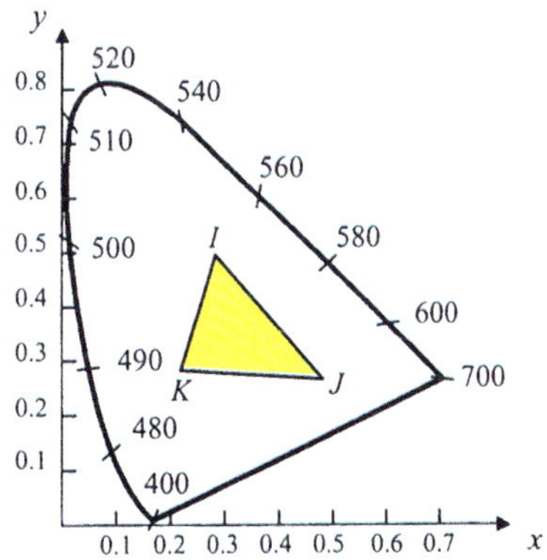

Figure 3.5. *By mixing the colors I and J, all colors can be produced on line IJ. By mixing the colors I, J, and K, all colors within the triangle IJK can be produced (after [Fol et al 94]).*

purple colors). These are the *dominant wavelength* λ_d , used for the description of hue, and the *excitation purity* p_e, for saturation.

The dominant wavelength of a chromaticity F can be constructed as point of intersection S of a ray arising in W and running through F with the spectral color transmission. With colors that do not have such a point of intersection, the purple boundary intersects them. The resulting wavelength is called the *complementary wavelength* λ_c . The spectral color component is defined as WF:WS.

Note that there exist differences of chromaticities, which are not perceived as equal, but are, however, equidistant in the chromaticity diagram. For example, more small color nuances can be differentiated in the green domain than in the blue domain. Examinations by MacAdam have revealed large discrepancies in the colors perceived as equal by the human visual system for differing reference color stimuli. From this, MacAdam obtained the distributions for 25 stimuli, which appear in the chromaticity diagram as ellipses (*MacAdam ellipses*) (see Fig. 3.6). The relationship of the area between the smallest and the largest ellipse amounts to approximately 0.94:69.4.

3.2 PHYSICS AND TECHNICS-BASED COLOR SPACES

Spaces that have a direct technical reference are presented below. As with the *RGB* color space, overlapping with physiological color spaces also occurs.

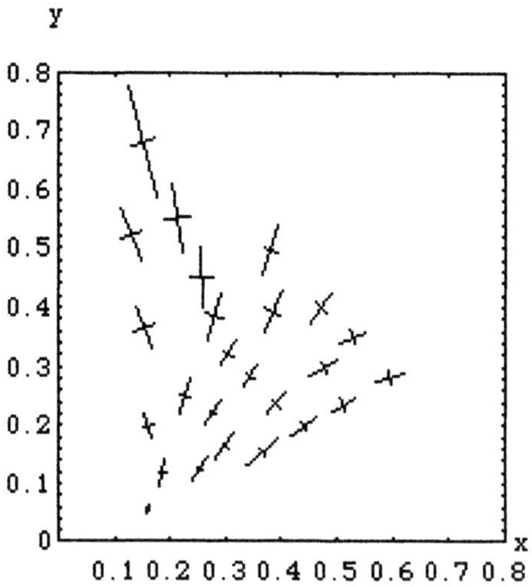

Figure 3.6. *In this chromaticity diagram for the 2° CIE standard observer, the MacAdam ellipses are represented by their two half-axes. The half-axes are scaled by a factor of 10. The data was adopted from [WysSti82] for this representation.*

3.2.1 *RGB* Color Spaces

The most commonly employed color space in computer technology is the *RGB color space*, which is based on the additive mixture of three primary colors **R**, **G**, and **B** (see Section 3.1). The international standardized wavelengths of the primary colors **R**ed, **G**reen, and **B**lue were already given in Table 3.1. It should be noted that the terms *red*, *green*, and *blue* were introduced solely for the purpose of standardization to provide descriptions for the primary colors. Visible colors and wavelengths are not equivalent. In order to avoid possible confusion, the notations *L*, *M*, *S* may be used for light containing long, middle, and short wavelengths instead of the notations *R*, *G*, *B*. However, the usual notations are *R*, *G*, and *B* and they will be used in the following.

The primary colors are for the most part the "reference colors" of the imaging sensors. They form the base vectors of a three-dimensional orthogonal (color)-vector space, where the zero-vector represents black (see Fig. 3.7). The origin is also described as *black point*. Any color can therefore be viewed as a linear combination of the base vectors in the *RGB* space. In one such accepted *RGB* color space, a color image is mathematically treated as a vector function with three components. The three vector components are determined by the measured intensities of visible light in the long-wave, middle-wave, and short-wave area. For a (three-channel) *digital color image* **C**, three vector components *R*, *G*, *B* are to be indicated for each image pixel (*x*,*y*):

$$\mathbf{C}(x,y) = \left(R(x,y), G(x,y), B(x,y)\right)^T = (R,G,B)^T.$$

These values are referred to as *tristimulus values*. The colors that are represented by explicit value combinations of the vector components *R*, *G*, *B*, are relative, device-dependent entities. All vectors $(R,G,B)^T$ with integer components $0 \le R,G,B \le G_{\max}$ characterize one color in the *RGB* color space. $G_{\max}+1$ indicates the largest permitted value in each vector component. Using permeable filters in the generation of a color image in the *RGB* color space, so-called red, green, and blue extracts are generated in the long wave, middle wave and short wave area of visible light. If one refrains from using the filter, each of the three scannings is identical with the digitalization of a gray-level image. The rational numbers

$$r = \frac{R}{R+G+B}, \quad g = \frac{G}{R+G+B}, \text{ and } b = \frac{B}{R+G+B} \tag{3.1}$$

are the color value components that are normalized with respect to the intensity (see Section 3.1.3).

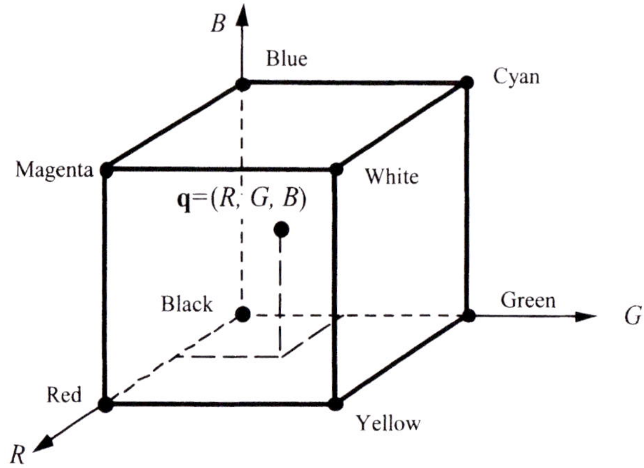

Figure 3.7. *In the RGB color space, every vector* $\mathbf{q} = (R,G,B)^T$ *inside the color cube represents exactly one color, where* $0 \le R,G,B \le G_{\max}$ *and R, G, B are integers.*

The primary colors red $(G_{\max},0,0)^T$, green $(0,G_{\max},0)^T$, blue $(0,0,G_{\max})^T$, and the complementary colors yellow $(G_{\max},G_{\max},0)^T$, magenta $(G_{\max},0,G_{\max})^T$, cyan $(0,G_{\max},G_{\max})^T$, as well as the achromatic colors white $(G_{\max},G_{\max},G_{\max})^T$ and black $(0,0,0)^T$, represent the boundaries of the color cube, which is formed through the possible value combinations of R, G, B. All color vectors $(R,G,B)^T$ with $0 \le R,G,B \le G_{\max}$ each characterize a color in the *RGB* color space. This color cube is represented in Fig. 3.7. All achromatic colors (gray tones) lie on the principal diagonal $(u,u,u)^T$, with $0 \le u \le G_{\max}$.

The *RGB* color space is the most applied computer-internal representation of color images. Its wide distribution is, among other things, traced back to the well-standardized three primary colors. Almost all visible colors can be represented by a linear combination of the three vectors (see Section 3.1.3). For identical objects, differing color values are generated with different cameras or scanners since their primary colors in general do not match. The process of adjusting color values between different devices (e.g., camera*RGB*, monitor*RGB*, and printer*RGB*) is called *color management*. See [Har01] and [GioMad98] for an introduction to color management.

A special case of the *RGB* color space is the primary color system $R_N G_N B_N$ for television receivers (*receiver primary color system*), which refers to the established phosphors in the American standard NTSC (National Television System Committee). Values deviating from this are valid for the phosphors through the television standards PAL (Phase Alternation Line) and SECAM (Séquentiel Couleur à Mémoire). The *RGB* color space, which was determined by the CIE, is transformed [Pra91] into the NTSC-primary color system $R_N G_N B_N$ through

$$\begin{pmatrix} R_N \\ G_N \\ B_N \end{pmatrix} = \begin{pmatrix} 0.842 & 0.156 & 0.091 \\ -0.129 & 1.320 & -0.203 \\ 0.008 & -0.069 & 0.897 \end{pmatrix} \cdot \begin{pmatrix} R \\ G \\ B \end{pmatrix} \quad (3.2)$$

In 1996, the International Color Consortium (ICC) proposed a standard color space *sRGB* for the Internet [Sto et al. 96]. The standard considers cathode ray tube (CRT) monitors and D_{65} daylight. The reference viewing environment parameters can be found in [Sto et al. 96]. *sRGB* tristimulus values are simply linear combinations of the CIE *XYZ* values. They are also known as Rec. 709 *RGB* values and they can be computed using the following relationship:

$$\begin{pmatrix} R_{sRGB} \\ G_{sRGB} \\ B_{sRGB} \end{pmatrix} = \begin{pmatrix} 3.2410 & -1.5374 & -0.4986 \\ -0.9692 & 1.8760 & 0.0416 \\ 0.05556 & -0.2040 & 1.0570 \end{pmatrix} \cdot \begin{pmatrix} X \\ Y \\ Z \end{pmatrix} \quad (3.3)$$

In a second processing step (which modifies the value of the gamma correction factor used in the ITU-R BT.709 standard), these values are transformed to nonlinear s*R'G'B'* values as follows:

If $R_{sRGB}, G_{sRGB}, B_{sRGB} \leq 0.0034$ **then**

$$sR' = 12.92 R_{sRGB}$$
$$sG' = 12.92 G_{sRGB}$$
$$sB' = 12.92 B_{sRGB}$$

else if $R_{sRGB}, G_{sRGB}, B_{sRGB} > 0.0034$ **then**

$$sR' = 1.055 R_{sRGB}^{1.0/2.4} - 0.055$$
$$sG' = 1.055 G_{sRGB}^{1.0/2.4} - 0.055\,.$$
$$sB' = 1.055 B_{sRGB}^{1.0/2.4} - 0.055$$

end {if}

The effect of the above equations is to closely fit a straightforward gamma 2.2 curve with a slight offset to allow for invertability in integer math [Sto et al. 96]. Finally, the nonlinear *sR'G'B'* values are converted to digital code values by

$$
\begin{aligned}
R &= G_{\max} \cdot sR' \\
G &= G_{\max} \cdot sG' \quad . \\
B &= G_{\max} \cdot sB'
\end{aligned}
$$

This transformation proposes using a black digital count of 0 and a white digital count of 255 for 24-bit (8-bits/channel) encoding (which is different from digital broadcast television using a black digital count of 16 and a white digital count of 235 in order to provide a larger encoded color gamut).

Note that the Internet standard *sRGB* is made for CRT monitors under daylight viewing conditions. However, many people work under nondaylight conditions and/or they use notebooks with liquid crystal displays (LCDs). Recently, several modifications of the *sRGB* are proposed for extended color gamuts (e.g., e-sRGB, scRGB, ROMM RGB, RIMM RGB). For an overview on *RGB* color spaces see, for example, [Süs et al. 99] and [Spa et al. 00].

3.2.2 *CMY(K)* Color Space

In the printout of a color image with a color printer, the *CMY(K)* color space, with the subtractive primary colors cyan, magenta, and yellow as well as possibly an additional black (*K*arbon), is used. A subtractive color space forms the basis for the printing process. Since cyan, magenta, and yellow are the complementary colors to red, green, and blue, the *RGB* color space and the *CMY* color space can be transferred through

$$
\begin{pmatrix} R \\ G \\ B \end{pmatrix} = \begin{pmatrix} G_{\max} \\ G_{\max} \\ G_{\max} \end{pmatrix} - \begin{pmatrix} C \\ M \\ Y \end{pmatrix} \text{ and } \begin{pmatrix} C \\ M \\ Y \end{pmatrix} = \begin{pmatrix} G_{\max} \\ G_{\max} \\ G_{\max} \end{pmatrix} - \begin{pmatrix} R \\ G \\ B \end{pmatrix} \tag{3.4}
$$

into one another, where $G_{\max} + 1$ again denotes the greatest representable value in each color channel. As in the *RGB* color space, a color cube represents the *CMY* color space (see Fig. 3.8).

Frequently, additional black ink is used in printing. There are essentially three reasons for this: First, the printing, one on top of the other, of the colors cyan, magenta, and yellow leads to a larger emission of fluid and with that come longer drying times. Second, it can occur, on the basis of mechanical tolerances, that the three color inks do not print exactly on the same position. Because of this, colored edges develop along black borders. Third, the black ink is less expensive to manufacture than the three color inks. Toward the end of the *CMY* color space, black is used in place of the same portions of *C*, *M*, and *Y* in accordance with the relations

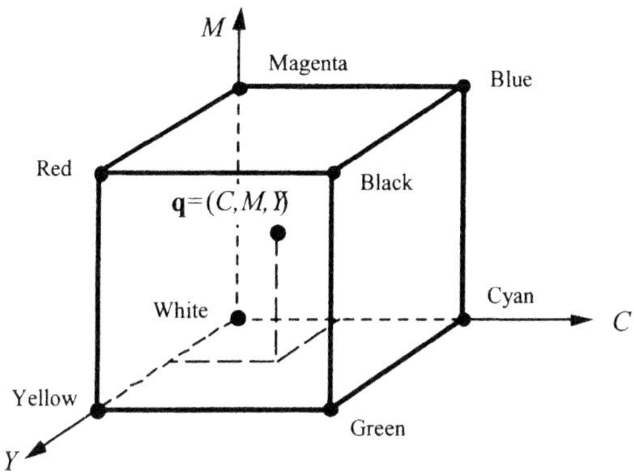

Figure 3.8. *The CMY color space for color printing.*

$$K = min(C, M, Y),$$
$$C = C - K,$$
$$M = M - K, \text{ and}$$
$$Y = Y - K.$$

In the *CMY* color space it is assumed that filters with nonoverlapping spectral absorption curves exist for the three subtractive primary colors. In practice, the dyes and printer inks used in the printing technology frequently have absorption curves that severely overlap. Through this, nonlinear dependencies between the printing colors develop in the color mixture. In addition, the printed result is influenced by the various representations of the printed color depending on the number of ink dots per inch and the composition of the paper as well as the material to be printed [Poy97]. These complex correlations will not be covered further at this point.

3.2.3 *YIQ* Color Space

In the development of the NTSC television system used in the United States, a color coordinate system with the coordinates *Y*, *I*, and *Q* was defined for transmission purposes. To transmit a color signal efficiently, the $R_N G_N B_N$ signal was more conveniently coded from a linear transformation. The luminance signal is coded in the *Y*-component. The additional portions *I* (*in-phase*) and *Q* (*quadrature*) contain the entire chromaticity information that is also denoted as *chrominance signal* in television technology.

I and *Q* are transmitted by a much shorter waveband since the *Y* signal contains by far the largest part of the information. The *Y* signal contains no color information so that the *YIQ* system remains compatible with the black-white system. By using only the Y signal in a black-and-white television, gray-level images can be displayed, which would not be possible by a direct transmission of the $R_N G_N B_N$ signal.

The values in the $R_N G_N B_N$ color space can be transformed with

$$\begin{pmatrix} Y \\ I \\ Q \end{pmatrix} = \begin{pmatrix} 0.299 & 0.587 & 0.114 \\ 0.596 & -0.274 & -0.322 \\ 0.211 & -0.523 & 0.312 \end{pmatrix} \cdot \begin{pmatrix} R_N \\ G_N \\ B_N \end{pmatrix} \tag{3.5}$$

into the values in the YIQ color space.

3.2.4 *YUV* Color Space

The color television systems PAL and SECAM, developed in Germany and France, use the YUV color space for transmission. The Y component is identical with the one of the YIQ color space. The values in the $R_N G_N B_N$ color space can be transformed with

$$\begin{pmatrix} Y \\ U \\ V \end{pmatrix} = \begin{pmatrix} 0.299 & 0.587 & 0.114 \\ -0.418 & -0.289 & 0.437 \\ 0.615 & -0.515 & -0.100 \end{pmatrix} \cdot \begin{pmatrix} R_N \\ G_N \\ B_N \end{pmatrix} \tag{3.6}$$

into the values in the YUV color space [Pra91]. On account of the low information content, the U and V signals, which are usually related to the Y signal, are reduced by half (two successive image pixels each having a separate Y portion, but with a common color type) and by a quarter for simple demands.

The I and Q signals of the YIQ color space are determined from the U and V signals of the YUV color space by a simple rotation in the color coordinate system. The following applies [Pra91]:

$$I = -U \cdot \sin(33°) + V \cdot \cos(33°),$$

$$Q = U \cdot \cos(33°) + V \cdot \sin(33°).$$

Presentations in the YIQ and YUV color space are very suitable for image compression since luminance and chrominance can be coded with different numbers of bits, which is not possible when using RGB values.

In the literature, YUV also indicates a color space, in which U corresponds to the color difference red-blue and V to the color difference green-magenta. Y corresponds to the equally weighted (arithmetical) averages of red, green, and blue. This color space is, for example, employed in highlight analysis of color images (see Section 8.1.4). We will denote this color space with $(YUV)'$ for a better distinction. A linear correlation exists between the $(YUV)'$ color space and the RGB system, which is given by the transformation

$$(Y,U,V)' = (R,G,B)\begin{pmatrix} \frac{1}{3} & \frac{1}{2} & \frac{-1}{2\sqrt{3}} \\ \frac{1}{3} & 0 & \frac{1}{\sqrt{3}} \\ \frac{1}{3} & \frac{-1}{2} & \frac{-1}{2\sqrt{3}} \end{pmatrix}. \tag{3.7}$$

Brightness normalization can be defined by

$$u = \frac{U}{R+G+B} \quad \text{and} \quad v = \frac{V}{R+G+B}. \tag{3.8}$$

If u and v form the axes of a Cartesian coordinate system, then red, green, and blue stretch an equilateral triangle in which black lies in the origin (see Fig. 3.9).

3.2.5 YC_BC_R Color Space

In the area of digital video, which is increasingly gaining importance, the internationally standardized YC_BC_R color space is employed for the representation of color vectors. This color space differs from the color space used in analog video recording, which will not be gone into at this point. A detailed description of digital and analog video technology is included in [Poy96]. The values in the $R_NG_NB_N$ color space can be transformed into the values in the YC_BC_R color space [Poy96]:

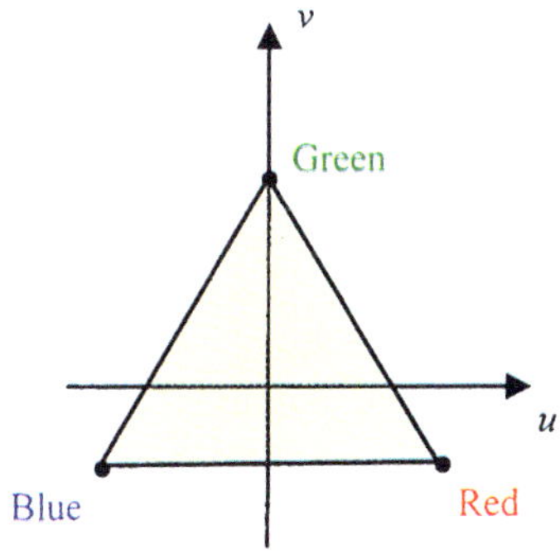

Figure 3.9. *The uv-plane of the $(YUV)'$-model.*

$$\begin{pmatrix} Y \\ C_B \\ C_R \end{pmatrix} = \begin{pmatrix} 16 \\ 128 \\ 128 \end{pmatrix} + \frac{1}{256} \begin{pmatrix} 65.738 & 129.057 & 25.064 \\ -37.945 & -74.494 & 112.439 \\ 112.439 & -94.154 & -18.285 \end{pmatrix} \cdot \begin{pmatrix} R_N \\ G_N \\ B_N \end{pmatrix}$$

From this transformation it is assumed that the *RGB* data has already undergone a gamma correction. The quantities for the *Y* components refer to the fixed values for phosphors in the reference Rec. ITU-R BT.601-4 of the NTSC system. The back transformation from the YC_BC_R color space into the $R_NG_NB_N$ color space is (except for a few rounding errors) given by

$$\begin{pmatrix} R_N \\ G_N \\ B_N \end{pmatrix} = \frac{1}{256} \begin{pmatrix} 298.082 & 0.0 & 408.583 \\ 298.082 & -100.291 & -208.120 \\ 298.082 & 516.411 & 0.0 \end{pmatrix} \cdot \begin{pmatrix} Y-16 \\ C_B-128 \\ C_R-128 \end{pmatrix}$$

[Poy96]. The YC_BC_R color space was developed for representations in the television format common until now. It does not apply to the HDTV (*high definition television*) format.

3.2.6 Kodak PhotoCD YC_1C_2 Color Space

The Kodak Company developed the YC_1C_2 color space for their PhotoCD system. The YC_1C_2 color space is quite similar to the YC_BC_R color space. The differences are that the color gamut of the YC_1C_2 color space was adapted as closely as possible to the color gamut of photographic film, while the color gamut of the YC_BC_R color space orientates itself to the phosphors of the NTSC system. The values in the $R_NG_NB_N$ color space can be obtained from the PhotoCD YC_1C_2 color space with the equation

$$\begin{pmatrix} R_N \\ G_N \\ B_N \end{pmatrix} = 255 \cdot \begin{pmatrix} 0.00549804 & 0.0 & 0.0051681 \\ 0.00549804 & -0.0015446 & -0.0026325 \\ 0.00549804 & 0.0079533 & 0.0 \end{pmatrix} \cdot \begin{pmatrix} Y \\ C_1-156 \\ C_2-137 \end{pmatrix}$$

[Poy96]. It should be taken into consideration that on the PhotoCD, the C_1 and C_2 components are subsampled both horizontally and vertically by a factor of 2 for compression purposes. This is, however, an integral part of the PhotoCD system's compression and not an element of the YC_1C_2 color space.

3.2.7 $I_1I_2I_3$ Color Space

The $I_1I_2I_3$ color space depicts a feature model rather than a color space. In an examination of eight randomly chosen color images and eleven color spaces, Ohta, Kanade, and Sakai [Oht et al. 80] achieved the best segmenting results in color images by using the $I_1I_2I_3$ color space. Due to these good results this color space is frequently used in color image processing. The three components are defined by

$$I_1 = \frac{R+G+B}{3},\ I_2 = \frac{R-B}{2}\ \text{, and }\ I_3 = \frac{2G-R-B}{4}. \tag{3.9}$$

Note the similarity between the $I_1I_2I_3$ color space and the opponent color space.

3.3 UNIFORM COLOR SPACES

A *uniform color space* is a color space in which same-size changes in the color coordinates also correspond to same-size recognizable changes in the visible color tones and color saturation. This is not the case with physically and technically oriented color spaces, which were described in the previous section. In order to reduce such variations, several revised color spaces were defined. However, no color space exists that allows an undistorted or unbiased representation. Therefore, "uniform" means in this sense "more or less uniform." The international lighting commission CIE recommended the $L^*a^*b^*$ *color space CIE 1976* and the $L^*u^*v^*$ *color space CIE 1976* as an approximation of uniform color spaces. The German institute of standardization DIN (Deutsches Institut für Normierung) also adopted these color spaces. These are usually abbreviated as CIELAB and CIELUV.

Both color spaces are derived from the *XYZ* standard color system (*CIE XYZ primary system*). The calibration of the measured camera data in the *XYZ* standard color space presents a complex problem, which is the subject of Section 4.5.4. In the following section, the CIELAB color space and the CIELUV color space are described. Uniform color spaces are interesting for several areas of color image processing applications, especially if very similar colors have to be compared. Both these color spaces are device-independent color spaces. However, they are computationally intensive to transform to other color spaces. This is a drawback if video real-time processing is required, that is 24 frames per second.

3.3.1 CIELAB Color Space

The CIELAB color space was developed in order to obtain an easy-to-calculate color measurement that is in accord with the Munsell color order system (see Section 3.6.1). The following conversion is valid for $X/X_n, Y/Y_n, Z/Z_n >$

0.008856. It is further expanded for $X/X_n, Y/Y_n, Z/Z_n \leq 0.008856$ (see CIE, DIN 5033, and DIN 6174). The following applies:

$$\begin{aligned} X^* &= \sqrt[3]{X/X_n} && \text{for } X/X_n > 0.008856 \\ X^* &= 7.787 \cdot (X/X_n) + 0.138 && \text{for } X/X_n \leq 0.008856 \end{aligned} \tag{3.10a}$$

$$\begin{aligned} Y^* &= \sqrt[3]{Y/Y_n} && \text{for } Y/Y_n > 0.008856 \\ Y^* &= 7.787 \cdot (Y/Y_n) + 0.138 && \text{for } Y/Y_n \leq 0.008856 \end{aligned} \tag{3.10b}$$

$$\begin{aligned} Z^* &= \sqrt[3]{Z/Z_n} && \text{for } Z/Z_n > 0.008856 \\ Z^* &= 7.787 \cdot (Z/Z_n) + 0.138 && \text{for } Z/Z_n \leq 0.008856 \end{aligned} \tag{3.10c}$$

$$\begin{aligned} L^* &= 116 \cdot Y^* - 16, \\ a^* &= 500 \cdot (X^* - Y^*), \\ b^* &= 200 \cdot (Y^* - Z^*). \end{aligned} \tag{3.11}$$

On this occasion (X_n, Y_n, Z_n) describes the white reference point, which represents a perfectly matte white body under an illuminant (e.g., *D*65, *A* or *C*; see Section 4.2.2). *X*, *Y*, and *Z* are the values in the standard color system (see Section 3.1.2). L^* is also indicated as *L*-lightness* (pronounced *L-star-lightness*) and is scaled approximately equidistant according to sensitivity. The functional relations between *Y* and L^* are represented for $Y_n = 100$ in Fig. 3.10.

It should be understood that equally distant differences between *Y* pairs are more strongly noticeable in the dark area than in the lighter area. In contrast to the

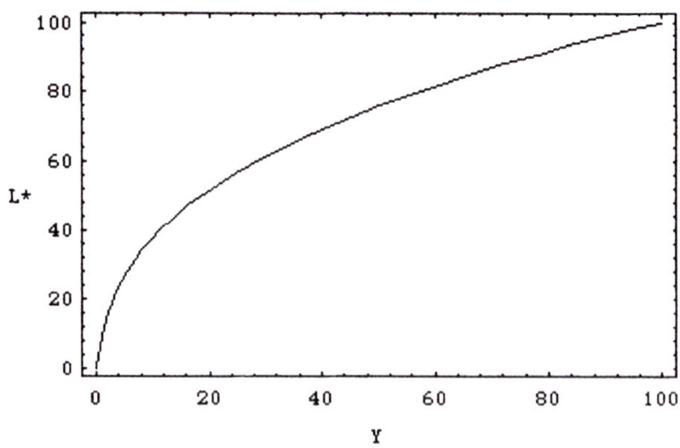

Figure 3.10. *L^*-lightness for* $Y_n = 100$.

standard CIE chromaticity diagram and the (u', v') chromaticity diagram, chromaticities cannot be clearly assigned to the chromaticity points, the (a^*, b^*) chromaticity diagrams, which belong to the CIELAB color space since their position depends on L^*. By constant L^*, chromaticity points that lie on a straight line in the standard color table become curved.

For the improvement of the uniformity of the color appearance, modifications to the CIELAB color space were continually developed. A recent example of this is called the LLAB color space, the description of which is referred to in [Luo96]. Furthermore, several approximations of the CIELAB color space were suggested in order to accelerate the computationally intensive conversion of data in the *RGB* color space into the CIELAB color space [ConFli97]. Especially mathematically efficient is the single calculation of a number of $L^*a^*b^*$ values and their storage in a lookup table. If all $L^*a^*b^*$ values for all possible *RGB* values were stored, then the lookup table, with a quantization of eight bits, would have to contain about 16 million entries. However, the number of entries in the table can be considerably reduced by an approximation of the $L^*a^*b^*$ values. The maximum error between the correct values and the approximate values that arises from this can be determined from Eq. (3.17) described later in Section 3.5.3. In a lookup table with 2000 real-valued entries, the error amounts to $\Delta E^*_{ab} = 1.96$ units, and with 8000 entries, it decreases to $\Delta E^*_{ab} = 0.6$ units. The CIELAB color space (see Fig. 3.11) is used for the description of body colors (for nonluminous materials), whereas the CIELUV color space described in the following (see Fig. 3.12) is usually used for the description of colors of light.

3.3.2 CIELUV Color Space

The chromaticity diagram of the CIELUV color space is indicated as the *CIE 1976 uniform chromaticity scale diagram* (or *CIE 1976 UCS diagram*). In accordance with DIN 5033, the term (u', v')-space is also used. The CIELUV color space is defined using the symbols taken from the previous section:

$$\begin{aligned} L^* &= 116 \cdot \sqrt[3]{Y/Y_n} - 16 \quad &\text{for } Y/Y_n > 0.008856 \\ X^* &= 903.3 \cdot (Y/Y_n) \quad &\text{for } Y/Y_n \leq 0.008856 \end{aligned} \tag{3.12a}$$

$$\begin{aligned} u^* &= 13 \cdot L^* \cdot (u' - u'_n)\,, \\ v^* &= 13 \cdot L^* \cdot (v' - v'_n) \end{aligned} \tag{3.12b}$$

with

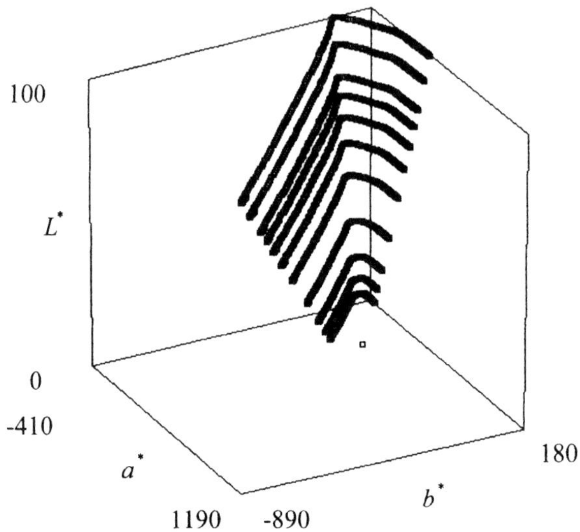

Figure 3.11. *Illustration of the 2° spectral color transmission for Y = 0, 2, 3, 5, 10, 20, 30, 40, 50, 60, 80, and 100 in the CIELAB color space. A point results for Y = 0 (black). Illuminant C is used as white reference.*

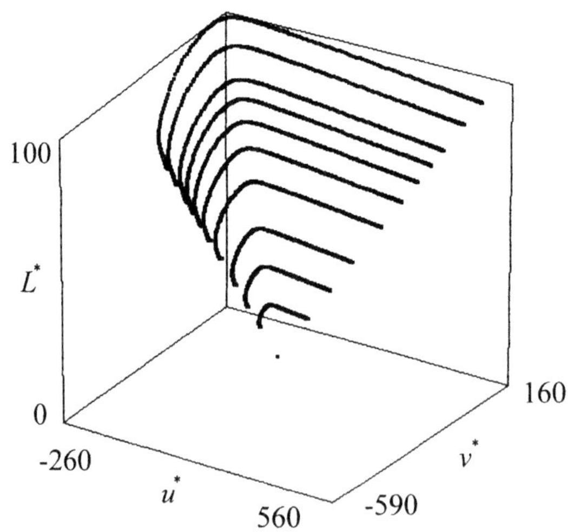

Figure 3.12. *Depiction of the 2° spectral color transmission for Y = 0, 2, 3, 5, 10, 20, 30, 40, 50, 60, 80 and 100 in the CIELUV color space. For Y = 0 (black) a point is shown. As white reference the standard illuminant C was assumed.*

$$u' = \frac{4 \cdot X}{X + 15 \cdot Y + 3 \cdot Z} = \frac{4 \cdot x}{3 - 2 \cdot x + 12 \cdot y},$$

$$u'_n = \frac{4 \cdot X_n}{X_n + 15 \cdot Y_n + 3 \cdot Z_n} = \frac{4 \cdot x_n}{3 - 2 \cdot x_n + 12 \cdot y_n},$$

and

$$v' = \frac{9 \cdot Y}{X + 15 \cdot Y + 3 \cdot Z} = \frac{9 \cdot y}{3 - 2 \cdot x + 12 \cdot y},$$

$$v'_n = \frac{9 \cdot Y_n}{X_n + 15 \cdot Y_n + 3 \cdot Z_n} = \frac{9 \cdot y_n}{3 - 2 \cdot x_n + 12 \cdot y_n}.$$

The L^*-component is identical to the L^*-component in the CIELAB color space. In contrast to the CIELAB color space, straight lines are again mapped onto straight lines in the CIELUV color space, which is a great advantage to additive mixture calculations.

The sizes of wavelengths with the same hue and the sizes that correspond to the spectral color proportion are indicated as *hue angles* h^*_{ab}, h^*_{uv} and *chroma* C^*_{ab}, C^*_{uv}. The following applies:

$$h^*_{ab} = \arctan(b^* / a^*), \qquad h^*_{uv} = \arctan(v^* / u^*),$$
$$C^*_{ab} = \sqrt{a^* \cdot a^* + b^* \cdot b^*}, \quad C^*_{uv} = \sqrt{u^* \cdot u^* + v^* \cdot v^*}. \tag{3.13}$$

For the CIELUV color space the relation of multicoloredness to brightness is defined as a measure of saturation by

$$S^*_{uv} = C^*_{uv} / L^* = 13\sqrt{(u'-u'_n) \cdot (u'-u'_n) + (v'-v'_n) \cdot (v'-v'_n)}. \tag{3.14}$$

3.4 PERCEPTION-BASED COLOR SPACES

Color spaces that are based intuitively on human color perception are of interest for the fields of computer vision and computer graphics. With the *HSI* and the *HSV* color spaces, the wish for user-friendly input and description of color values is in the foreground. A color can be more easily described intuitively (above all by untrained users) by values for hue, color saturation, and intensity than from vector components in the *RGB* or *CMYK* color space.

3.4.1 *HSI* Color Space

In the *HSI* color space *hue*, *saturation*, and *intensity* are used as coordinate axes. Fig. 3.13 shows a possible representation of the *HSI* color space. This color space is well suited for the processing of color images and for visually defining interpretable local characteristics. A color $\mathbf{q} = (R, G, B)^T$ is given in the *RGB* color space. The hue H of the color $\mathbf{q}$ characterizes the dominant color contained in $\mathbf{q}$. Red is specified as a "reference color." Because of that, $H = 0°$ and $H = 360°$ correspond to the color red. Formally, H is given by

$$H = \begin{cases} \delta & \text{if } B \leq G \\ 360° - \delta & \text{if } B > G \end{cases} \tag{3.15a}$$

with

$$\delta = \arccos\left(\frac{(R-G)+(R-B)}{2\sqrt{(R-G)^2 + (R-B)\cdot(G-B}} \right).$$

The saturation S of the color $\mathbf{q}$ is a measurement of color purity. This parameter is dependent on the number of wavelengths that contribute to the color perception. The wider the range of the wavelengths, the lower the purity of the color. The more narrow the range of the wavelengths, the higher the purity of the color. The extreme case $S = 1$ is true for a pure color and the extreme case $S = 0$ for an achromatic color. S is given by

$$S = 1 - 3 \cdot \frac{\min\,(R, G, B)}{R + G + B}. \tag{3.15b}$$

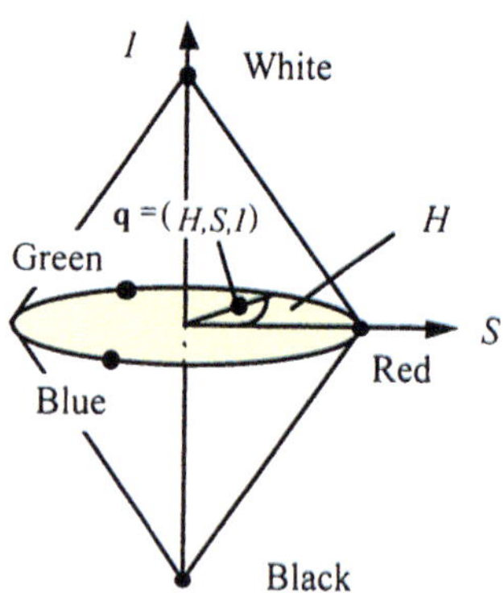

Figure 3.13. *The HSI-color space.*

The *intensity* I of the color $\mathbf{q}$ corresponds to the relative brightness (in the sense of a gray-level image). The extreme case $I = 0$ corresponds to the color black. The intensity is defined in accordance with

$$I = \frac{R+G+B}{3}. \tag{3.15c}$$

For the color $\mathbf{q} = (R, G, B)^T$ in the *RGB* color space, a representation $(H, S, I)^T$ of this color is given in the *HSI* color space. This conversion is clearly reversible (except for inaccuracies in rounding and some singularities). The back transformation is given in Fig. 3.14.

---

if $(S = 0)$ **then** {gray tone}
    $R = G = B = I$;
else
    if $(0 \le H \le 120)$ **then** {B is minimum}
        $B = (1 - S) \cdot I$;
        $H = \frac{1}{\sqrt{3}} \cdot \tan(H - 60)$;
        $G = (1.5 + 1.5 \cdot H) \cdot I - (0.5 + 1.5 \cdot H) \cdot B$;
        $R = 3 \cdot I - G - B$
    else
      if $(120 \le H \le 240)$ **then** {R is minimum}
          $R = (1 - S) \cdot I$;
          $H = \frac{1}{\sqrt{3}} \cdot \tan(H - 180)$;
          $B = (1.5 + 1.5 \cdot H) \cdot I - (0.5 + 1.5 \cdot H) \cdot R$;
          $G = 3 \cdot I - B - R$
      else {G ist minimum}
          $G = (1 - S) \cdot I$;
          $H = \frac{1}{\sqrt{3}} \cdot \tan(H - 300)$;
          $R = (1.5 + 1.5 \cdot H) \cdot I - (0.5 + 1.5 \cdot H) \cdot G$;
          $B = 3 \cdot I - R - G$
        end {if}
    end {if}
end {if}

Figure 3.14. *Conversion of color images from the HSI representation into a RGB representation (according to [Fre88]).*

One of the advantages of the *HSI* color space is the separation of chromatic and achromatic information. The existence of singularities is a disadvantage for the *HSI* color space. Furthermore, it must be observed that the information content and the reliability of the calculation of hue and saturation depend on the luminosity (see [Fre88] and [Fre90]). In achromatic colors, neither hue nor saturation is defined. The characteristic nonlinearity of the cameras in general can affect the *HSI* conversion unfavorably (see Section 4.4.1).

Transformations between the color spaces can be significantly accelerated when using hardware. Image processing boards are available for PCs and workstations that transfer a video image (in NTSC or PAL format) or an *RGB* image into an *HSI* image in real time. The back transformation of *HSI* into the *RGB* color space can be derived from Eq. (3.15abc). The algorithm is given in pseudocode in Fig. 3.14.

3.4.2 *HSV* Color Space

The *HSV* color space, which is also called the *HSB* color space, is particularly common in the field of computer graphics. As in the *HSI* color space, hue, saturation, and brightness value are used as coordinate axes. By projecting the *RGB* unit cube along the diagonals of white to black, a hexacone results that forms the topside of the *HSV* pyramid. The hue H is indicated as an angle around the vertical axle. As in the *HSI* color space, red is determined with $H = 0°$ or $H = 360°$, green with $H = 120°$, and so on (see Fig. 3.15).

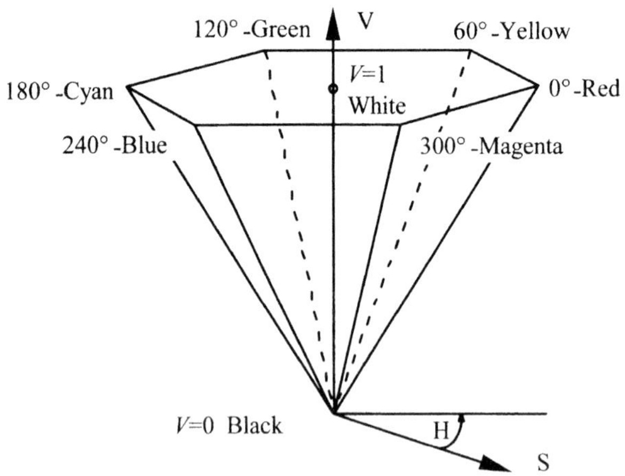

Figure 3.15. *Hexacone representation of the HSV color space.*

The saturation S is a number between 0 on the central axis (the V-axis) and 1 on the sides of the pyramid. For $S = 0$, H is undefined. The brightness value V (or B) lies between 0 on the apex of the pyramid and 1 on the base. The point on the apex of the pyramid with $V = 0$ is black. At this point, the values of H and S have no significance. The lightest colors lie on the topside of the pyramid; however, not all colors with the same brightness are visible on the plane $V = 1$. The pseudocode for the conversion of a color image from the RGB color space into the HSV color space is indicated in Fig. 3.16, where again $G_{\max} + 1$ denotes the largest possible value in each color channel. A representation of the back transformation is not given here but can be found in [Fol et al. 94]. Some image processing programs (e.g., Adobe PhotoshopTM) contain modules for transforming images between the RGB and the HSV representation (there called HSB).

```
max = Max(R,G,B);
min = Min(R,G,B);
V = max / Gmax ;                                   {brightness value}
if (max ≠ 0) then S = (max − min) / max            {saturation value}
             else S = 0 ;
                H = UNDEFINED
end {if};
if (S > 0) then
   DR = (max − R)/(max − min);
   DG = (max − G)/(max − min);
   DB = (max − B)/(max − min);
   if (max = R) then if (min = G) then H = 5 + DB       {color between
                     else H = 1 − DG                magenta and yellow}
   else if (max = G) then if (min = B) then H = 1 + DR   {color between
                          else H = 3 − DB              yellow and cyan}
   else if (min = R) then H = 3 + DG                   {color between
                     else H = 5 − DR ;                      cyan and
   end {if}                                                 magenta}
   if (H < 6) then H = H · 60
              else H = 0 ;
end {if}
```

Figure 3.16. *Conversion of color images from the RGB representation into an HSV representation.*

With both the *HSV* and the *HSI* color spaces described in the previous paragraph, there exists the problem, apart from the singularities in the color space already specified, that a straight line in the *RGB* space is not generally mapped onto a straight line in the two other color models. Here in particular it is to be noted in the cases of interpolations in the color spaces and transformations between the color spaces. An advantage of the *HSV* color space lies in the fact that it intuitively corresponds to the color system of a painter when mixing the colors and its operation is very easy to learn. In digital color image processing, the *HSV* color space is of only secondary importance. It is used for the easily operated manipulation of a color image's color values (e.g., with Adobe PhotoshopTM).

3.4.3 Opponent Color Spaces

In the opponent color theory or four-color theory, antagonistic neural processes, with the opponent colors red-green and blue-yellow and likewise an antagonistically organized light-dark system, are named as the basis for human color vision (see Section 2.3). Fig. 3.17 shows one presentation of the opponent color space. The receptive field of an opponent color cell is divided into a center system and a peripheral system (see Section 2.3). In color image processing, both systems are also modeled by two-dimensional, rotationally symmetrical Gaussian functions. The opponent color space is used, for example, in color stereo analysis [BroYan89] (see Chapter 9), in color image segmentation [Hol82] (see Chapter 7), and in an approximation of color constancy (see Chapter 8).

3.5 COLOR DIFFERENCE FORMULAS

In virtually all areas of color image processing it is necessary to compare colors within an image or between several images. The difference between two colors in a color space must be determined and described. In the following, some measurements of color difference are given as examples for the *RGB* color space, the *HSI* color space, the CIELAB color space, and the CIELUV color space.

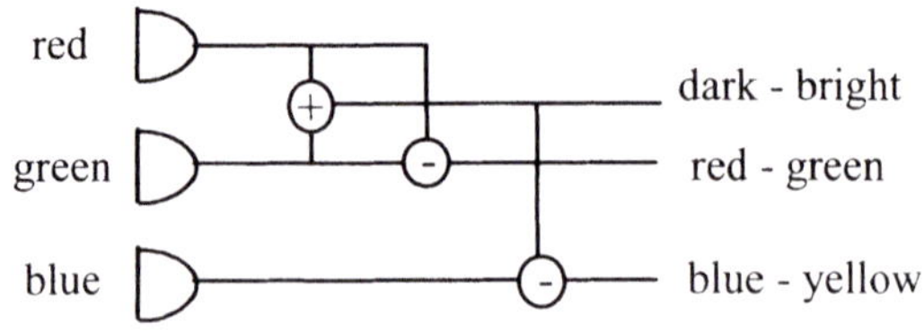

Figure 3.17. *A representation of the opponent color space (after [BroYan89]).*

3.5.1 Color Difference Formulas in the *RGB* Color Space

If the *RGB* space is regarded as Euclidean space, then known distance measurements can be adopted for the calculation of a color difference. A metric (e.g., the Euclidean metric) is to be selected. The color distance can be used in accordance with Eq. (3.1) for the color value quantities, standardized by intensity, as well as for the nonstandardized values. The color distance measurement can be used both for the color proportional values $(r,g,b)^T$ standardized in accordance with Eq. (3.1) by intensity, and for the nonstandardized values. In the *RGB* color space, the Euclidean distance between two color vectors $\mathbf{F}_1$ and $\mathbf{F}_2$ represents the angle between the two vectors (see Fig. 3.18). These distance measurements are widely used within the area of color image processing on account of their simple predictability. However, there is no connection between these distance measurements and human color perception.

3.5.2 Color Difference Formulas in the *HSI* Color Space

The formulas valid in the *RGB* color space are not suitable for a difference measurement in the *HSI* color space. For example, the color difference Δ_{HSI} can be used there [TseCha92]. For two color values $\mathbf{F}_1 = (H_1, S_1, I_1)^T$ and $\mathbf{F}_2 = (H_2, S_2, I_2)^T$ in the *HSI* color space, the color difference between them is determined by

$$\Delta_{HSI}(\mathbf{F}_1, \mathbf{F}_2) = \sqrt{(\Delta_I)^2 + (\Delta_C)^2} \tag{3.16}$$

with

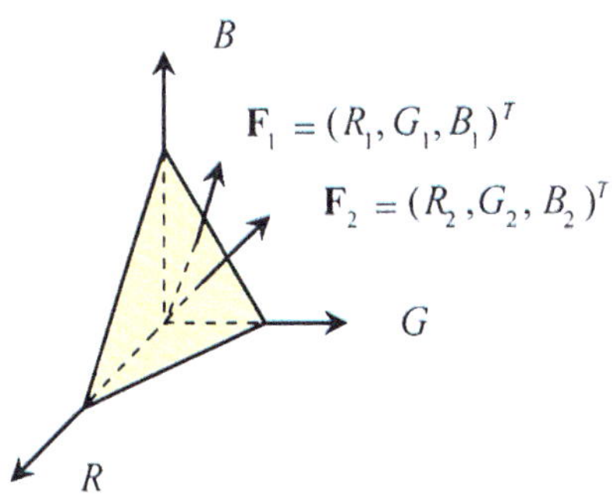

Figure 3.18. *Representation of two color vectors $\mathbf{F}_1$ and $\mathbf{F}_2$ in the RGB space.*

$$\Delta_I = |I_1 - I_2| \text{ and } \Delta_C = \sqrt{S_1^2 + S_2^2 - 2S_1S_2\cos\theta}\ ,$$

whereby applies

$$\theta = \begin{cases} |H_1 - H_2| & \text{if } |H_1 - H_2| \le \pi \\ 2\pi - |H_1 - H_2| & \text{if } |H_1 - H_2| > \pi \end{cases}.$$

3.5.3 Color Difference Formulas in the CIELAB and CIELUV Color Spaces

According to Schrödinger [Sch20], it is generally accepted that the area of color appearance possesses a Riemann metric. The uniform color spaces represented can be used for this, using the Riemann metric as an adequate aid due to the approximately equal distance. However, this is reasonably possible only if the color-mapping behavior in the image-formation process can be related to the standard color system. This is the subject of Section 4.5.4 (calibration in the standard color system).

In the CIELAB and CIELUV color spaces the Euclidian distance is used for determining color distance (according to DIN 6174). The following applies:

$$\Delta E_{ab}^* = \sqrt{\Delta L^* \cdot \Delta L^* + \Delta a^* \cdot \Delta a^* + \Delta b^* \cdot \Delta b^*}$$

and (3.17)

$$\Delta E_{uv}^* = \sqrt{\Delta L^* \cdot \Delta L^* + \Delta u^* \cdot \Delta u^* + \Delta v^* \cdot \Delta v^*}\ .$$

The color difference ΔE_{ab}^* is used, for example, in object classification [Tom90] and the color difference ΔE_{uv}^* in color calibration. The formula for the CIELAB space is to be applied especially to body colors.

For typical applications in the printing industry, Stamm [Sta81] gives justifiable CIELAB color differences from a mean of six units with a standard deviation of around three to four units. In [Hal89], color surfaces and their color differences are represented in CIELAB units. Through this a rough impression for the estimation of CIELAB units can be won. Color distortions are possible, however, through the reproduction of color on paper. The data concerning the number of CIELUV units ΔE_{ab}^* and ΔE_{uv}^*, which by "juxtaposition" comparison of single colors are still distinguishable to the human eye under controlled viewing conditions (*JND, just noticeable difference*), varies between one unit [Sto et al. 88] and five units [Dal88].

Note that the color difference formulas described here are valid in general only for small color differences [Ric81]. In addition, the color differences are no longer valid under changed lighting conditions. The color difference measurements are designed above all for the dye industry, so that a transfer to color image processing is not automatically useful. Furthermore, their applicability in color image processing has not yet been sufficiently researched. These characteristics are sometimes not given adequate attention in publications dealing with color image processing.

Color differences can also be described by means of L^*-lightness, hue angle, and chroma. The following applies:

$$\Delta E^*_{ab} = \sqrt{\Delta L^* \cdot \Delta L^* + \Delta h^*_{ab} \cdot \Delta h^*_{ab} + \Delta C^*_{ab} \cdot \Delta C^*_{ab}}$$

and (3.18)

$$\Delta E^*_{uv} = \sqrt{\Delta L^* \cdot \Delta L^* + \Delta h^*_{uv} \cdot \Delta h^*_{uv} + \Delta C^*_{uv} \cdot \Delta C^*_{uv}} \,.$$

Likewise, it can be useful, for more exact specification, to examine a difference triplet $(\Delta L^*, \Delta C^*, \Delta H^*)$ of L^*-lightness, chroma, and hue angle in the respective color space. The hue angle difference is calculated from ΔE^*_{ab} and ΔE^*_{uv} with

$$\Delta H^*_{ab} = \sqrt{\Delta E^*_{ab} \cdot \Delta E^*_{ab} - \Delta L^* \cdot \Delta L^* - \Delta C^*_{ab} \cdot \Delta C^*_{ab}}$$

and (3.19)

$$\Delta H^*_{uv} = \sqrt{\Delta E^*_{uv} \cdot \Delta E^*_{uv} - \Delta L^* \cdot \Delta L^* - \Delta C^*_{uv} \cdot \Delta C^*_{uv}} \,.$$

The difference of the hue angle has a plus sign if the angle increases; otherwise it has a minus sign. A number of additional color difference formulas can be found, for example, in [Ber94], [Hun91], [Ric81] and [WysSti82]. A recent one for CIELAB is the CIE 2000 color difference formula CIDE2000 that can be found in [Luo et al. 01] together with a comparison of different color formulas for the CIELAB.

3.6 COLOR ORDERING SYSTEMS

The colors to characterize were described in the preceding sections through a set of numbers. This numeric systematization is often not suitable for the daily contact with colors, for by changing the number values every person makes an entirely individual conception of color. This fact has no great importance during the automatic segmentation of a color image in regions. In contrast, the classification of materials or the recognition of objects in color images can be supported considerably by the use of reference systems. Apart from the wealth of trade-specific color references there exist also a number of nationally and internationally standardized catalogs called *color ordering systems*. In color image processing they are used above all in colorimetric and photometric calibration. In the following sections, the Munsell Color System, the Macbeth ColorChecker, and the DIN colormap are described.

3.6.1 Munsell Color System

The Munsell Color System, developed by A.H. Munsell in 1905, found frequent distribution. Each of the over 1000 defined matte color planes are described by the three ordering attributes *hue H*, lightness *value V*, and *chroma C*. The hues are arranged in a circle that is separated into 100 perceptually equidistant steps with five principal hues and five intermediate-positioned hue steps. The principal and intermediate hues are indicated by red, yellow-red, yellow, green-yellow, green, blue-green, blue, purple-blue, purple, and purple-red. The lightness value is subdivided according to sensitivity into 11 equidistant gray shades. In this case, "absolute white" is indicated as "10/" and "absolute black" as "0/". The value "5/" lies, according to sensitivity, exactly in between and is indicated as a "middle gray."

The chroma has a resolution dependent on the hue. Its notation indicates the degree of divergence of a given hue from a neutral gray of the same value. the value "/0" indicates an absolute desaturation of color. The maxima lie by 10 to 14 and are even higher in some colors. A chromatic color is clearly referenced by the notation "*H V/C*" and an achromatic color by "*N V/*." Therefore, for example, *N* 8/ corresponds to a light gray, 5*YR* 7/12 a "rich" orange, and 5*R* 8/4 a "soft" pink. A transformation of *RGB* values into the Munsell notation is explained in [Tom87].

3.6.2 Macbeth ColorChecker

The Macbeth ColorChecker is frequently used in English-speaking countries for the calibration of color image processing systems (see Section 4.5.2). It concerns a reference map, which contains 24 matte color surfaces, used for the purpose of color evaluation. The 24 color surfaces comprise the primary colors of the additive and subtractive color mixture (see Section 2.2) as well as six achromatic colors (gray shades). The remaining colors adapt to our environment. For example, a leaf color, two skin colors, and several sky colors are included on the chart (see Fig. 3.19). The CIE coordinates *x*, *y*, *Y* and the Munsell notation are given for each color surface are presented in Table 3.2, where the *xy* coordinates relate to the standard illuminant *C* (see Fig. 4.2.2). The spectral reflection factors were represented graphically in [McC et al. 76] and in [Mey88].

Figure 3.19. *Color image of the MacbethTM ColorChecker.*

3YR 3.7/3.2 Dark Skin	2.2YR 6.47/4.1 Light Skin	4.3PB 4.95/5.5 Blue Sky	6.7GY 4.2/4.1 Foliage	9.7PB 5.4/6.7 Blue Flower	2.5BG 7/6 Bluish Green
5YR 6/11 Orange	7.5PB 4/10 Purplish Blue	2.5R 5/10 Moderate Red	5P 3/7 Purple	5GY 7.1/9.1 Yellow Green	10YR 7/10 Orange Yellow
7.5PB 2.9/12.7 Blue	0.25G 5.4/8.65 Green	5R 4/12 Red	5Y 8/11.1 Yellow	2.5RP 5/12 Magenta	5B 5.08/8.0 Cyan
N 9.5/ White	N 8/ Gray	N 6.5/ Gray	N 5/ Gray	N 3.5/ Gray	N 2/ Black

Table 3.2. *The color or gray-level denotations and the Munsell denotations for the corresponding standardized color patches of the Macbeth ColorChecker.*

3.6.3 DIN Color Map

The German ordering system for body colors is the DIN colormap in accordance with DIN 6164. *Hue*, *saturation*, and *lightness* form the ordering features in the DIN color system. These terms were chosen because they correspond to a very natural description of the sensitivity of a color impression.

Hue	Number
Yellow	1
Red	7
Blue	16
Green	22

Table 3.3. *DIN6164 hue numbers.*

The color identification results from the hue number *T*, the degree of saturation *S*, and the degree of darkness *D* with the notation $T : S : D$. The hues are subdivided into 24 parts on a circle (see Table 3.3). For achromatic colors, an *N* is noted for the hue number. In contrast to the Munsell system, the saturation lines, in relation to the standard color table, remain equal with the lightness and lie on straight lines that pass through the origin. The fundamental characteristic of this system is the equal distance according to the sensitivity of the individually defined color rows.

In addition to the DIN colormap, other DIN color models exist that were developed and standardized for the color quality control of color television transmissions and color film scanning (see DIN 6169).

3.7 FURTHER READING

A representation of color spaces from a computer graphics point of view is found in [Fol et al. 94] and a representation from the image processing point of view in [Pra91], [SanHor98], and [PlaVen00]. We recommend [Ber94] for an introduction to color metrics and color differences. The topic of color management is discussed in [Har01] and [GioMad98]. A very comprehensive description (976 pages) of color spaces and color difference measurements is given in [WysSti82].

Nowadays, the data for the CIE chromaticity diagram no longer needs to be copied from the data sheets of the lighting commission CIE. It is available at http://www-cvrl.ucsd.edu/ on the Internet. Beyond that, the homepage of the lighting commission CIE, http://www.cie.co.at/cie/home.html, contains additional interesting references to further data material. Information about *sRGB* is available at http://www.srgb.com. We refer to the newsgroup news:sci.engr.color for current discussions on color-related topics.

3.8 REFERENCES

[Ber94] A. Berger-Schunn. Practical Color Measurement: A Primer for the Beginner, a Reminder for the Expert. Wiley, New York, 1994.

[BroYan89] D.C. Brockelbank, Y.H. Yang. An experimental investigation in the use of color in computational stereopsis. *IEEE Transactions on Systems, Man, and Cybernetics* **19** (1989), pp. 1365-1383.

[ConFli97] C. Connolly, T. Fliess. A study of efficiency and accuracy in the transformation from RGB to CIELAB color space. *IEEE Transactions on Image Processing* **6** (1997), pp. 1046-1048.

[Dal88] C.J. Dalton. The measurement of the colorimetric fidelity of television cameras. *J. Inst. of Electronic and Radio Engineers* **58** (1988), pp.181-186.

[DIN 5033] Deutsche Normen. DIN 5033: *Farbmessung*, Teil 1-9, March 1979. [DIN 6174] Deutsche Normen. DIN 6174: Farbmetrische Bestimmung von Farbabständen bei Körperfarben nach der CIELAB-Formel. January 1979.

[Fol et al. 94] J.D. Foley, A. van Dam, S.K. Feiner, J.F. Hughes, R.L. Phillips. *Introduction to Computer Graphics*. Addison-Wesley, Reading, 1994.

[Fre88] H. Frey. *Digitale Bildverarbeitung in Farbräumen*. Doktorarbeit (Ph.D. Thesis), University Ulm, Germany, 1988.

[GioMad98] E.J. Giorgianni, T.E. Madden. *Digital Color Management: Encoding Solutions*. Prentice Hall, Englewood Cliffs, New Jersey, 1998.

[Hal89] R. Hall. Illumination and Color in Computer Generated Imagery. Springer, Berlin, Germany, 1989.

[Har01] J. Hardeberg. Acquisition and Reproduction of Color Images: Colorimetric and Multispectral Approaches. Universal, Parkland, Florida, 2001.

[Hol82] K. Holla. Opponent colors as a 2-dimenional feature within a model of the first stages of the human visual system. *Proc. 6th Int. Conference on Pattern Recognition*, Munich, Germany, 1982, pp. 561-563.

[Hun91] R.W.G. Hunt. *Measuring Colour* Ellis Horwood, Chichester, Sussex, England, 1991.

[Luo96] M.R. Luo. The LLAB model for colour appearance and colour difference evaluation. *Proc. SPIE* ***2658***, Color Imaging: Device-Independent Color, Color Hard Copy and Graphic Arts, San Jose, California, 1996, pp. 261-269.

[Luo et al. 01] M.R. Luo, G. Cui, B. Rigg. The development of the CIE 2000 colour-difference formula: CIDE2000. *Color Research & Application* **26** (2001), pp. 340-350.

[McC et al. 76] C.S. McCamy, H. Marcus, J.G. Davidson. A color-rendition chart. *J. Applied Photographic Engineering* **2** (1976), pp. 95-99.

[Mey88] G.W. Meyer. Wavelength selection for synthetic image generation. *Computer Vision, Graphics, and Image Processing* **41** (1988), pp. 57-79.

[MonFai97] E.D. Montag, M.D. Fairchild. Psychophysical evaluation of gamut mapping techniques using simple rendered images and artificial gamut boundaries. *IEEE Transactions on Image Processing* **6** (1997), pp. 977-989.

[Oht et al. 80] Y.-I. Ohta, T. Kanade, T. Sakai. Color information for region segmentation. *Computer Graphics and Image Processing* **13** (1980), pp. 222-241.

[PlaVen00] K.N. Plataniotis, A.N. Venetsanopoulos. *Color Image Processing and Applications*. Springer, Berlin, Germany, 2000.

[Poy96] C.A. Poynton:. *A Technical Introduction to Digital Video*. Wiley, New York, 1996.

[Poy97] C. Poynton. Poynton's color FAQ. WWW-Note at http://www.poynton.com/ColorFAQ.html, 1997.

[Pra91] W.K. Pratt. *Digital Image Processing*, 2nd ed., Wiley, New York, 1991.

[Ric81] M. Richter. *Einführung in die Farbmetrik*. 2nd ed., Walter de Gruyter, Berlin, 1981.

[SanHor98] S.J. Sangwine, R.E.N. Horne. *The Colour Image Processing Handbook*. Kluwer, Boston, 1998.

[Sch20] E. Schrödinger. Grundlagen einer Theorie der Farbmetrik im Tagessehen. *Ann. Physik (IV)* **63**, 1920, pp. 397-456, 489-520.

[Spa et al. 00] K.E. Spaulding, G.J. Woolfe, E.J. Giorgianni. Image states and standard color encodings (RIMM/ROMM RGB). *Proc. IS&T 8th Color Imaging Conference*, 2000, pp. 288-294.

[Sta81] S. Stamm. An investigation of color tolerance. *Technical Association of the Graphic Arts Conference*, 1981, pp. 157-173.

[Sto et al. 88] M.C. Stone, W.B. Cowan, J.C. Beatty. Color gamut mapping and the printing of digital color images. *ACM Transactions on Graphics* **7** (1988), pp. 249-292.

[Sto et al. 96] M. Stokes, M. Anderson, S. Chandrasekar, R. Motta. A standard default color space for the internet: sRGB, version 1.10, 1996 at http://www.color.org/sRGB.html.

[Süs et al. 99] S. Süsstrunk, R. Buckley, S. Sven. Standard RGB color spaces. *Proc. of IS&T/SID's 7th Color Imaging Conference*, 1999, pp. 127-134.

[Tom87] S. Tominaga. Expansion of color images using three perceptual attributes. *Pattern Recognition Letters* **6** (1987), pp. 77-85.

[Tom90] S. Tominaga. A color classification method for color images using a uniform color space. *Proc. 10th Int. Conference on Pattern Recognition*, Atlantic City, New Jersey, 1990, Vol. I, pp. 803-807.

[TruKul96] H.J. Trussell, M.S. Kulkarni. Sampling and processing of color signals. *IEEE Transactions on Image Processing* **5** (1996), pp. 677-681.

[TseCha92] D.-C. Tseng, C.-H. Chang. Color segmentation using perceptual attributes. *Proc. 11th Int. Conference on Pattern Recognition*, The Hague, the Netherlands, 1992, Vol. III, pp. 228-231.

[WysSti82] G. Wyszecki, W.S. Stiles. *Color Science: Concepts and Methods, Quantitative Data and Formulae*, 2nd ed., Wiley, New York, 1982 (also available in paperback, printed 2000).

4 COLOR IMAGE FORMATION

The acquisition of digital color images is the first step in a color image processing pipeline. The physical and geometrical parameters, which have substantially affected the formation of the image function, must be determined for the subsequent processing and evaluation of the images. On the one hand, the resulting function values of the digital image function depend on the material and reflectance characteristics of the object surfaces in the acquired scene. On the other hand, the sensor characteristics of the camera, the optical characteristics of the lenses, the scanning of the image signal, the photometric characteristics of the lighting source, and, last but not least, the geometrical regularities, to which the image acquisition process is subject, directly influence the resulting image function.

Color and photometric influence factors form the emphasis of this chapter. For this reason some terms from the area of colorimetrics are introduced. The geometrical characteristics of the camera model, which are needed, for example, for color stereo analysis (Chapter 9), do not differ from those of three-dimensional gray-level image analysis. A detailed representation of camera geometry and two well-known procedures for geometrical calibration of the camera system are presented in [Kle et al. 98]. First, an introduction to common image acquisition devices follows.

4.1 TECHNICAL DESIGN OF ELECTRONIC COLOR CAMERAS

The camera used for image acquisition has an important influence on the generated image. Its selection should always be suited to the problem to be solved. For this reason demands on image quality, size, and weight of the camera and the resulting purchase price must be biased against each other. In the following section the three most commonly used sensor concepts in modern cameras are introduced. In addition, various camera concepts for the generation of color images are described.

Color images can be produced with black-and-white cameras using color filters, inexpensive one-chip CCD color cameras, and high-quality expensive three-chip CCD color cameras. In addition to this, color image formation is also possible with scanners. However, scanners will not be covered further here since the geometrical and photometrical parameters that influence the generation of the image function cannot be easily determined with scanned images. Thus, the latter image class is only conditionally suitable for three-dimensional color image processing. Furthermore, chromatic distortions in the digitalized images arise frequently by errors in the mechanical adjustment and by the chromatic aberration (see Section 4.3.2) of the lens system of the scanner. The results of a subsequent technique for two-dimensional color image processing (e.g., edge detection in color images) are unfavorably affected by this (see [KhoZin96]). The principal structure of semiconductor cameras and the advantages and disadvantages of the individual concepts are summarized in [Len89]. A detailed representation of the technical design for color television is given in [Eva94].

4.1.1 Image Sensors

The structure of a modern camera suitable for image processing is based almost entirely on semiconductor sensors, the *CCD chips* (*charge coupled device*). Here the interline transfer is frequently used as a sensor concept (see Fig. 4.1). Furthermore, frame transfer and full frame transfer sensors are also commonly used (see Fig. 4.2). *Interline transfer sensors* contain columnarly arranged photosensitive sensor elements. Each column is connected to a vertical CCD shift - register by a transfer gate. In each clock phase a half-image is selected whereby two of the sensor elements, each belonging to different half-images, are assigned to the cells of the transport register.

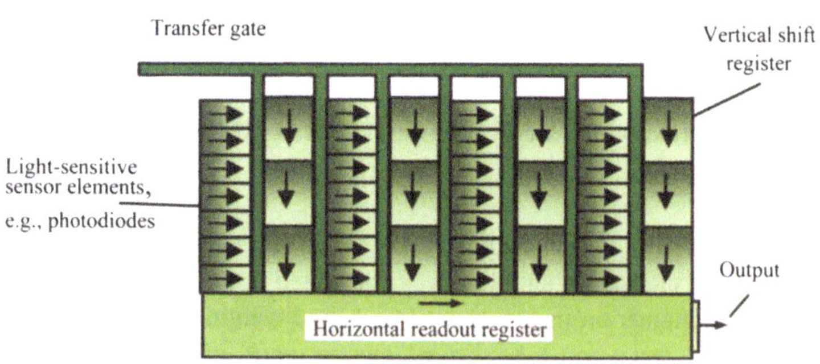

Figure 4.1. *The interline transfer concept (adopted from [Kle et al. 96]).*

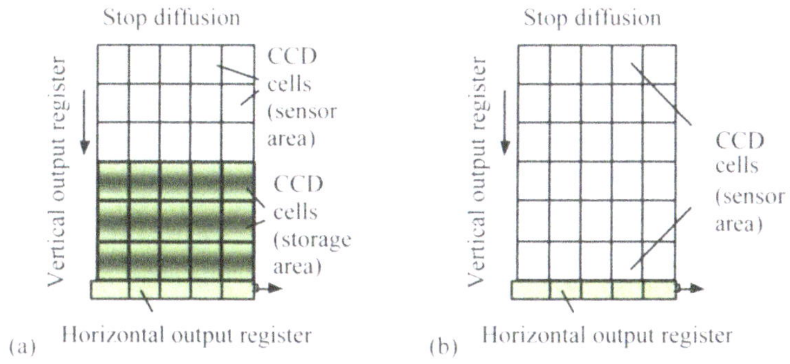

Figure 4.2. *(a) Frame transfer concept. (b) Full frame transfer concept.*

This concept offers the advantage that only half as many CCD cells are necessary as pixels present in the resulting image. However, there are just as many sensor elements present as there are pixels. Thus, the formation of a full image lasts two clock phases (interlaced mode), which can lead to an impairment of image quality due to the time difference between the acquisition of odd and even image scanlines. This occurs only with very-fast-moving objects. The interline transfer concept is implemented in black-and-white cameras as well as in color cameras. The utilization of the sensor elements and also the image produced differs according to the type of camera.

On the frame transfer sensor, the CCD registers are arranged vertically next to each other. The upper photosensitive sensor area and the same-sized lower storage area, protected from light, can be distinguished in the frame transfer image sensor (see Fig. 4.2a). The entire charge is removed during one readout cycle from the image area and the storage area. Thus, the frame transfer sensor, in contrast to the interline transfer, needs only half as many photosensitive sensor elements as would be necessary for a complete television image. For this reason the frame transfer sensor contains only the number of lines that are necessary for a half-image. Image integration results twice as fast as with the other readout concepts, likewise with half-image frequency. Line interlacing between the half-images is achieved by a corresponding control of the CCD register.

The full frame transfer sensor, unlike the frame transfer and the interline transfer sensor, does not contain a storage section. The total sensor area is light sensitive (see Fig. 4.2b). Full frame sensors always have to be incorporated with a shutter camera. The external shutter determines the integration time. Full frame transfer sensors are mostly used in time-critical applications and also in high-resolution cameras (e.g., 4000×4000 pixels). In the following, various camera concepts for the generation of color images are presented.

4.1.2 Multispectral Imaging Using Black-and-White Cameras with Color Filters

One possibility of generating color images is to place *color filters* (see Section 4.2.1) in front of the lens of a black-and-white camera. The spectral transmissions are provided timewise successively. A static scene is a prerequisite here since the channels of the image are acquired one after another. The local dissolution of the images is not affected with this method. The simple and controllable color mapping is advantageous since the number as well as the kind of filters (and thus the spectral transmission factors) are freely selectable. Therefore, color images with more than three color channels can also be produced when needed, whereby color image segmentation (see Chapter 7) sometimes can be improved. Perez [Per95] proposes for this the use of a six-channel image with six different spectral transmissions. A liquid crystal tunable filter (LCTF) may be applied to reduce mechanical vibrations caused by filter changes [Miy et al. 02]. An introduction to multispectral imaging and its possible application in face recognition is given in Chapter 12.

In generating three-channel color images, this method has the advantage that the spectral transmission factors of the filter used, corresponding to the data of the filter manufacturer, are known for each color channel. This data is usually not available in color cameras. Furthermore, color images with nonoverlapping spectral sensitivities can be produced, which is not possible with any of the commercial color cameras (without using color filters).

Special attention is to be directed to white balancing and color aberrations with this method. Color aberrations can arise, for example, from a chromatic aberration of the lens. Chromatic aberration of the lens is the reason that longer wavelengths focus at a larger distance than short wavelengths and produce an enlarged image (see Section 4.3.2). This physical effect must be taken into consideration when using color filters in image acquisition. Furthermore, the camera has a different sensitivity when using color filters in the different spectral bands. Here an alignment must likewise be implemented (see Section 4.2.1).

4.1.3 One-Chip CCD Color Camera

A color camera is necessary if the employment of the filter method is too complex or if dynamic scenes are to be digitized. One-chip and three-chip CCD cameras can be differentiated according to their structure. The spectral sensitivities of the CCD chips are, however, not variably selectable. The photosensitive cells of the more inexpensive one-chip camera are covered with a color filter mask. Essentially two variants can be distinguished. Red, green, and blue filters are applied to the CCD chip and arranged into the form of a *mosaic filter* (see Fig. 4.3).

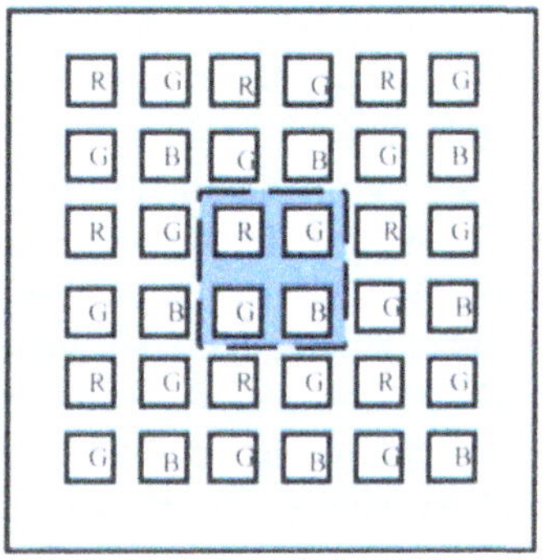

Figure 4.3. *Color mask composition of a one-chip CCD camera. The four centered CCD elements form one macropixel by reducing the local resolution.*

For example, four elements, of which two green and one element each of red and blue are sensitive, form a 2×2 macropixel. The characteristic of human color perception, that the brightness sensitivity of daylight vision is maximal at 560 nm, is considered by using two green filters. The image resolution in the vertical and horizontal directions is halved during this arrangement. Thus, a chip with, for example, 2000×2000 sensor elements delivers an image with only 1000×1000 pixels. Different arrangements of the red, green, and blue filters are common within a mosaic filter. Furthermore, Sony developed a four-color filter with red, green, blue, and an additional emerald where a special image processor is computing the *RGB* output signal.

Another distribution of sensor elements, which was used, for example, in the Spindler & Hoyer camera 310 C, is based on the arrangement of *color strip filters*. Here the sensor surface is covered by periodically repeating vertical red, green, and blue strips. Thus, the vertical resolution remains the same and the horizontal resolution is reduced. In addition to cameras constructed with additive filters, there are also models with subtractive filters, such as the Spindler & Hoyer camera 230 C, which functions with a *complementary mosaic sample*. Filters of the colors magenta, cyan, green, and yellow are contained in one macropixel. The camera does not provide separate *RGB* signals. The sensitivity of the camera is increased by the complementary sample; however, the color reproduction decreases slightly.

One-chip cameras have the disadvantage that the transmission behavior of the filter masks is not always identical for different sensor elements due to a manufacturing process that is too imprecise. This results in a lesser color fidelity. Furthermore, color moiré errors arise at a signal frequency that is too high since not all CCD elements that form a macropixel are illuminated uniformly. One-chip camera technologies are commonly used in video cameras of the consumer branch.

4.1.4 Three-Chip CCD Color Cameras

Three-chip color cameras are used for the generation of high-quality color images. Three-chip CCD cameras split the light beam with semitransparent prisms (*beam-splitter prisms*) and steer the separate spectral intervals toward three locally separated CCD chips (see Fig. 4.4). In addition to the higher price of the three-chip camera, which can also be explained by the very exact chip fitting of less than one micrometer, the digitalization units are more complex due to the three-way A/D transformation and parallel storage. One advantage of the three-chip CCD camera over the one-chip CCD camera exists in the higher color quality of the generated images. The higher color quality is obtained, among other things, by the fact that the filters used in three-chip CCD cameras exhibit a substantially more homogeneous transmission behavior than is usually the case with one-chip CCD cameras. Disadvantages of the three-chip CCD camera are its larger dimensions and its greater weight. Previous attempts to construct three-chip cameras more compactly are based above all on a reduction of image resolution, in which, instead of the 2/3-inch or 1/2-inchCCD sensors, three 1/3-inch CCD sensors are employed.

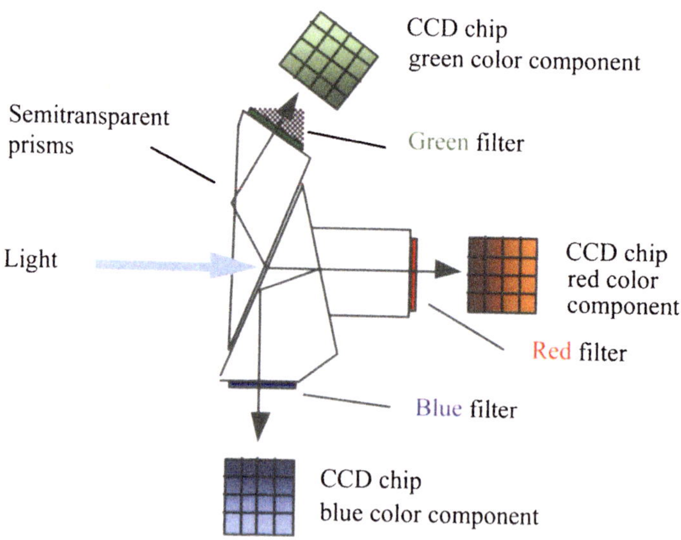

Figure 4.4. *Light beam splitting by applying semitransparent prisms in a standard three-chip CCD camera (adopted from [Kle et al. 96]).*

4.1.5 Digital Cameras

Recently, digital cameras have received even greater attention. In this case, one-chip or three-chip CCD cameras deliver (and/or store) the image not as an analog signal, as previously, but rather as a digital signal. The structure of the image acquisition sensors is identical to that of the analog camera. Digital cameras are offered for the acquisition of image sequences (*digital video*) as well as for the acquisition of single still images (*digital photography*). An interesting technical introduction into the area of digital video can be found in [Poy96] and an introduction to the area of digital photography can be found in [Has et al. 95].

While in the past quadratic image formats dominated in digital image processing, digital cameras supply right-angled image formats of differing sizes (e.g., the Nikon Coolpix 5400 uses, for example, the formats 2592×1944, 2592×1728, and 1280×960 pixels, the Kodak DX4900 the format 2448×1632 pixels, and the Canon EOS-1Ds formats up to size 4064×2704 pixels). The sampling frequency of the digital camera must be taken into consideration for the computation of pixel size in geometric stereo analysis. Furthermore, attention must be given to the programs used for the processing of color images to ensure their compatibility with the available image format. The manufacturer's software delivered with each digital camera allows only very simple image manipulations.

One disadvantage of digital photo cameras existed until recently in their memory limitations. Usually only a few images could be stored in uncompressed data format. In order to increase the number of storable images, digital cameras allow image coding according to the standard *JPEG* set by the Joint Photographic Experts Group. The JPEG coding is based substantially on a discrete cosine transformation (DCT). The lossy compressed images are suitable, however, for simple image presentations (e.g., on the internet). Their use is limited for applications of scene analysis, shape reconstruction, and for high-quality print products. So far, not all digital cameras allow storing data in raw format and the gamma correction cannot be switched off with most digital cameras, which limits its applicableness for scene analysis in computer vision.

With digital video cameras, the image formats are based predominantly on the values given by the television standards PAL and NTSC. The use of larger image formats, as intended for ATV, AVCHD, PALplus, or HDTV, is not finalized yet. The format 720p60 is 1280×720 pixels progressive encoding with 60 frames per second (60 Hz). The format 1080i50 is 1920×1080 pixels, interlaced encoding with 50 fields (25 frames) per second. The rate can sometimes be derived from the context, in which case it can usually be assumed to be either 50 or 60, except for 1080p which is often used to denote either 1080p24, 1080p25 or 1080p30 at present but will also denote 1080p50 and 1080p60 in the future. In the early days of digital video, as with analog video systems, a cassette was inserted into the digital camcorder (also called *digicam*). In Hitachi's first digital camcorders without tape, images were compressed to 1/100th of their original size and stored on a 260 MB card. Therefore, an image sequence of a maximum length of 20 minutes could be stored. Due to the large compression rate of the images, nothing more than VHS quality could be attained. This has changed, and newer systems record onto conventional DVDs or memory sticks. There are still many limitations and challenges in digital video, such as cost, size, and energy consumption (which reduces the maximum possible video recording time).

4.2 STANDARD COLOR FILTERS AND STANDARD ILLUMINANTS

Various color filters and reference lighting types are used for purposes of color measurement and colorimetric calibration of an image acquisition system. The systems used most frequently are described in the following. A detailed representation of this topic can be found in [Ber94] and [WysSti82].

4.2.1 Standard Color Filters

For the generation of color images with black-and-white cameras, or the generation of color images with more than three color channels, glass filters as well as gelatin filters are applied. During the production of gelatin filters organic coloring materials are added to the gelatin and the mixture is applied to a glass base. After drying, the mixture is removed from the glass base and can be cut to the necessary size. Gelatin filters are substantially less expensive than glass filters and are therefore used frequently in experimental laboratories during color image formation (e.g., at Carnegie Mellon University or the University of California at Irvine).

Eastman Kodak offers gelatin filters in a large selection of various spectral characteristics. The individual filters are named after Frederick Wratten, a British inventor, and are referenced by numbers. Commonly used are the Kodak Wratten Filter #25 in the long-wave (so-called red) band, filter #47B in the middle-wave (so-called green) band, and filter #58 in the short wave (so-called blue) band. The spectral transmissions are represented in Fig. 4.5 for these three filters. Additional spectral transmissions are given, for example, for Wratten Filters #22, #24, #29, #44, #50, #90, and #99 in [Vri et al. 95]. The disadvantage of the gelatin filter as opposed to the glass filters exists in the substantially greater instability of the gelatin filter in higher humidity or higher temperatures.

It should be noted that during the generation of color images with black-and-white cameras, the sensitivity of the camera could vary significantly when using color filters in different spectral areas. Differently scaled intensity values in the individual spectral responses are due to this. The illustration in Fig. 4.5b shows an example of the spectral sensitivity of a CCD camera when using color filters. For the adjustment of the spectral responses it is not recommended to multiply the

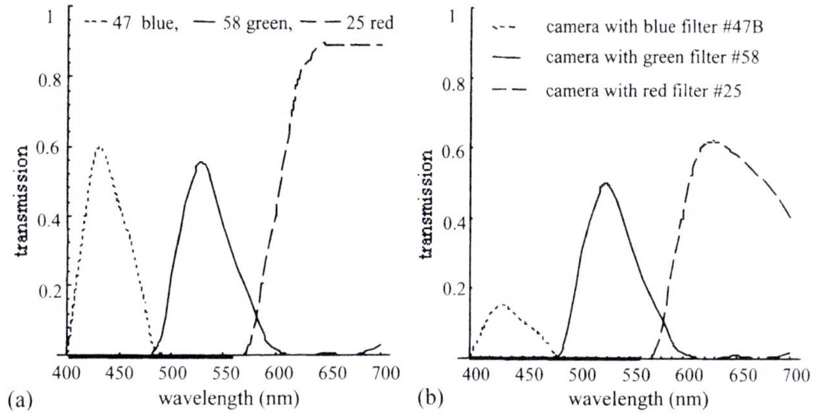

Figure 4.5. *(a) Spectral transmission of the Wratten gelatin filters #25, #58, and #47B. (b) Example of the spectral sensitivity of a CCD camera using the above mentioned filters (adapted from [Nov et al. 90]).*

measured intensity values within the blue range by a scaling factor (e.g., 3.5). The noise is equally strengthened by this (and differently from the other spectral responses). Substantially more favorable is the adjustment of the aperture for each filter or each spectral response (see [Nov et al. 90]). Note that when inserting several color filters into a path of rays, the laws of subtractive and not additive color mixture apply (see Section 2.2).

Glass filters are also employed for the measurement of camera linearity. Here *neutral filter glasses* are introduced into the ray of light from the light source and a white surface is illuminated or supported. The resulting intensity values in the camera are measured and compared. Neutral filter glasses reduce the transmission of the incidental light by a given factor. This transmission factor is set by the selection of the glass type and glass thickness. The dependency of the transmission on glass density d is defined by

$$d = \frac{\lg\left(\frac{1}{\tau(\lambda)}\right) - \lg\left(\frac{1}{R(\lambda)}\right)}{\lg\left(\frac{1}{\tau_R(\lambda)}\right)},$$

where $\tau(\lambda)$ describes the desired degree of spectral transmission, $\tau_R(\lambda)$ is the degree of pure transmission for the reference density $d = 1$ mm, and $R(\lambda)$ is the reflection factor of the glass. The degree of pure transmission is the transmission of the filter glass without consideration for the reflection losses. In the neutral filter set of the Schott company, which provides filters with transmission factors between 10^{-5} and 0.708 (for λ = 546 nm), the reflection factor lies at approximately 0.92. Another common description of the transmission properties of filter glasses is given on the basis of *optical density* $D(\lambda)$. The optical densities of the filter set lie between 0.15 and 5.0. The relation between optical density and transmission factor is given by

$$D(\lambda) = \lg\left(\frac{1}{\tau(\lambda)}\right).$$

The measurement of camera linearity is described in Section 4.4.2.

4.2.2 Standard Illuminants

The spectral distribution of the illumination (or radiation) existing during image acquisition has an influence on the generated color image. This is known from the area of photography. The color of the same dress that was photographed in the light of the midday sun appears totally different in the photo that was taken at sunset or under artificial neon lighting. For color comparison and color measurements, a number of color measurement numbers must be given for each color according to the spectral distribution of the illumination (or radiation). In order to reduce this variety, some spectral distributions were standardized internationally.

Note that there are no standardized lighting sources, but instead *standard illuminants*. That is in the corresponding standards different spectral distributions were set tabularly, independent of whether lighting sources (lamps) exist that correspond to these distributions. Several terms from the area of colorimetrics are necessary for the explanation of the standard illuminants.

For the color measurement of body colors, *standard illuminants* were set in 1931 by the International Commission on Illumination CIE (and were subsequently adopted in the German standard DIN 5033). The *body color* denotes the color of a nonluminous object (i.e., a body that is visible only under an illumination). The spectral distribution of the illumination directly influences, as already mentioned, the perceived or measured body color of the illuminated object. Therefore, the color values of a body color are defined only in relation to the spectral power distribution of the illumination.

The illumination describes a ray of a certain relative spectral distribution in the spectral transmission that has an influence on the color of the object (see DIN 5033). In colorimetrics, only the *relative spectral distribution* S_λ is used since only relative values are needed. The value of the spectral distribution is set at 560 nm, being equal to 100, and is divided by this value. The spectral power distribution depends on the wavelength and the color temperature.

For the definition of *color temperature*, the terms *black body* or *Planckian radiator* must be introduced. These radiators have no technical relevance, but rather are included merely as reference size. For them the spectral power distribution can be specified according to the Planckian radiation law. There are highly heatable black cavities (called *black bodies*) whose radiation can escape by a comparatively small opening. If the black body is heated, then the color of the radiation is changed depending on the temperature. In lower temperatures they appear red; as the temperature rises, they appear first yellow and then white.

Therefore, at rising temperatures the black body goes through a clear color sequence. The color of any (technical) radiator can now be described by the temperature T_f of the black radiator, by which this equienergetic color radiates to the technical radiator. This temperature T_f of the black radiator is called the *color temperature* of the radiator concerned. Note that here only color similarity is the subject. The color temperature does not relate to the spectral distribution of the radiator concerned here. The measurement for the color temperature is the absolute temperature K. It corresponds to the temperature of + 273 (°C).

The standard illuminant *A* was established as representative of an artificial illumination with tungsten light. It corresponds to the illuminant of a Planckian radiator with the color temperature 2856 K. As representative of average daylight, with the most similar color temperature 6500 K, this standard illuminant was set as *D*65. Earlier, the standard illuminant *C* (the most similar color temperature, likewise at 6500 K) was used for "daylight." However, on account of its lack of long-wave UV radiation, it was replaced by *D*65. The spectral power distributions represented in Fig. 4.6 for the standard illuminants *A*, *D*65, and *C* were calculated by interpolation of the values given in tabular form in DIN 5033, part 7.

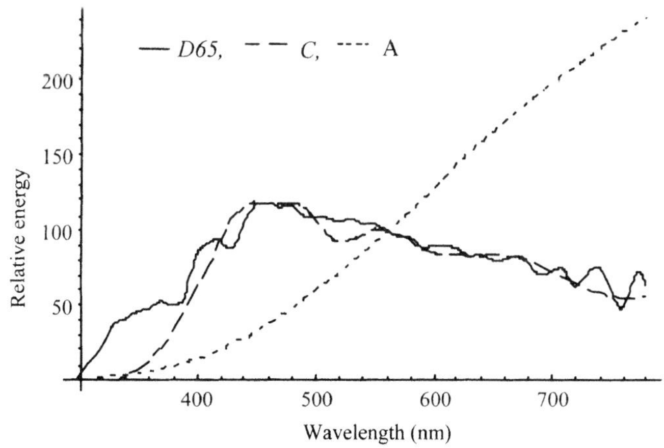

Figure 4.6. *Spectral power distributions of the standard illuminants A, D65, and C.*

Again, the disadvantage of the standard illuminant *D*65 is that it cannot be reproduced by any technical lighting source. At the present time, filtered short-arc xenon lamps are the nearest alternative to *D*65. In DIN 5033, the illuminants *B* (sunlight), *G* (tungsten light), *P* (petroleum and candle light), as well as *D*50, *D*55, and *D*75 (average daylight with the most similar color temperatures 5000 K, 5500 K, and 7500 K, respectively) are specified. However, they are not regarded as standard illuminants. Standard illuminants also play a role in the calculation of color constancy (see Section 8.3).

4.3 PHOTOMETRIC SENSOR MODEL

The following *sensor model* is employed in general to describe the generation of a color image:

$$\begin{aligned}\rho_k &= \int_{\lambda_l}^{\lambda_h} E(\lambda)\cdot R(\lambda)\cdot S_k(\lambda)\, d\lambda \\ &= \int_{\lambda_l}^{\lambda_h} C(\lambda)\cdot S_k(\lambda)\, d\lambda \qquad \text{with } C(\lambda) = E(\lambda)\cdot R(\lambda)\,.\end{aligned} \tag{4.1}$$

Thereby $E(\lambda)$ is the *spectral power distribution* of the illumination, $R(\lambda)$ is the *spectral reflectance factor* of the surface material, $S_k(\lambda)$ is the *spectral sensitivity* (or *spectral response curve*) of the subsensor *k*, and ρ_k its *response*. The interval of integration describes the measurement range. The product of the functions $E(\lambda)$ and $R(\lambda)$ is called the *color signal* $C(\lambda)$. $C(\lambda)$ describes the radiation reflecting from the object that reaches the sensor. Note that the color signal $C(\lambda)$ should not be mistaken for the color image **C**.

Only a finite number of measurements can be established in a real measurement of the spectral quantities $E(\lambda)$, $R(\lambda)$, $C(\lambda)$, and $S_k(\lambda)$. The width of the wavelength interval $\Delta\lambda$ lies in general at roughly 2 to 10 nm. The smaller the interval, the smaller the amount of energy that reaches the sensor, so that natural, technical measurement borders exist at the bottom. Thus, the equation

$$\begin{aligned}\rho_k &= \sum_{i=1}^{n} E(\lambda(i))\cdot R(\lambda(i))\cdot S_k(\lambda(i)) \\ &= \sum_{i=1}^{n} C(\lambda(i))\cdot S_k(\lambda(i)),\end{aligned}$$

with $C(\lambda(i)) = E(\lambda(i))\cdot R(\lambda(i))$ is possible. The function $\lambda(i)$ represents the *i*th value of the *n* measurements on the wavelength interval $\Delta\lambda$ belonging to it. $E(\lambda)$, $R(\lambda)$, $C(\lambda)$, and $S_k(\lambda)$ can also be represented by vectors. In the case of an *integral equation* there are infinitelydimensional vectors and in a *sum equation* finitelydimensional vectors. The above equation can be abbreviated as

$$\rho_k = \mathbf{c}^T \cdot \mathbf{s}_k ,$$

whereby "·" stands for the inner vector product and **c** and $\mathbf{s}_k$ are n-dimensional column vectors. Equation (4.1) can be described as

$$\rho_k = \mathbf{e}^T \cdot diag(\mathbf{R}(\lambda(i))) \cdot \mathbf{s}_k = \mathbf{e}^T \cdot \mathbf{R} \cdot \mathbf{s}_k ,$$

whereby the spectral reflectance factor is represented as diagonal matrix **R** of the dimension $n \times n$ with the elements $\mathbf{R}(\lambda(i))$. The n-dimensional column vector representing the spectral power distribution of the illumination is indicated by **e**. Instead of representing the spectral reflectance factor as a diagonal matrix, $\mathbf{E}(\lambda(i))$ can be represented as a diagonal matrix of dimension $n \times n$. Then the following applies:

$$\rho_k = \mathbf{r}^T \cdot diag(\mathbf{E}(\lambda(i))) \cdot \mathbf{s}_k = \mathbf{r}^T \cdot \mathbf{E} \cdot \mathbf{s}_k .$$

The n-dimensional reflectance column vector is indicated by **r**. By combining the m sensor responses as one m-dimensional row vector $\boldsymbol{\rho}$ and the column vectors $\mathbf{s}_k$ as a single $n \times n$ -matrix **S**, the following results:

$$\boldsymbol{\rho} = \mathbf{c}^T \cdot \mathbf{S} \text{ and } \boldsymbol{\rho} = \mathbf{e}^T \cdot \mathbf{R} \cdot \mathbf{S} \text{ and } \boldsymbol{\rho} = \mathbf{r}^T \cdot \mathbf{E} \cdot \mathbf{S} \qquad (4.2)$$

4.3.1 Attenuation, Clipping, and Blooming

Attenuation of the color signal arises when more photons fall on the photodiode than it can process. This attenuation (or more exactly nonlinearity) can falsify the signal by exceeding the *dynamic range* of the processing unit in the phases of its analog processing, as well as in the analog/digital transformation. Special attention is to be given to this problem in color image processing since attenuation does not occur at the same time in the different color channels. This can lead to strong color distortions. Attenuation is usually accomplished in the analog section of the camera by a *pre-knee-circuit*. Fig. 4.7 shows the characteristics of the pre-knee-circuit of the Sony three-chip camera DXC-930P. The signal is attenuated linearly from a certain input signal. The attenuation factor is 1:78. In color image processing, circuits of this type tend to distort the color since the camera signal loses its (up to then) global linearity. In addition, a *clipping* of the input signal arises when the analog signal exceeds the highest processable voltage during the A/D conversion.

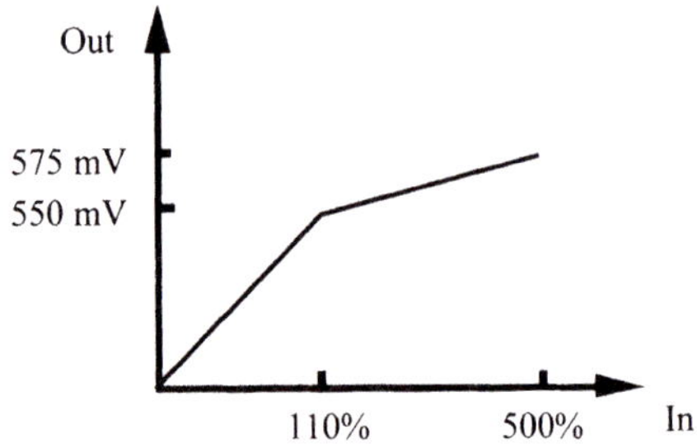

Figure 4.7. *Example of a pre-knee-circuit to preserve the dynamic area (reprinted from [Kle et al. 96] with permission from Vieweg).*

The clipping range can be controlled if a clipping level (white level) can be specified for the A/D conversion. One disadvantage here is that quantization errors can occur. The green channel is saturated, for example, from a certain intensity. If the maximum intensity is not yet reached in the red and blue components, then the values in these components can rise further. Resulting from this is a projection of the colors lying outside the dynamic area onto the clipping level of the blue component through which the most important attribute of the color, namely the hue, changes. A further increase in intensity can cause two or all colors to lie in the saturation region. This leads to special problems for image processing techniques that are based on color comparisons.

The following example, in which the color values above the value 1.0 are clipped, will clarify this. The entry *RGB* vector (0.8, 0.4, 0.2) is not first subjected to the clipping process. Let us change the iris setting of the camera in a way that the intensity is doubled. Now, only the red channel is concerned. In the case of a further duplication of the intensity, it concerns the green channel as well as the red channel. The change of hue H and saturation S (for definitions see Section 3.4.1) is to be attributed to clipping due to the independence from the intensity. Table 4.1 shows the distortion in H and S.

	***RGB* Original**			***RGB* Clipping**				
Intensity Factor	R	G	B	R	G	B	H	S in %
× 1	0.8	0.4	0.2	0.8	0.4	0.2	19.1	57.1
× 2	1.6	0.8	0.4	1.0	0.8	0.4	40.9	45.5
× 4	3.2	1.6	0.8	1.0	1.0	0.8	60.0	14.3

Table 4.1. *Example of hue and saturation distortion caused by a clipping of the signal. The clipping level lies at 1.0 (reprinted from [Kle et al. 96] with permission from Vieweg).*

Figure 4.8. *Blooming as an effect of intensive illumination (reprinted from [Kle et al. 96] with permission from Vieweg).*

It is not possible to identify a pixel impaired by attenuation or clipping in the color channels clearly, but the color components can be compared with the known saturation values and pixels with doubtful color. A color most similar to the original color can be subsequently approximated by a correction technique only if there is certain knowledge of the colors and their distribution in the image.

If the measured intensity in the sensor element is greater than the clipping level by several factors, then the CCD cell is no longer able to store more charge per time unit. The additional charge is spread into the neighboring CCD cells. This effect is called *blooming*. Blooming becomes particularly apparent, for example, during highlight analysis in color images (see Section 8.1). Blooming in gray-level images, however, also directly influences the results of shade analysis or photometric stereo analysis (see Section 10.2). In interline transfer sensors, blooming starts, according to [Len89], at roughly 600% overload. MOS-XY sensors (see Section 4.1.1) show an even lesser sensitivity toward overloading. Figure 4.8 shows a highlight on a rectangular flat metallic surface and the results of overloading.

4.3.2 Chromatic Aberration

The sharpness of a color image is impaired by the fact that the focusing depends both on the effective focal length of the lens and on the wavelength of the incidental light. For each of the three sensitivity areas R, G, and B of the camera, there is a focal length that is best for the focusing (and vice versa). This depends on the *chromatic aberration* of the lens. *Color longitudinal aberration* (or *axial chromatic aberration*) causes blurring. The size of the "blur circle" is determined by the distance between the focus rays when they intersect the image plane. Additionally, the *color transverse aberration* (or *lateral chromatic aberration*) causes shifting and magnification of the mapped object point. An illustration of these circumstances is to be found in Fig. 4.9.

Due to the chromatic aberration, some areas in the color image are shown more focused than others according to the wavelength of the incidental light. At first sight this effect appears more insignificant than serious for the processing result of a color image. This is surely correct if the techniques used are sufficiently robust. However, if one takes into consideration, for example, a procedure for cluster analysis in a color model, then the chromatic aberration greatly affects the processing result. This applies equally to the detection and localization of color edges in the image. By the chromatic aberration, a shift of the edges of one to two pixels can occur if the edge detection is accomplished separately in the individual color channels. Substantial inaccuracies arise in the results if a procedure for computing *depth from focus* is applied to a color image. The computed depth depends then not only on the distance between object and camera lens, but also on the wavelength of the incidental light due to the chromatic aberration. Independently of whether one of the procedures specified above is used, an improvement of the accuracy of the results is always desirable if it can be attained without large expenditure.

An obvious consideration for the avoidance of the chromatic aberration exists in the use of corrected lenses. Note that the correction process consists of a relation between the correction of the chromatic aberration and the correction of the spherical aberration (opening error) as well as the coma (asymmetry error) of the lens system. Neither of the latter aberration errors are dependent on the wavelength and, therefore, they are not further explained here (a detailed description can be found in [Hec02]). The complex computations necessary for the design of a corrected lens system are usually not accomplished with inexpensive objectives.

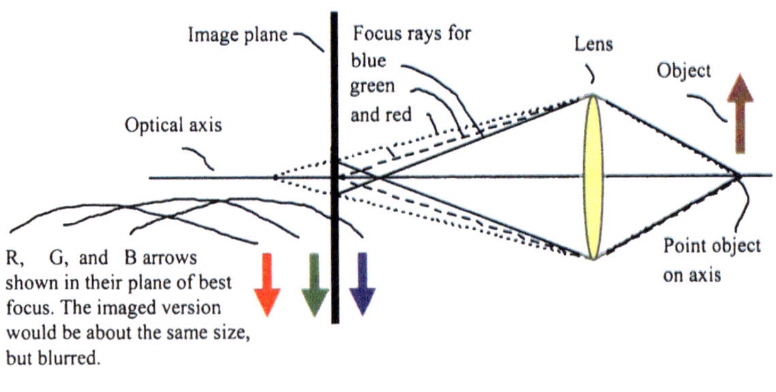

Figure 4.9. *Geometric optical representation of chromatic aberration caused by a thin lens. Longer wavelengths focus long, and are magnified more than the shorter wavelengths (after [BouWol92]).*

The design of multilens systems (especially of zoom-lenses) is not a long-known technology, but rather the subject of modern research [BouWol92]. Furthermore, Willson and Shafer [WilSha91] have determined in an examination that somehow "corrected" lens systems (called *achromates*) are not corrected for all wavelengths. Significant errors caused by a chromatic aberration can also arise with these lens systems. An additional problem is that lens connections are not standardized. Thus, there is no guarantee that high-quality lenses are available for each color camera. Finally, it should be mentioned that color constancy could also be achieved by analysis of the chromatic aberration (see [FunHo89]).

4.3.3 Correction of the Chromatic Aberration

One possibility of correcting the chromatic aberration consists of translating the camera and a subsequent refocusing. This concerns an active process that was developed at Carnegie Mellon University [WilSha91]. Precise mechanics are necessary for its implementation. For example, for the correction of lateral chromatic aberration, a translation of the camera must take place in the order of magnitude from 0.05 to 0.1 mm. This is possible only at a large expenditure of laboratory conditions.

Another possibility for correcting the chromatic aberration consists of an additional treatment of the image signal [BouWol92]. The technique is employed in computer graphics for *image warping*. Here no mathematically complex algorithms need be used, but rather the warping can take place one-dimensionally along the lines and the columns (line for line and column for column). The main idea of the procedure is based on a rescanning of the approximated image signal. For this, the model of a *PSF* (*point spread function*) with a pixel is used in order to attain a functional description of the input signal. This function is "warped" and "denoised" using an output PSF.

The goal of image warping is a correction of the color lateral aberration. For this, a test image (e.g., a chessboard) focused on one (e.g., the blue) color channel is acquired in the three color channels R, G, and B. Subsequently, in all three images the edges are determined with subpixel accuracy (i.e., more exact than the resolution of the camera). Since the test image is monochromatic in our case, the edge positions should be identical in all three color channels. This is not the case due to the chromatic aberration. The size of the deviation is given by the shift of the edge elements in the individual color channels. Using the procedure described above, the image signal in the two other color channels (in this example, R and G) can again be scanned according to the shift, line for line and column for column. This produces a corrected color image. In order to keep the influence of the color noise on the localization of the edge positions as small as possible, it is recommended to average the function values in the color channels over at least 10 images [BouWol92].

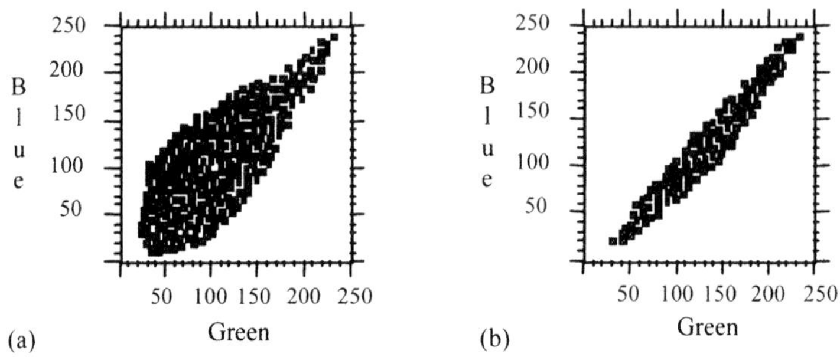

Figure 4.10. *Example of a 2D histogram (blue on green) in the color space. (a) Histogram for an uncorrected image with chromatic aberration and (b) histogram of the image corrected by image warping.*

The effect of chromatic aberration by image warping is shown in Fig. 4.10. The 2D histogram (for example, blue on green) of the monochromatic test image should be a straight line in the ideal case. To simplify the representation, a gray-level coding of the frequency values is omitted. Each arising value is shown black (independently of its frequency).However, due to the chromatic aberration, a point cloud results, and the more strongly it is strewn, the greater the aberration (see Fig. 4.10a). This aberration is reduced by rescanning the image signal (see Fig. 4.10b) and the result is substantially better suited to the ideal case.

Finally, it should be noted that especially with simple (uncorrected) lenses a better correction of the chromatic aberration can be performed with the active technique from Willson and Shafer [WilSha91]. For this, high precision mechanics are necessary, which can be carried out only under laboratory conditions (with considerable additional hardware costs). The passive technique of image warping can be used, however, in every camera configuration. In addition, hardly a difference between the results of the two procedures can be determined with lenses of higher quality [BouWol92].

4.4 PHOTOMETRIC AND COLORIMETRIC CALIBRATION

In many techniques of digital image processing a linear response of the imaging sensors is presupposed. This applies especially to shading-based techniques such as shape-from-shading or photometric stereo. In these techniques it is assumed that the gray levels (or the color values) in the image are in *direct linear relation* to image irradiances, or rather, the radiation hitting the sensor. Furthermore, many techniques of color image processing, of which obviously no physical radiation model forms the basis, also often require linearity of the sensor response. This is based on the fact that also simple conversions between color spaces implicitly presuppose linearity. In the following, some reasons for the nonlinearity of cameras are described and in connection a technique for the linearization of the signal of CCD cameras is introduced.

4.4.1 Nonlinearities of Camera Signals

In principle, CCD chip signals are highly linear since photons are directly transformed into charge in a CCD cell (optoelectric transformation principle). Due to differing boundary conditions the signal supplied by the CCD camera does not exhibit any linearity. Two of these influences are described in the following. The first influence of the linear camera signal results from the desire to display images in good quality on the monitor. For this it must be ensured that the intensity that is received by the camera is transformed by the monitor into a proportional intensity. For a conventional monitor like a cathode-ray tube (CRT), the intensity I produced at the front of the display is approximately proportional to the applied voltage U raised to a power factor γ,

$$I \propto U^{\gamma} ,$$

where γ is known as *gamma value* or *gamma*. In order to achieve better display fidelity rendition quality on the monitor, a circuit for the gamma correction is integrated into the camera, which the *linear* CCD camera signal adapts to the inherently *nonlinear* monitor technique. The adjustment of the values of an intensity signal with the maximum *max* results in

$$I_{Out} = \max^{(1-1/\gamma)} \cdot I_{In}^{1/\gamma} .$$

The gamma value is dependent on the monitor and the video standard of the television system. Usually the gamma values of monitors lie between 2.0 and 3.0. In Table 4.2, the gamma values for the three most common television systems are represented. These are the PAL system, developed in Germany; the SECAM system, developed in France; and the NTSC system used in North America and Japan. Further details are specified in the CCIR standard.

The gamma correction must be disabled so that a reflection-based technique (such as shape-from-shading or photometric stereo) can be applied to the image. Furthermore, color constancy or color normalization can be obtained much more simply when the gamma correction is disabled. So far, only a few digital cameras offer the feature of disabling gamma correction.

	PAL	NTSC	SECAM
γ-value	2.8	2.2	2.8

Table 4.2. *Gamma values for television systems.*

The second reason for the occurrence of nonlinearities in the camera signal is based on the fact (as already mentioned in Section 4.3.1) that the camera signal is clipped whenever the analog signal exceeds the highest processable voltage. The attenuated signal must be postprocessed if the pre-knee-circuit cannot be disabled. For this, either the camera signal is linearized globally or all signal values that exceed the clipping level must be disregarded in succeeding processing tasks.

In the first case, the noise is greatly strengthened; in the second case, a more reduced, more linear, and more dynamic work area results for the camera. If the gray levels of the individual color channels, determined by a measurement in the nonattenuated area, are plotted graphically, then the typical process represented in Fig. 4.11 results.

In Fig. 4.12, the relation between intensity value and reflection is plotted for $\gamma = 1$ (after inverse gamma correction has been applied). The figure illustrates that an additional white balance and the subtraction of the black levels is necessary. This will be addressed in Section 4.4.3.

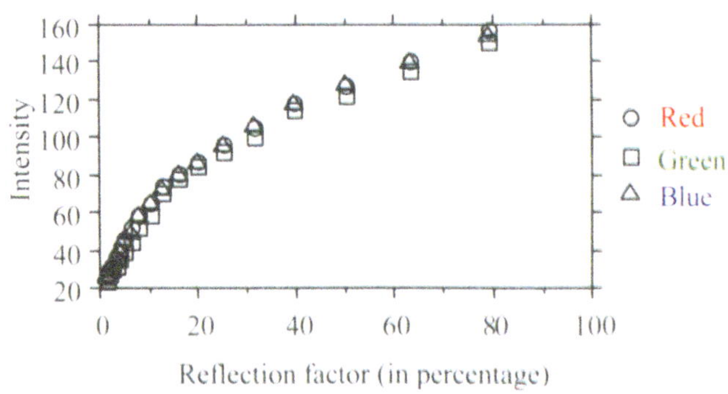

Figure 4.11. *Graph of the gamma corrected intensities in the RGB channels depending on reflection factor.*

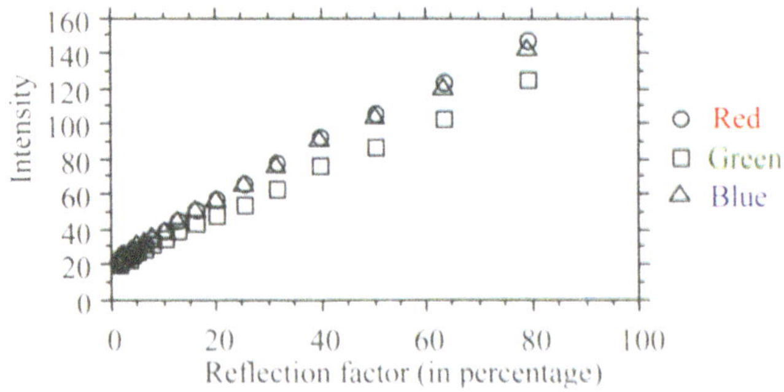

Figure 4.12. *Graph of the intensities in the RGB channels depending on reflection factor after inverse gamma correction ($\gamma = 1$).*

4.4.2 Measurement of Camera Linearity

The preparation of models of known reflection factors for the measurement of camera linearity can take place, among other things, using glass filters or standardized gray patches. The method using glass filters is more complex and also more exact than the method based on gray patches. In practice, however, gray patches are used for determining linearity simply because of their easy procurement.

For the measurement of camera linearity with glass filters, *neutral filter glasses* (see Section 4.2.1) are brought into the path of rays of the light source used and a white surface is radiated or supported. The radiated surface is captured by the camera and the resulting intensity values are compared with the values predicted by the known transmission factors of the neutral filter. Camera linearity can be measured on the basis of the deviations and agreements between the values.

Somewhat simpler is the use of grayscale reference cards for determining camera linearity. The most well-known are the Kodak grayscale card, the 31-level Macbeth grayscale card, and the six-level gray patches of the Macbeth ColorChecker (see Section 3.6.2). While the Macbeth gray patches are matte, the Kodak gray patches are glossy. The Kodak grayscale card contains 20 gray patches of an optical density between 0.0. and 1.9, as indicated by the manufacturer. The relation between the reflection factor R and the optical density D is defined by

$$R = 10^{-D}.$$

The filter method is generally more exact, since on the one hand the gray patches are subjected to a natural aging process and on the other hand the glass filters are manufactured in a precisely controlled production process, so that a set of transmission and tolerance values are available for these filters. A logarithmical representation is needed for the estimation of the γ value (see Section 4.4.1). With the determined γ value a transformation table can be subsequently established for all intensity values. By checking in this *lookup table,* the linearization of the individual intensity values can be implemented. Several video cards permit these operations in terms of hardware so that an automatic linearization is possible.

4.4.3 White Balance and Black-Level Determination

The absolute intensity values of the color channels depend largely on the spectral sensitivity of the filter-coated CCD cells. This becomes especially clear when a black-and-white camera is used in combination with color filters. On the one hand, the sensitivity of the CCD chip is strongly different within the spectral interval of interest, and on the other hand, the maximum transmission factors of the filters do not agree. For achromatic objects the falsifications are the most remarkable. A false-color image with differing red, green, and blue pixel levels (*color imbalance*) occurs. Many cameras offer the possibility of implementing a *white balance* with hardware. For this, the white balance function of the camera is activated directly after startup of the camera and a format-filling capture of a white object is produced. For the generation of an "ideal" white, barium sulfate must be used. In practice, however, good results can be obtained using a white sheet of paper that is as homogeneous as possible. With cameras that do not allow for the possibility of automatic white balance, a correction can be accomplished by an adapted aperture selection and therefore without a decrease in the signal-to-noise ratio (see [Nov et al. 90] and Section 4.2.1).

Apart from the white balance it is necessary to consider the *black level* of the image acquisition system and the gray level of the background lighting. The black level is that gray level which the A/D converter converts the black current to. The black current is generated without incidence of light. The amount of the black level can most simply be determined by image acquisition with an attached objective cover. It is to be considered that both are added additively to the image signal; however, they are to be viewed separately since the black level is independent of the aperture and the background lighting is dependent on the aperture.

The black level is partly determined directly in the camera by reading nonilluminated CCD elements located at border of the sensor. If the A/D converter can be parameterized with the black level, then the black level treatment can take place during image acquisition. In Fig. 4.13, the intensities are graphically represented in the long-wave spectral responses (red channel) depending on the reflection factor of an acquisition.

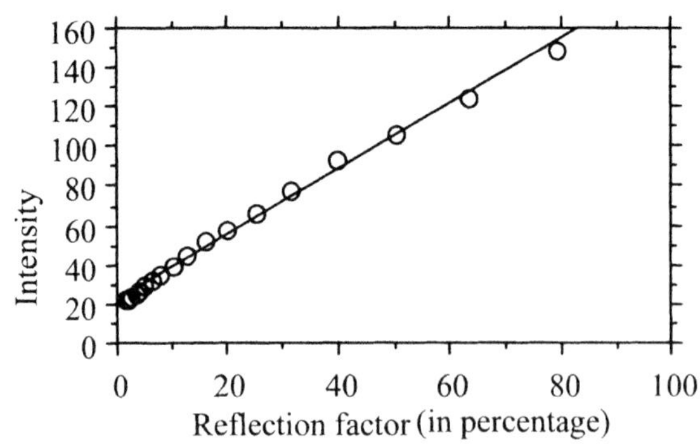

Figure 4.13. *Graph of intensities in long-wave spectral responses depending on reflection factor or an acquisition with (per hardware) switched-off gamma correction. The regression line shows that certain nonlinearities remain. This line can be consulted for the determination of the black level.*

4.4.4 Transformation into the Standard Color System *XYZ*

For tasks of color measurement, the color image supplied by the camera is to be transformed into a standard color image. The difficulty of the conversion lies in the fact that the camera does not possess the same sensitivity curves as the human eye. Furthermore, color standardization occurs over the virtual standard color matching functions $\bar{x}(\lambda), \bar{y}(\lambda), \bar{z}(\lambda) \neq X, Y, Z$ (see Section 3.1), which are derived by color comparison. The following comparison of the formulas clarifies the differences, whereby the symbols from Section 4.3 are used. The following applies:

$$\rho_1 = \int C(\lambda) \cdot S_1(\lambda)\, d\lambda, \tag{4.3}$$

$$X = \int C(\lambda) \cdot \bar{x}(\lambda)\, d\lambda, \tag{4.4}$$

$$\rho_2 = \int C(\lambda) \cdot S_2(\lambda)\, d\lambda, \tag{4.5}$$

$$Y = \int C(\lambda) \cdot \bar{y}(\lambda)\, d\lambda, \tag{4.6}$$

$$\rho_3 = \int C(\lambda) \cdot S_3(\lambda)\, d\lambda, \tag{4.7}$$

$$Z = \int C(\lambda) \cdot \bar{z}(\lambda)\, d\lambda, \tag{4.8}$$

By using the vector equations from Section 4.3, the following results:

$$\boldsymbol{\rho}_1 = \mathbf{c} \cdot \mathbf{s}_1, \quad X = \mathbf{c} \cdot \bar{\mathbf{x}},$$
$$\boldsymbol{\rho}_2 = \mathbf{c} \cdot \mathbf{s}_2, \quad Y = \mathbf{c} \cdot \bar{\mathbf{y}},$$
$$\boldsymbol{\rho}_3 = \mathbf{c} \cdot \mathbf{s}_3, \quad Z = \mathbf{c} \cdot \bar{\mathbf{z}}.$$

The task of conversion into the standard color system consists of determining the tristimulus values *XYZ* of the unknown color signal $C(\lambda)$ from a sensor response triple. One possibility of achieving this is to first completely reconstruct the color signal $C(\lambda)$ and to subsequently insert the color signal into Eqs. (4.4), (4.6), and (4.8). By doing this, the problems that arise due to metamers would be completely eliminated.

The goal to be attained in color calibration now consists of representing sensor metamers on eye metamers. For this the tristimulus values *XYZ* are used since those colors that appear similar to the human eye are represented on them. The dependency on the illumination used should be noted, since the color signal $C(\lambda)$ is influenced by the spectral reflectance factor $R(\lambda)$ of the object as well as the spectral distribution of the light source $E(\lambda)$. Two body colors that can be differentiated under one lighting condition may appear identical to the human eye under another lighting condition and vice versa. This problem is referred to as *eye-versus-camera metamer*.

The tristimulus values are then correctly determined when the sensor is adapted to the color sensitivity of the human observer with a suitable transformation. The problem of determining the tristimulus values *XYZ* according to the method outlined is *ill posed*, since three measured values stand opposite an infinite-dimensional color signal vector. Apart from that, the representation associated with it is not bijective. An infinite number of spectral color signals are represented by integration on a sensor response triple. The sensor standards described in the previous sections illustrate a precondition for color calibration. As previously suggested, each technique that approximates the color signal from a finite number of measured values can also be used for the estimation of tristimulus values and in addition for color measurement. In the following, a technique by R.L. Lee [Lee88] for color calibration using the Macbeth ColorChecker is introduced.

The problem of color calibration can be simplified by approximating the spectral reflectance factors $R(\lambda)$ and by adopting a known spectral distribution of lighting $E(\lambda)$ that does not change between calibration and usage. To do this, first the sensor sensitivities, for which a first estimate must be known, must be approximated with help of the Macbeth-ColorChecker (see Section 3.5.2). The quality of the calculated 2°-tristimulus values and the spectral reflectance factors is comparable to measurements using a spectroradiometer [Lee88].

The spectral reflectance factors $R_i(\lambda)$ (see Section 3.5.2) and the spectral distribution of the lighting are known for the 24 color and gray surfaces of the Macbeth ColorChecker. Therefore, the *RGB* signal to be expected can be calculated with the first estimate of the spectral sensor sensitivity (matrix $\mathbf{S}_1$). Thus, the following equation results:

$$\mathbf{P}_{Macbeth} = \mathbf{R}^T_{Macbeth} \cdot \mathbf{E} \cdot \mathbf{S}_1 ,$$

whereby the spectral reflectance factors of the Macbeth ColorChecker (column vectors) combined to an $n \times 24$-matrix $\mathbf{R}_{Macbeth}$ and the sensor responses for the color surfaces to a 24×3-matrix $\mathbf{P}_{Macbeth}$. The above equation is an expansion of Eq. (4.2) on 24 *RGB* triples. The importance of $\mathbf{E}$ and n is adopted from Section 4.3. These 24 calculated triples can be linked to the measured triples by a transformation matrix $\mathbf{S}_2$. It follows that

$$\mathbf{P}_{Macbeth} = \mathbf{R}^T_{Macbeth} \cdot \mathbf{E} \cdot \mathbf{S}_1 \cdot \mathbf{S}_2 .$$

An overdetermined linear equation system is to be solved. For this, for example, the determination of the *pseudoinverse* or a *singular value decomposition* is possible. An overview of optimization techniques and techniques for determining the pseudoinverses and their implementation in various programming languages (FORTRAN, PASCAL, C) is presented in [Pre et al. 92]. The improved approximation matrix **S** of the sensor sensitivities is formed by the product of the first estimate $\mathbf{S}_1$ and the transformation matrix $\mathbf{S}_2$. In the ensuing step, the spectral reflectance factors $R_i(\lambda)$ of the Macbeth-ColorChecker are approximated with the help of a main component analysis by three basic vectors plus the vector of the average spectral reflectance factor. The basic vectors can be represented by an $n \times 3$-matrix $\mathbf{R}_{basis}$ and the average reflection with the n-dimensional column vector $\mathbf{r}_{mean}$. Thus, each n-dimensional reflection vector **r** is to be represented by three weighting factors. The problem to be solved is reduced therefore to the calculation of a three-dimensional weighting column vector **b**, consisting of the weighting factors:

$$\mathbf{r} = \mathbf{r}_{mean} + \mathbf{R}_{basis} \cdot \mathbf{b} \,. \tag{4.9}$$

The weighting factors are linear functions of the measured *RGB*-system, sensor sensitivity, and lighting. The basic vectors can be calculated by insertion into Eq. (4.2) by

$$\mathbf{b}^T = (\rho - \mathbf{r}_{mean}^T \cdot \mathbf{E} \cdot \mathbf{S}) \cdot (\mathbf{R}_{basis}^T \cdot \mathbf{E} \cdot \mathbf{S})^{-1} \,.$$

Therefore, the spectral reflectance factor and the tristimulus values are able to be determined for the *RGB*-sensor response ρ by insertion into Eq. (4.9). With this approximation small errors still arise in the tristimulus values and in the spectral reflectance factor. An improved solution can be reached if the *RGB* triple ρ is multiplied by an empirically determined correction factor for each color surface and for each color channel. The determination of correction factors will not be elaborated upon here. For this, please refer to [Lee88].

The previous calculations refer to the color surfaces of the calibration chart. Colors to be examined are determined from them by interpolation of the correction factors. This occurs under the assumption that the interpolation scheme, which produces a new *RGB* triple ρ from the ones available, should also produce the corresponding correction factors. Furthermore, it is assumed that small changes of the *RGB* values of the color chart are connected only by small changes of the spectral reflectance factors (*smoothness constraint*). For a new *RGB* triple, that distance to each color of the calibration chart is determined, with which a weighting vector is erected, whose elements drop exponentially with the Euclidian *RGB* color distance. Since the weighting formula is not normalized to the value domain [0, 1], the correction matrix (consisting of the correction factors for each of the 24 color surfaces) must be adapted by a further linear transformation to avoid errors. This transformation is formed by the inverse of that matrix, which consists of line vectors of the weighting vectors for the calibration chart. Hereby all entries of the original correction matrix are maintained for the colors of the chart.

Real measurements with a film projector as light source show the exactness of this technique. The *root mean square error* amounts to approximately 8% according to [Lee88], if the samples are fully saturated and lie within the convex hull of the calibration colors represented in the 2° standard color table. Outside this domain, only limited reasonable results can be calculated. Within the reconstructed reflectance, factors (referring to the color type) can be designated as almost metameric. Therefore, an estimate of the tristimulus values *XYZ* is possible to a limited extent. For a calibration of the *RGB* values with reference to inhomogeneous illumination, please refer to [ChaRei96].

4.5 FURTHER READING

A technical introduction to the area of digital video can be found in [Poy96], and an introduction into the area of digital photography can be found, for example, in [Kas et al. 97]. For a detailed representation of color television systems please refer to [Eva94], and for an introduction to optical physics please refer to [Hec02]. A representation of color measurement and standard illuminators can be found in [Ber94] and [WysSti82]. The spectral distributions of standard illuminants established by the International Commission on Illumination (CIE) are available on the Internet (http://www.cie.co.at/cie/) and no longer need to be written out from tables. Reviews of digital cameras can be found on the Internet (e.g., at http://www.pcphotoreview.com/, which also provides a gallery of images acquired with various digital cameras and a list of Internet addresses of manufacturers of digital cameras).

4.6 REFERENCES

[Ber94] A. Berger-Schunn. Practical Color Measurement: A Primer for the Beginner, a Reminder for the Expert. Wiley, New York, 1994.

[BouWol92] T.E. Boult, G. Wolberg. Correcting chromatic aberrations using image warping. *Proc. Image Understanding Workshop*, San Diego, California, 1992, pp. 363-377.

[ChaRei96] Y.-C. Chang, J.F. Reid. RGB calibration for color image analysis in machine vision. *IEEE Transactions on Image Processing* **5** (1996), pp. 1414-1422.

[Eva94] B. Evans. Understanding Digital TV: The Route to HDTV. IEEE Press, 1994.

[FunHo89] B.V. Funt, J. Ho. Color from black and white. *Int. J. of Computer Vision* **3** (1989), pp. 109-117.

[Has et al. 95] Y. Hashimoto, M. Yamamoto, T. Asaida. Cameras and display systems. *Proc. IEEE* **83** (1995), pp. 1032-1043.

[Hec02] E. Hecht. *Optics*. 4th ed., Addison Wesley, Reading, Massachusetts, 2002.

[Kas et al. 97] A. Kasai, R. Sparkman, E. Hurley. *Essentials of Digital Photography*. New Riders, 1997.

[KhoZin96] A. Khotanzad, E. Zink. Color paper map segmentation using eigenvector line-fitting. *Proc. IEEE Southwest Symposium on Image Analysis and Interpretation*, San Antonio, Texas, 1996, pp. 190-194.

[Kle et al. 96] R. Klette, A. Koschan, K. Schlüns. *Computer Vision: Räumliche Information aus digitalen Bildern*. Vieweg, Braunschweig/Wiesbaden, Germany, 1996.

[Kle et al. 98] R. Klette, K. Schlüns, A. Koschan. *Computer Vision: Three-Dimensional Data from Images*. Springer, Singapore, 1998.

[Lee88] R.L. Lee. Colorimetric calibration of a video digitizing system: algorithm and applications. *Color Research and Application* **13** (1988), pp. 180-186.

[Len89] R. Lenz. Image data acquisition with CCD cameras. In: A. Gruen, H. Kahmen (eds.), *Optical 3-D Measurement Techniques*. Wichmann, Karlsruhe, Germany, 1989, pp. 22-34.

[Miy et al. 01] K. Miyazawa, K. Kurashiki, M. Hauta-Kasari, S. Toyooka. Broad-band color filters with arbitrary spectral transmittance using a liquid crystal tunable filter (LCTF). *Proc. SPIE* **4421** *9th Congress of the International Colour Association*, Rochester, New York, 2001, pp 753-756.

[Nov et al. 90] C. Novak, S.A. Shafer, S.A Willson. Obtaining accurate color images for machine vision research. *Proc. SPIE* **1250**, Perceiving, Measuring and Using Color, 1990, pp. 54-68.

[Per95] F.A. Perez. Hue segmentation, color circuitry, and the mantis shrimp. Ph.D. Thesis, California Institute of Technology, Pasadena, California, 1995.

[Poy96] C.A. Poynton. *A Technical Introduction to Digital Video*. Wiley, New York, 1996.

[Pre et al. 92] W.H. Press, S.A. Teukolsky, W.T. Vetterling, B.P. Flannery. *Numerical Recipes in C (FORTRAN, PASCAL)*. 2nd ed., Cambridge University Press, Cambridge, 1992.

[SPIE2658] *Proc. SPIE Symposium on Electronic Imaging: Science & Technology* **2658**: Color imaging: Device-independent color, color hard copy and graphic arts. San Jose, California, 1996.

[Vri et al. 95] M. Vriesenga, G. Healey, J. Sklansky, K. Peleg. Colored illumination for enhancing discriminability in machine vision. *J. Visual Communication and Image Representation* **6** (1995), pp. 244-255.

[WilSha91] R.G. Willson, S.A. Shafer. Active lens control for high precision computer imaging. *Proc. Int. Conference on Robotics and Automation*, Sacramento, 1991, pp. 2063-2070.

[WysSti82] R.G. Wyszecki, W.S. Stiles. *Color Science: Concepts and Methods, Quantitative Data and Formulae*. 2nd ed., Wiley, New York, 1982.

5 COLOR IMAGE ENHANCEMENT

The general interest in digital techniques for the improvement of image quality has continually increased in the previous years. This can be traced back to, among other things, the ever-more-frequent use of color image processing methods in medical research, in the printing industry, and in aerial image analysis. A further increase in the importance of color image enhancement techniques can be noted by the archiving of digital images and in the area of digital video. Here existing programs for digital image processing, such as XV or Adobe PhotoshopTM, already offer a number of interactive tools. The standards imposed on the techniques differ from one area of application to another.

Image acquisition definitely has the biggest influence on image quality. Therefore, image quality can also be enhanced when an especially high-quality camera and/or an especially high-quality sensor is used (see [Hsi et al. 92]). It can also be enhanced when the color of the lighting source is optimally adapted to the sensitivity of the camera (see [Vri et al. 92] and [Vri et al. 95]). However, the examination of hardware characteristics is not the subject of this chapter. Here we are concerned the question of how the quality of an already-existing color image can be later enhanced by suitable techniques. Before an expansion of the known techniques from gray-level image processing to color image enhancement is introduced in the following sections, an explanation of the terminology is necessary. W.K. Pratt [Pra91] combines the three areas "image enhancement," "image restoration," and "geometrical image modification" into the generic term *image improvement*.

1. *Image enhancement* includes those techniques that enhance the recognizability of objects in an image. The goal can be to enhance the view for the human observer as well as the transformation of an image into a more-suitable format for computer-aided processing [Pra91]. These techniques consist of (but are not limited to) contrast enhancement, filtering in the spatial domain, noise suppression, image smoothing, and image sharpening [HarSha91].
2. *Image restoration* is a process through which a degraded image is restored as well as possible. A perfect image restoration is only possible when the

degradation function is mathematically invertible. Typical techniques of image restoration are inverse filtering, Wiener filtering, and filtering using smallest squares [HarSha91].

3. *Geometrical image modification* includes image enlargement, image reduction, image rotation, and nonlinear image distortion.

Only color techniques that can be summarized by the term "image enhancement" will be introduced in the following sections. One technique for the restoration of color images can be found, for example, in [Zhu et al. 92]. Chellappa's compilation [Che93] is recommended for an extensive overview of techniques for image enhancement in gray-level images. The main theme of this chapter concerns the examination of techniques for the area of color image processing.

5.1 FALSE COLORS AND PSEUDOCOLORS

According to the definition given above, one goal of image enhancement is to enhance the visual recognizability of objects in the image for the human observer. Since the human eye can distinguish more colors than gray scales, image enhancement is therefore also reached by a pseudocolor coding of a gray-level image. Depending on the derived results either significant regions in the gray-level image are colored (e.g., toxicated areas in an aerial image) or the entire gray-level scale is spread and transformed into a color representation. Likewise, selected regions in a color image can be colored and in this way be emphasized. This topic will not be addressed further here. An introduction to this topic can be found in [GonWin87] or in [Pra91].

Another possibility of image enhancement exists in *false color coding*. In this procedure, the original color image, which is represented by its three primary colors or by a set of multispectral images, is mapped pixel by pixel into a color space. This color space is defined by three representational coordinates (e.g., *RGB*) that are described as a linear or a nonlinear function of the original color pixel. The general goal is the formation of an image in which the objects do not have the colors that one would expect (see [Pra91]). Therefore, a red river or a blue banana attracts the attention of the observer more than the natural color representation of these objects. In contrast to pseudocolor coding, in false color coding of an image individual domains are not dyed (e.g., the river or the banana), but rather a (linear or nonlinear) transformation is implemented on all color values. This transformation can be achieved, for example, simply by manipulation of the video lookup table.

The differing sensitivity of the human eye with regard to various wavelengths is frequently an additional motivation for false color coding. In this way, the transformation of an object's color into a domain with higher sensitivity (e.g., into the middlewave, so-called green domain) makes the recognizability of an object easier. Furthermore, in multispectral images the sensitivity of a sensor frequently lies outside the visible wave domain (e.g., in the infrared domain). In this case a transformation (false color coding) of the pixel values is necessary for a visualization.

Pseudocolor coding can be applied, for example, to an x-ray image to aid luggage inspection. Fig. 5.1a shows an x-ray image of a bag containing a low-density knife. The knife is not easy to recognize by a screener at an airport. The visibility of the threat object in the image is significantly enhanced by pseudocolor coding (see Fig. 5.1b), especially when considering the fatigue of a screener after several hours of duty. A study of several different pseudocolor codes applied to improving luggage inspection is discussed in detail in Chapter 14.

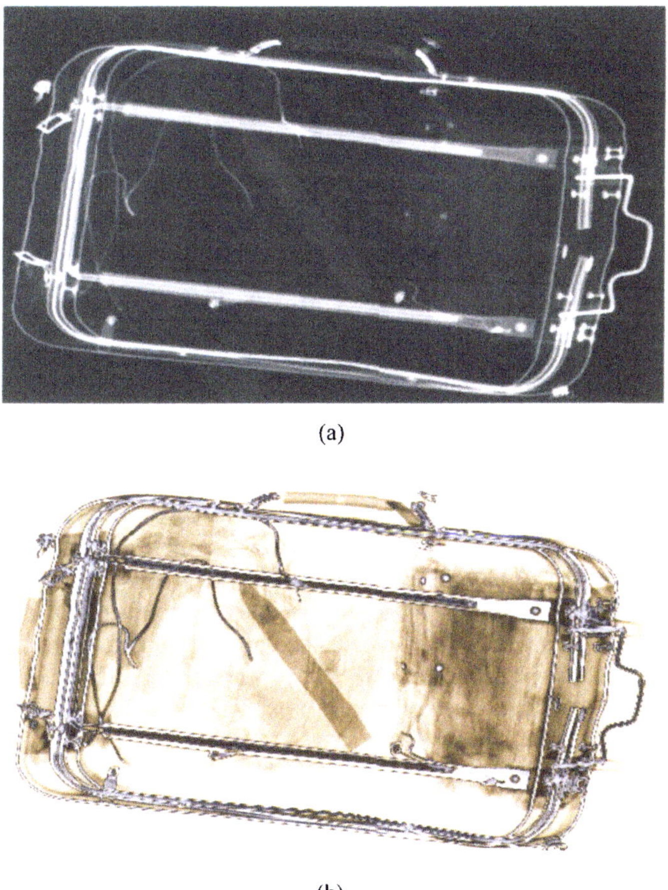

(a)

(b)

Figure 5.1. *Bag containing a low-density knife. (a) X-ray image. (b) Enhanced color-coded version of this image.*

5.2 ENHANCEMENT OF REAL COLOR IMAGES

The quality enhancement of real color images places far greater demands on the techniques than is the case with gray-level images. This lies first in the fact that vectors instead of scalars are used; second, the complexity of the image perception must be taken into consideration (see Chapter 2). A monochromatic-based procedure in color image enhancement is still widely used, although one can expect better results with vector-valued formulations that take into consideration the connection between vector components. In the monochromatic-based procedures, the components of the color vectors for each pixel are transformed independently of each other into another set of components.

The principal course of this formulation is as follows. In each of these new color coordinates a separate (monochrome) algorithm is used for image enhancement. The "enhanced" image coordinates T_1', T_2', and T_3' attained in this manner are transformed back (as a rule) into the *RGB* color space for representation. Since the components of a pixel T_k, $k = 1,2,3$ are processed independently of each other, it must be guaranteed that the "enhanced" coordinates T_k' lie within the representation domain of the *RGB* system. The color coordinate system T_k, $k = 1,2,3$ for the enhancement technique is individually selected according to the requirements for the problem. Note that color distortions can appear on the basis of quantization errors in the individual components [RodYan94]. Apart from this, there is no correlation between this method and human color perception. Therefore, the success of this method (i.e., the enhancement of image quality to be achieved by this method) cannot be predetermined. For this reason, a number of more recent techniques also include color perception-based aspects in predominantly vector-valued formulations.

In the following, several problems in the monochromatic-based procedure are described and the advantages of vector-valued formulations over monochromatic-based formulations are shown. These are introduced in the following sections by examples of noise suppression and contrast enhancement in color images. Here it is always assumed that the color images are present in full 24-bit quantization that the *RGB* output delivers to a CCD camera. The enhancement of subsampled color signals that are used for compression purposes in digital video recording and in the Kodak PhotoCD is not the topic of this chapter. For techniques concerning the treatment of this formulation of the question please refer to [SchSte97].

5.3 NOISE REMOVAL IN COLOR IMAGES

Digital color images are degraded during their formation by various kinds of noise (see Section 1.2.6). This noise can be completely eliminated only when its cause is entirely known, can be modeled mathematically, and is invertible. However, this knowledge is generally not available. In addition, few investigations into the modeling and treatment of noise in vector-valued color images exist so far, as previously mentioned. Nevertheless, an attempt should be made to at least partially reduce the noise. Methods that are already known for noise suppression in gray-level images are expanded and used for color images. These techniques are introduced here only when it is necessary to understand their expansion to color images. A detailed presentation of the gray-level technique is found in [GonWin87] and [KleZam96]. In the following, several monochromatic-based and vector-valued techniques for noise suppression in color images are introduced.

5.3.1 Box-Filter

A simple way of smoothing a gray-level image involves replacing each pixel by the mean value of its neighboring pixels. This linear operator is called *box-filter,* since the image is scanned as through an ideal column and the form of the operator is similar to that of a box (see Fig. 5.2). The attempt to transfer this filter for use on vector-valued color images demonstrates several principal problems of color image processing. In a monochromatic-based procedure, the box-filter would be used separately on each individual vector component. However, this would lead to color changes in the resulting image. This problem can be partly overcome by first transforming the *RGB* image into the *HSI* color space, and then applying the box-filter to the chromaticity (hue and saturation) of the image.

Let chromaticity be defined as a complex function with $j = \sqrt{-1}$, where the hue $t(x,y)$ is noted as phase and the saturation $s(x,y)$ is noted as absolute value (see [Fre88]). Chromaticity $b(x,y)$ is given by

$$b(x,y) = s(x,y) \cdot e^{jt(x,y)} . \qquad (5.1)$$

The real part ($\Re$) and the imaginary part ($\Im$) of $b(x,y)$ can be computed due to Euler's formula by

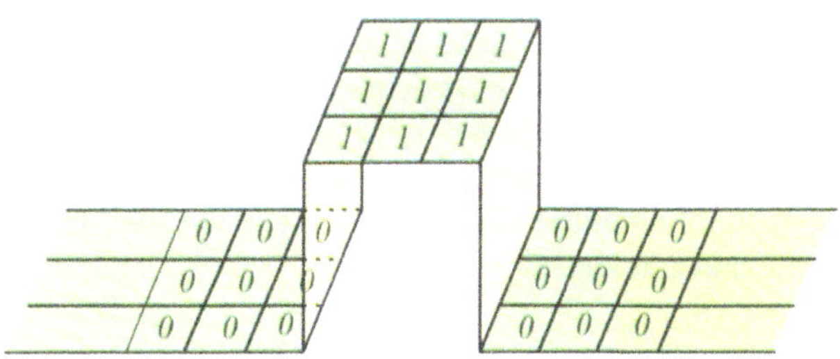

Figure 5.2. *Three-dimensional representation of the coefficients of a two-dimensional box-filter for gray-level images.*

$$\Re(b(x,y)) = s(x,y)\cdot\cos(t(x,y)), \tag{5.2}$$

$$\Im(b(x,y)) = s(x,y)\cdot\sin(t(x,y)), \tag{5.3}$$

Chromaticity $b(x,y)$ is defined by

$$b(x,y) = \Re(b(x,y)) + j\cdot\Im(b(x,y)) \text{ or} \tag{5.4}$$

$$b(x,y) = s(x,y)\cdot\cos(t(x,y)) + j\cdot s(x,y)\cdot\sin(t(x,y)). \tag{5.5}$$

Considering the abovenamed complex-valued definition of chromaticity, the box-filter is now applied to the chromaticity. In this connection, the chromaticities within the operator window are vectorally added. For an operator size of $n \times m$ pixels, where n and m are assumed to be odd numbers, the mean chromaticity $b_\mu(x,y)$ results at position (*x*,*y*) of the image as:

$$b_\mu(x,y) = \frac{1}{n\cdot m}\sum_{k=-(n-1)/2}^{(n-1)/2}\;\sum_{l=-(m-1)/2}^{(m-1)/2}\Re(b(x-k,y-l)) + j\cdot\frac{1}{n\cdot m}\sum_{k=-(n-1)/2}^{(n-1)/2}\;\sum_{l=-(m-1)/2}^{(m-1)/2}\Im(b(x-k,y-l)).$$

Using the above formula, the color distortions in the results reduce in contrast to a separate use of the box-filter in each vector component in the *RGB* space. However, by calculation of a mean chromaticity at color transitions, colors lying in between can also occur. Averaging at a red-green transition yields, for example, a yellow transition that is perceived as unnatural. The same effect results likewise when a Gauss filter is applied to the chromaticity.

The example of the box-filter shows the difficulties with the design of filters for color image processing. So far, few signal-theoretical investigations for vector-valued color signals exist. Simple terms from scalar mathematics, for example, that of the average value, cannot be directly transferred to vector-valued functions. How can one describe the mean value of a quantity of vectors? Is a however-defined "averaged" vector a suitable candidate for the description of a less "noisy" color signal? These questions cannot be answered satisfactorily at this time. However, the computation of mean values is important for the treatment of color images. Here the use of nonlinear filters appears more suitable, such as the median filter, which is discussed in the following section.

5.3.2 Median Filter

The median filter is often applied to gray-level images due to its property of edge-preserving smoothing. The median filter is a nonlinear operator that arranges the pixels in a local window according to the size of their intensity values and replaces the value of the pixel in the resulting image by the middle value in this order. The extension of the concept of scalar median filtering to color image processing is not a simple procedure (see [Val et al. 91]). One essential difficulty in defining the median in a set of vectors is the lack of a "natural" concept of ranking vectors. The problems occurring here are outlined based on the following example.

Consider the following example: Three color pixels in the *RGB* color space are defined by the three color vectors $\mathbf{p}_1 = (10,40,50)^T$, $\mathbf{p}_2 = (80,50,10)^T$, and $\mathbf{p}_3 = (50,100,150)^T$. If the median filter is applied separately to each vector component, the resulting vector is $\mathbf{p}' = (50,50,50)^T$. This vector does not exist in the input data and it represents an achromatic gray level. Furthermore, the separate median filtering may cause different shifts of the local maxima in the particular color components. This problem is illustrated in the following on the basis of a one-dimensional linear vector signal with two entries (see Fig. 5.3).

The original signal represents three segments: left band bounded-right at n_1, middle band bounded-left by n_1 and bounded-right by n_2, and right band bounded-left by n_2. Both entries are contaminated by impulse noise: one at position $n_1 - 3$ and the other at $n_2 + 1$ (see Fig. 5.3a). The result of separate filtering with a window size of five pixels is shown in Fig. 5.3b. The impulse at $n_1 - 1$ in entry 1 is reduced and the distinction, which separates two segments in the original image, is shifted by one pixel position to the left. A shift also occurs at n_2 in entry 2. From this observation, one may conclude that individual processing does not remove the impulse noise. Instead, it moves the noise position to affect its neighboring values in the image [Zhe et al. 93].

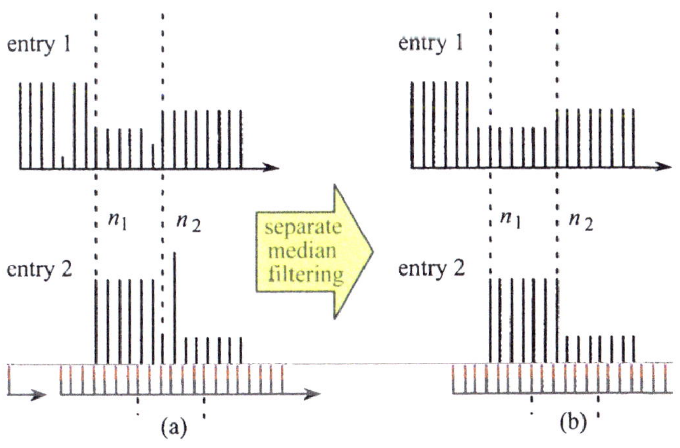

Figure 5.3. *The edge jitter problem caused by separate median filtering (after [Zhe et al. 93]).*

The example specified above shows that both a color distortion and the loss of the characteristic of edge preservation may occur when the median filter is applied separately to each single component of the color vectors. If all three color components are regarded at the same time, then an order must be defined for the sequence of color vectors that defines, for example, when a vector is larger or smaller than another vector. Several techniques have been proposed for median filtering in color images, for example:

1. An adaptive scalar median filter [Val et al. 91]
2. A vector median filter (with weighting [Wic et al. 92] or without weighting [Arg et al. 91])
3. A reduced vector median filter [RegTes97]
4. A median filter applied to chromaticity in the HSI space [Fre88]
5. A median filter based on conditional ordering in the HSV space [Var et al. 01]
6. Vector directional filters (see [PlaVen00], [TraVen93], [Tra et al. 96])

Furthermore, vector median filters can be connected to morphological operations considering a lexicographic order. Detailed information about the mathematical theory on this connection may be found in [Cas et al. 00]. The six techniques mentioned above are discussed in the following.

Adaptive Scalar Median

Valavanis et al. [Val et al. 91] presented a color-oriented modification of the median filter using scalar medians. The algorithm is quite heuristic and requires a more exact explanation. For this some notations are specified. Consider a color image **C** in the *RGB* color space. It holds that

$$\mathbf{C} = \begin{pmatrix} R \\ G \\ B \end{pmatrix} \quad \text{or rather} \quad C(x,y) = \begin{pmatrix} R(x,y) \\ G(x,y) \\ B(x,y) \end{pmatrix},$$

where $\mathbf{C}(x,y)$ is the pixel vector at location (x,y) in color image **C**. Furthermore, denote

$$\mathbf{C}(x_R, y_R) = med_R\left\{\mathbf{C}(x,y) \mid x,y \in W\right\},$$

$$\mathbf{C}(x_G, y_G) = med_G\left\{\mathbf{C}(x,y) \mid x,y \in W\right\}, \text{ and}$$

$$\mathbf{C}(x_B, y_B) = med_B\left\{\mathbf{C}(x,y) \mid x,y \in W\right\},$$

where W is a window of odd size. $\mathbf{C}(x_R, y_R)$, $\mathbf{C}(x_G, y_G)$, and $\mathbf{C}(x_B, y_B)$ are the coordinates of the median values for the respective color channel within the window W and $med_R\{\ \}$, $med_G\{\ \}$, and $med_B\{\ \}$ are the median operators for the computation of the resulting RGB values. Since only one component of the vector is regarded at a time, three vectors result. Each of those vectors contains a median element and two assigned elements. The three vectors can be combined to a "median matrix" $\mathbf{M}$ given by

$$\mathbf{M} = \left(\mathbf{C}(x_R, y_R), \mathbf{C}(x_G, y_G), \mathbf{C}(x_B, y_B)\right)$$

$$= \begin{pmatrix} R(x_R, y_R) & R(x_G, y_G) & R(x_B, y_B) \\ G(x_R, y_R) & G(x_G, y_G) & G(x_B, y_B) \\ B(x_R, y_R) & B(x_G, y_G) & B(x_B, y_B) \end{pmatrix}.$$

Every column in $\mathbf{M}$ represents a real pixel vector in the original image. A monochromatic-based (separate) median filtering in each vector component yields a new "median" as $\left(R(x_R, y_R), G(x_R, y_R), B(x_B, y_B)\right)^T$, consisting of the diagonal elements of $\mathbf{M}$. This vector does not necessarily exist in the input image and its use may cause color distortions in the filtered image. Consider again the example mentioned above with three color vectors $\mathbf{p}_1 = (10,40,50)^T$, $\mathbf{p}_2 = (80,50,10)^T$, and $\mathbf{p}_3 = (50,100,150)^T$. The resulting median matrix is

$$\begin{pmatrix} 50 & 80 & 10 \\ 100 & 50 & 40 \\ 150 & 10 & 50 \end{pmatrix}.$$

The result $(50, 50, 50)^T$ of a separate median filtering is certainly the worst choice for a median candidate. Altogether 27 combinations can be computed from 3×3 matrix elements ($(50, 100, 150)^T$, $(50, 100, 10)^T$, $(50, 100, 50)^T$, etc.). Since color distortions in the image have a direct influence on the color perception of the viewer, perception-referred criteria for the selection of the "most suitable" candidate are consulted there. For this a representation in the intuitive perception-adapted *HSI* color space is used (see Section 3.4.1 for a transformation from *RGB* to *HSI*). The abstract quantitative values of a single vector pixel are named h, s, and i.

$$h = H(R,G,B),\ s = S(R,G,B),\ i = I(R,G,B). \tag{5.6}$$

The transformation of the *RGB* values into an *hsi* representation constitutes an ill-posed problem. Nonremovable singularities can occur; for example, no hue value and no saturation value can be determined for gray shades (achromatic colors) where the values are identical in the red, green, and blue channels. To decrease the probability for the occurrence of these disturbances, Zheng, Valavanis, and Gauch [Zhe et al. 93] suggest the following determinations for h, s, and i:

$$h = H(\bar{r}, \bar{g}, \bar{b}),\ s = S(\bar{r}, \bar{g}, \bar{b}) \text{ and } i = I(\bar{r}, \bar{g}, \bar{b}), \tag{5.7}$$

where $\bar{r}$, $\bar{g}$, and $\bar{b}$ denote the mean values for the red, green, and blue channel within a window. Note that the following holds:

$$H(\bar{r}, \bar{g}, \bar{b}) \neq \bar{h} = \iint h(x,y)\, dx\, dy,$$

$$S(\bar{r}, \bar{g}, \bar{b}) \neq \bar{s} = \iint s(x,y)\, dx\, dy, \text{ and}$$

$$I(\bar{r}, \bar{g}, \bar{b}) \neq \bar{i} = \iint i(x,y)\, dx\, dy.$$

In order to minimize the distortions in color appearance in a processed image, the following three criteria are proposed in [Zhe et al. 93]:

1. The hue changes should be minimized.
2. The shift of saturation should be as small as possible and it is better to increase than to decrease the saturation.
3. It is desirable to maximize the (relative) luminance contrast.

Since the change in each individual component has a completely different influence on the result, the three criteria are not examined at the same time, but successively, which is in some way similar to a conditional ordering. For this purpose, candidates are regarded according to the first criterion:

$$(r_l, g_m, b_n)^T = (r_i, g_j, b_k)^T \text{ if } min\{H(r_i, g_j, b_k) - h\}, \tag{5.8}$$

where $l,m,n \in \{i,j,k \mid i,j,k = 1,2,3\}$. The notation $min\{\cdot\}$ denotes the minimum value of the difference measurement. Since hue is judged separately from the other vector components, it is likely that there exist more than one possible candidate when the above criterion is applied. Several possible vector medians may possess the same hue and be similar to the overall average hue in the region. If more than one possible median remains, the second criterion is applied:

$$(r_x, g_y, b_z)^T = (r_l, g_m, b_n)^T \text{ if } min\{S(r_l, g_m, b_n) - s\}, \tag{5.9}$$

where $x,y,z \in \{l,m,n \mid l,m,n = 1,2,3\}$. Finally, the vector median with the largest intensity value is selected.

The search, described above, for a "heuristically perception-optimal" result based on the individual values of the median filters preserves, as far as possible, the characteristics of the median filtering and causes no large color distortions. The algorithm for computing the adaptive scalar median includes the following seven processing steps:

1. Process each color vector component separately.
2. Form the median matrix.
3. Remove the data with singularities.
4. Apply Eq. (5.7).
5. Determine candidates based on Eq. (5.8).
6. Reduce the number of candidates according to Eq. (5.9).
7. Insert the candidate with the highest intensity value into the output image.

Vector Median Filter

One important characteristic of the median filter is that it does not produce values for a pixel that do not exist in the original image. This characteristic is not always guaranteed during the adaptive scalar median filtering. One way to overcome this problem consists in the application of a vector median filter (see PitTsa91]). For a set of N vectors $x_1,\ldots,x_N$ within a right-angled window and any vector norm $\|\cdot\|_L$, the *vector median filter* VM is defined by the equation (see [Arg et al. 91])

$$VM\{\mathbf{x}_1,\mathbf{x}_2,\ldots,\mathbf{x}_N\}=\mathbf{x}_{VM}$$

where

$$\mathbf{x}_{VM}\in\{\mathbf{x}_1,\mathbf{x}_2,\ldots,\mathbf{x}_N\}$$

and

$$\sum_{i=1}^{N}\|\mathbf{x}_{VM}-\mathbf{x}_i\|_L\le\sum_{i=1}^{N}\|\mathbf{x}_j-\mathbf{x}_i\|_L\ ,\quad j=1,2,\ldots,N. \tag{5.10}$$

The result of this filter operation selects that vector in the window that minimizes the sum of the distances to the other N-1 vectors regarding the L-norm. Additionally, weightings can be specified for the vector median filter. In general, distance weights w_i, $i=1,\ldots,N$ and components weights v_i, $i=1,\ldots,N$ can be defined. The result of the weighted vector median filter is the vector $\mathbf{x}_{WVM}$ with

$$\mathbf{x}_{WVM}\in\{\mathbf{x}_1,\mathbf{x}_2,\ldots,\mathbf{x}_N\}$$

and

$$\sum_{i=1}^{N} w_i \left\| v_i \otimes (\mathbf{x}_{WVM} - \mathbf{x}_i) \right\|_L \leq \sum_{i=1}^{N} w_i \left\| v_i \otimes (\mathbf{x}_j - \mathbf{x}_i) \right\|_L , \quad j = 1,2,\ldots,N. \quad (5.11)$$

The pointwise multiplication is denoted with $\otimes$ (i.e., if $c = a \otimes b$, then $c_i = a_i \cdot b_i$ holds for all vector components). If several vectors fulfill Eqs. (5.10) and/or (5.11), then that vector that is closest to the vector in the center of the window is selected. The result of these filter operations is not unambiguous and beyond that also dependent on the selected L-norm. Additionally, the weightings have an influence on the result when the weighted filter is applied. Thus, weightings must be selected very carefully; otherwise, the characteristic of edge preservation can be lost. In each constellation it is guaranteed that the filtering produces no additional new color vector. An investigation of the influence of the weighting on the processing result can be found in [Wic et al. 92].

Reduced Vector Median Filter

The computation of the vector median for the entire color image is quite time-consuming. Regazzoni and Teschioni [RegTes97] suggest an approximation of the vector median filtering, which they call *reduced vector median filtering* (*RVMF*). For this they use "space-filling curves," as are also partly used with scanners, in order to map the three-dimensional color vectors into a one-dimensional space. In this one-dimensional space the median is then determined in a conventional way (as in gray-level images). A detailed presentation of the RVMF technique may be found in [RegTes97]. According to [RegTes97], the signal-to-noise ratios related to the nondistorted original images and the filtered distorted images are similar to those of the original vector median filtering. The signal-to-noise ratios for the reduced vector median filter are both for Gaussian noise and for impulse noise always worse than the values for the "original" vector median filter.

Median Filter Applied to the Chromaticity in the HSI Space

The difficulty of the definition of an order of rank between the color vectors arises also if a representation in the *HSI* color space is selected. Here the hue is indicated as angle and likewise the ranking cannot be easily specified. Frey [Fre88] suggested a procedure that works in the *HSI* model and already comes very close to a kind of median filter. He searches for the mean value in the chromaticity plane and thus guarantees that the value in the output image is identical to a value in the particular window in the input image. This procedure is a variant of the vector median described above and works exclusively on the chromaticity. The chromaticity is defined (as for the box-filter) as a complex function, where the hue $t(x, y)$ is noted as phase and the saturation $s(x, y)$ is noted as absolute value. The equations for the chromaticity image $b(x,y)$, the real part ($\Re$), and the imaginary part ($\Im$) of $b(x, y)$ were given in Eqs. (5.1) to (5.5).

That particular pixel in the window is looked for, for which the sum of the squared distances to all other pixels of the window is minimal. In the output image this chromaticity is then registered for the pixel. For a window of size $m \times n$ with $k = m \cdot n$ pixels, the squared Euclidean distance d_{ij}^2 of the pixel i to the pixel j is given by

$$d_{ij}^2 = \left(\Re(b_i) - \Re(b_j)\right)^2 + \left(\Im(b_i) - \Im(b_j)\right)^2$$

with $1 \le i, j \le k$. The sum of the squared distances d_i^2 of pixel i to all others is denoted as

$$d_i^2 = \sum_{j=1}^{k} d_{ij}^2 \quad \text{for } i \ne j.$$

The chromaticity of that pixel that fulfills $d^2 = \min\{d_i^2\}$ is selected for the resulting image. Just as with the vector median, this minimum is not always unambiguous. If several pixels fulfill the above condition, then the value that is most similar to the original value is selected for the chromaticity.

Median Filter Based on Conditional Ordering in the HSV Space

In *conditional ordering,* vectors are first ordered according to the ordered values of one of the components, such as the first component. Then, vectors having the same value for the first component are ordered according to the ordered values of another component (e.g., the second component) and so on. A color **c** in the *HSV* space is denoted by **c**(h,s,v) with the *hue* value $h \in [0,360)$, the *saturation* value $s \in [0,1]$, and the *value* $v \subset [0,1]$. Vardavoulia et al. [Var et al. 01] suggest the following ordering of *HSV* space vectors:

1. Vectors are sorted from those with smallest v to those with greatest v.
2. Vectors having the same value of v are sorted from those with greatest s to those with smallest s.
3. Vectors having the same values of s and v are sorted from those with smallest h to those with greatest h.

Using mathematical notation for these three ordering criteria, two operators $<_C$ and $=_C$ (see [Var et al. 01]) can be defined for two colors by

$$\mathbf{c}(h_i, s_i, v_i) <_c \mathbf{c}(h_k, s_k, vk) \Leftrightarrow$$
$$(v_i < v_k) \vee (v_i = v_k \wedge s_i > s_k) \vee$$
$$(v_i = v_k \wedge s_i = s_k \wedge h_i < h_k),$$

$$\mathbf{c}(h_i, s_i, v_i) =_c \mathbf{c}(h_k, s_k, vk) \Leftrightarrow$$
$$(v_i = v_k) \wedge (s_i = s_k) \wedge (h_i = h_k)^{\cdot}$$

The n color vectors $\mathbf{c}_1, \mathbf{c}_2, \ldots, \mathbf{c}_n$ representing the colors of the pixels inside a window are placed in ascending order, forming the set of ordered values $\{\mathbf{c}_1, \mathbf{c}_2, \ldots, \mathbf{c}_n\}$ in which $\mathbf{c}_1 <_c \mathbf{c}_2 <_c \ldots <_c \mathbf{c}_n$. The middle vector in this order is called the *vector median* and is denoted by $vmed_{HSV}$. In the case of a gray-level image, the median is identical to the conventional definition of a median.

Vector Directional Filters

Vector directional filters (VDFs) are a class of multivariate filters that are based on polar coordinates and vector ordering principles considering the angle between the color image vectors as ordering criterion (see [Tra et al. 96], [TraVen93], and [PlaVen00]). The class of VDFs operates on the direction of the image vectors with the objective of eliminating vectors with atypical directions (large chromaticity errors) in the vector space. A detailed investigation on the statistic characteristics of vector directional filters is to be found in [Tra et al. 96] and [PlaVen00]. Similar to the median filter applied to the chromaticity (see 2.4), the VDFs operate on the chromaticity components of a color. In other words, they are designed to detect chromaticity errors, but not intensity outliers.

Discussion

Since noise-removal techniques are designed to enhance image quality, their performance can be evaluated regarding at least three criteria. Criterion 1 considers the quality of the resulting color image based on its visual impression. Criterion 2 considers the quantity of the removed noise, and criterion 3 considers the computational cost of the technique. In Fig. 5.4 the results of filtering by applying the vector median and the adaptive scalar median to a test image disturbed by impulse noise are visualized. The color impulse noise is controlled via a noise rate and a noise impulse height. The noise rate indicates how many pixels of a color component are altered. The noise impulse height indicates the absolute value by which the pixel concerned is changed.

In the example in Fig. 5.4, the noise rate is equal to 11 and the noise impulse height is equal to 66. Thus, each eleventh pixel in a color component is altered by around ± 66 values. To enhance the visibility of the differences in the filtering results, Fig. 5.4 shows, in addition, the difference images (intensified by a factor of three) between the original image and both filtered images. It is to emphasize that a stronger noise reduction occurs when applying the vector median to the image than when applying the adaptive scalar median.

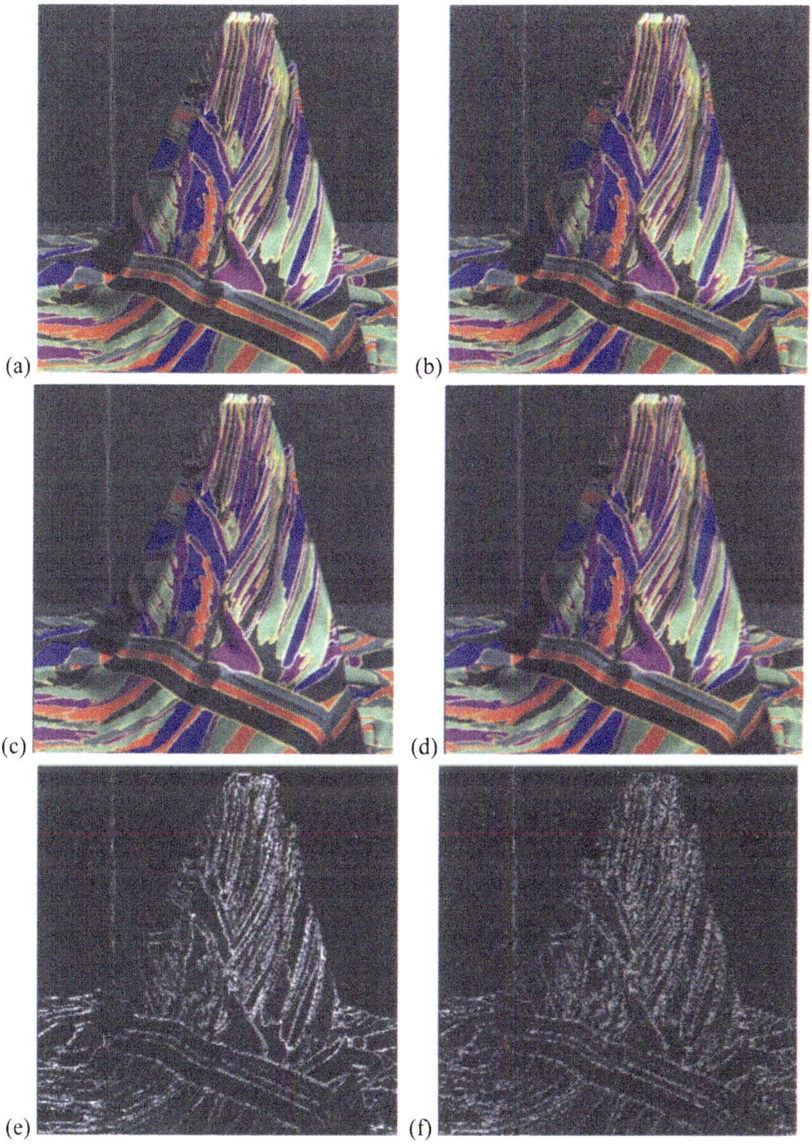

Figure 5.4. *Illustrations of (a) the original color image "Shawl," (b) the disorted color image, and the results of (c) vector median filtering, (d) adaptive scalar median filter (intensified by the factor three). Difference image between the original image and (e) the vector median filtered image, and (f) the adaptive scalar filtered image.*

This applies in general also to different noise rates. In Figs. 5.5 and 5.6 the interpolated standard deviations for the differences between original image and filtered image (by means of adaptive scalar median, vector median, and chromaticity median) are indicated for different noise rates. The noise impulse height always amounts to 66 values. Figure 5.5 shows the computed values for test image A and Fig. 5.6 shows the computed values for test image B. With all three techniques, the results improve with increasing noise rate. The worst results were obtained for both test images with the adaptive scalar median.

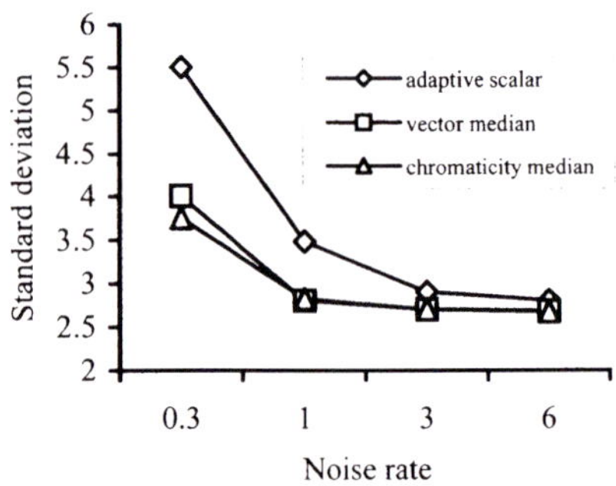

Figure 5.5. *Interpolated standard deviations for the differences between original image and filtered image for various noise rates and test image A. The noise impulse height always amounts to 66 values.*

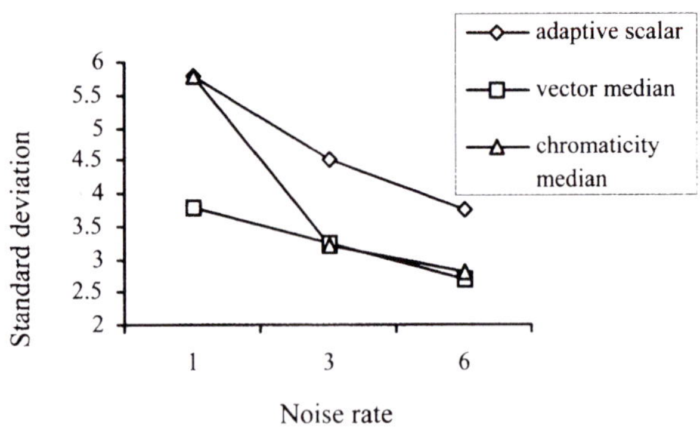

Figure 5.6. *Interpolated standard deviations for the differences between original image and filtered image for various noise rates and test image B. The noise impulse height always amounts to 66 values.*

However, these results are not representative. Already with two images better results are obtained in the case of a small noise rate, sometimes with the chromaticity median and sometimes with the vector median, depending on image content. Regarding criterion 1, the vector median performed best in this example, while no significant differences between VM and $vmed_{HSI}$ could be recognized in [Var et al. 01]. Regarding criterion 2, the adaptive scalar median performed worst in this example, while the performances of the other two filters were dependent on the noise. Due to [Var et al. 01] $vmed_{HSI}$ outperforms VM and VDFs based on a noise model in the *HSI* space, while better signal-to-noise ratios are reported for VDFs than for VM in [Tra et al. 96] and [TraVen93]. Regarding criterion 3, the vector median is the computationally most costly operator in this comparison. Its actual cost is dependent on the vector norm, and the cost can be reduced by applying, for example, a quasi-Euclidean norm [Bar et al. 00]. Since most of the other operators are designed for a special color space, their costs also include the transformations between the selected color spaces.

In summary, the vector median filter performs well in experiments in the *RGB* space. However, so far there is no single best technique for median filtering of color images. The performance of the operators depends on the image content and the kind of noise with which they are degraded. Moreover, the selection of a color space may also influence the selection of a median operator, since most of the operators are designed for a particular color space.

5.3.3 Morphological Filter

Another possibility of noise suppression is based on the application of morphological filters on image matrices. The elementary operations of mathematical morphology are *erosion* (or Minkowski subtraction) and *dilatation* (or Minkowski addition). Suppression of noise in gray-level images can be achieved by implementing these operations successively. Erosion followed by dilatation is indicated as closing and dilatation followed by erosion is indicated as opening. Crucial for the expected result is a suitable definition of the structuring element for erosion and dilatation. A detailed representation of morphological operations for gray-level images can be found in [Har et al. 87].

In a monochromatic-based expansion of morphological filtering of gray-level images to color images, the operations (analogous to their use in gray-level images) are carried out separately for each vector component of the color signal. If the structuring element is defined in this connection with definite size and definite form for all vector components, then color distortions can appear in the color image, produced by a combination of the individual results. The loss of details in morphologically filtered images, which is already recognizable in the filtering of gray-level images, increases even more in color images filtered in this manner.

In order to reduce this effect, Deng-Wong, Cheng, and Venetsanopoulos [Den et al. 92] propose to adaptively select the morphological filter depending on the local vicinity of a pixel in a color channel. In this connection, a structuring element that is best suited to the local geometrical characteristics is selected for each pixel in every component of the color vector. They achieve better detail and color fidelity with this rigid technique. The preservation of edges is less than with the vector median. In addition, the processing time for this monochromatic-based, adaptive, morphological technique, restricted by the search for the best structuring element for each pixel in the vector component, is roughly twice as long as with the vector median [Den et al. 92].

More suitable than the separate implementation of morphological filtering is a vector-valued morphological filtering. Vector median filters can be connected to morphological operations considering a lexicographic order. Detailed information about the mathematical theory on this connection may be found in [Cas et al. 00].

5.3.4 Filtering in the Frequency Domain

Techniques for noise suppression are used in the spatial domain as well as in the frequency domain of the image in gray-level image processing. In contrast, in color image processing, techniques are employed almost exclusively in the spatial domain. This is due to the fact that it is not yet well defined how a vector-valued color image should be represented in the frequency domain and how the achieved result is to be interpreted. In a monochromatic-based procedure, the discrete Fourier transformation (DFT) could be used separately on each vector component of the color signal. As a result, each pixel is represented in the frequency domain by six numbers of the three complex-valued Fourier transform. The connections between the individual vector components of the color signal are, however, completely lost with this way of looking at things.

A vector-valued approach for a Fourier transformation of a color image is proposed by Sangwine [San97]. He employs *quaternions*, introduced by Hamilton in mathematics, which are also used in robotics applications for treating coordinate systems with four variables. A quaternion number can be represented in the form

$$a + ib + jc + kd$$

whereby a, b, c, and d are real numbers and i, j, and k are complex operators. It holds that

$$i^2 = j^2 = k^2 = ijk = -1 .$$

The discrete quaternion Fourier transformation (DQFT) and its inverse are defined for a discrete field $f(m,n)$ of the dimension $M \times N$. With a scaling factor identical for the transformation and back transformation $S = 1/\sqrt{MN}$, the transformation is given by

$$F(u,v) = S \cdot \sum_{m=0}^{M-1} \sum_{n=0}^{N-1} \left(e^{j2\pi \frac{mu}{M}} \cdot f(m,n) \cdot e^{k2\pi \frac{nv}{N}} \right)$$

and the back transformation by

$$f(m,n) = S \cdot \sum_{u=0}^{M-1} \sum_{v=0}^{N-1} \left(e^{j2\pi \frac{mu}{M}} \cdot F(u,v) \cdot e^{k2\pi \frac{nv}{N}} \right).$$

The two-dimensional quaternion-valued functions can represent color images (e.g., in the *RGB* color space). For a quaternion-valued pixel of the form $a + ib + jc + kd$ that should be multiplied with both complex exponential functions, the three components of the affiliated color vector can be inserted into the three imaginary components and the real component is set to zero. Nevertheless, this assignment is not mathematically conclusive.

The DFT and the DQFT differ considerably in the way they represent information in the frequency domain. The DQFT separates the four possible combinations of horizontal and vertical cosine and sine components into the four components of the quaternion-valued spectral point, while the DFT combines several of them into a single real or imaginary component (see [San97]). It is not yet clear what effects this has on the analysis and interpretation of the information in the frequency domain [SanEll01]. Further investigations are necessary before the DQFT can be used in color image processing.

5.4 CONTRAST ENHANCEMENT IN COLOR IMAGES

The saturation and the lightness of a color image describe different types of information in the image. The saturation indicates whether a domain appears more or less chromatic in relation to the lightness. Frequently details in a color image with a low, relative lightness contrast can differ from the background on the basis of the differing color saturation. Unlike the increase of the relative contrast in gray-level images, an exclusive observation of the lightness of a color image is not sufficient for increasing its contrast. The differing meanings and interpretations of the term *contrast* were described in Section 1.2.5. In this section, techniques in changing the relative lightness contrast and the relative saturation contrast are described.

An enhancement of the image contrast can be achieved in isolated cases in which only the color saturation (the relative saturation contrast) is changed. Apart from the effects on the detectability of image details, the increase of the relative lightness contrast and the increase of the relative saturation contrast have differing aesthetic effects. The effect of saturated colors is frequently indicated as "stronger" and the effect of lighter colors as "friendlier" [Dav91]. Furthermore, it should be taken into consideration that a change of the perceived saturation can also be brought about by an increase in lightness (see [Dav91]).

Lightness as well as saturation can be changed for a (relative) contrast enhancement in a color image. In principle, treatment of image data is possible in the *RGB* color space. The transformation of the image data is presented in the *HSI* space for a separate observation of hue, saturation, and lightness. The treatment is carried out in this color space and the coordinates are, as usual, subsequently transferred back into the *RGB* color space.

5.4.1 Treatment of Color Saturation and Lightness

It is frequently advantageous for nearly real-time applications if a technique for image enhancement is carried out in the *RGB* color space. Computationally costly transformations between color spaces do not apply here and the problem of non-representability of colors in the one or the other color space does not need to be considered. A simple enhancement of the color saturation in the *RGB* color space is achieved by the transformation

$$\begin{pmatrix} R' \\ G' \\ B' \end{pmatrix} = \frac{\max\{R,G,B\}}{\max\{R,G,B\} - \min\{R,G,B\}} \begin{pmatrix} R - \min\{R,G,B\} \\ G - \min\{R,G,B\} \\ B - \min\{R,G,B\} \end{pmatrix}$$

on each pixel [Toe92]. Another possibility exists in the sole enhancement of the relative lightness contrast and ensuing back transformation into the *RGB* color space. Strickland, Kim, and McDonnell [Str et al. 87] propose for this the use of an observer-conformed lightness component L. It is

$$L(x,y) = 0.299 \cdot R(x,y) + 0.587 \cdot G(x,y) + 0.114 \cdot B(x,y), \tag{5.12}$$

whereby (x,y) are the pixel coordinates. The lightness component L is also called *luminance signal* in television technology and is equivalent to the Y component in the NTSC terminology (see Chapter 3). The luminance image produced in this manner $L(x,y)$ can be processed with a technique known from gray-level image enhancement. The result is an enhanced image $L'(x,y)$. The lightness scaling is described by a two-dimensional function $K(x,y)$, which is determined by

$$K(x,y) = \frac{L'(x,y)}{L(x,y)}.$$

The three components of a changed color vector can be calculated by multiplication of the original *RGB* data with $K(x,y)$. It holds that

$$\begin{pmatrix} R'(x,y) \\ G'(x,y) \\ B'(x,y) \end{pmatrix} = \begin{pmatrix} K(x,y) \cdot R(x,y) \\ K(x,y) \cdot G(x,y) \\ K(x,y) \cdot B(x,y) \end{pmatrix}.$$

Only the lightness of the image is changed by this scaling. Saturation S and hue H remain unchanged in the *HSI* or *HSL* color space. By an ensuing transformation of the data into the *YIQ* color space it is to be taken into account that not only the Y component is influenced by the scaling in the *RGB* color space, but also the I and Q components [YanRod96].

This simple procedure is monochromatic-based and does not consider the relation between the components of the color vectors. Furthermore, it is not always guaranteed that the resulting values R', G', B' do not exceed the allowed maximum value $G_{max} + 1$. This can be achieved by a restriction and/or a *clipping* of the function values. In gray-level images, an overmodulation of the lightness component is prevented by this clipping. By the treatment of vector-valued color images a color transition results from the clipping, for example, of only one color vector component (see Section 4.3.1).

In gray-level image processing, an equalization of the intensity histogram is frequently implemented for the increase of the relative intensity contrast. In analogy to that, the equalization of the one-dimensional histogram of the luminance values and an ensuing back transformation can be implemented into the *RGB* color space. However, it must be considered that, on the basis of quantization errors, the back transformation of the changed luminance values into the *RGB* color space can lead to color distortions (see [RodYan95]).

An expansion of this technique for equalizing a three-dimensional histogram in the *RGB* color space for color images is not possible without further details. Analogous to processing in gray-level images, a direct equalization of the common allocation function of the color components cannot be carried out in the *RGB* color space. For this, the three color channels must be uncorrelated, which (as a rule) they are not. This problem can be solved, for example, by first implementing a Karhunen-Loève transform and then stretching the histograms along the main components [SohSch78].

A less computationally costly possibility for solving this problem is suggested by Trahanias and Venetsanopoulos [TraVen92]. They search for the one symmetrical three-dimensional result histogram in the *RGB* color space that differs least from the input histogram in values. In Fig. 5.6, an example of an increase of the relative saturation contrast is given by three-dimensional histogram equalization. The manipulation of the three-dimensional histogram is very time consuming. A certain increase in speed can be achieved in this connection by a recursive calculation [Zha et al. 96].

From the perceptive-psychological view, it is frequently desirable that the vector components of the color image are not treated in the *RGB* color space, but rather instead in a color space that better conforms to the human observer. In this color space the image contrast can be influenced in various ways and means by the separate treatment of the color, saturation, and lightness components. A possibility for converting this concept exists in the spreading of color saturation under retention of the hue [Fre88]. In addition to this, a two-dimensional histogram of the chromaticity is constructed and for each hue in the histogram the smallest used saturation is sought. The entire saturation domain is now utilized in that all saturations of this hue are multiplied by a corresponding factor. So that the hues, which occupy only the unsaturated domain, are not spread too thickly, no spreading of the saturation is carried out below a given saturation border for this hue. The visual analysis of the image is enhanced by the increase of the color saturation in simultaneous retention of hues and lightnesses.

A modification of the histogram in the *RGB* color space would deliver substantially worse results here [Fre88]. All three color channels must be symmetrically spread in order to keep the hue constant. Additionally, the high values necessary as border values for the spreading in individual coordinates of the three-dimensional histogram of the *RGB* images are not necessarily caused by chromatic objects, but rather in many cases by the achromatic background. From this results only a minor increase of the color contrast. This disadvantage is avoided by the methods in the *HSI* space mentioned above, in which the saturations for differing hues are variously changed.

Another variant in the increase of the relative lightness contrast and the relative saturation contrast was proposed by Bockstein [Boc86]. He transformed the *RGB* data into the *HSL* space and subdivided the hue domain into equal-sized components (e.g., 96 subdivisions [Boc86]). In each of these hue domains, a histogram for the luminance and a histogram for the color saturation are constructed. An equalization of the luminance and color saturation is carried out afterward in each of these domains by retention of the hue. The outcome is again transformed back into the *RGB* space. With the rigid limiting of 96 hue domains, the actual number of hues appearing in the image is not taken into consideration.

A better differentiation of objects in the color image can be achieved in which the pixels in the image are first subdivided into chromatic and achromatic pixels. For this subdivision, the criterion that will be introduced in Section 7.2.2 can be implemented. For the achromatic pixels only the value domain of the lightness component is stretched. For the chromatic pixels the attempt is made, considering the Phong reflection model, to change the individual components of the vector-valued color signal in such a manner that the hue remains the same and the relative lightness contrast becomes maximum. This technique was successfully employed in microscopy images. A detailed representation can be found in [GupCha96]. The subdivision into chromatic and achromatic pixels is heuristic and in my opinion not conclusive. Apart from that, no comparison of the results with this and without this subdivision is contained in [GupCha96].

In the techniques previously introduced, saturation and lightness were always processed separately from each other. However, both components can also be set in relationship to each other. This can result from the modeling of lightness components under consideration of the variation of the saturation component in a pyramid with several resolution steps (see [Toe92]). The image data is first transformed into the *HSL* color space. Next, pyramids are formed for the saturation component *S* and for the luminance component *L*. Toet proposes to set the new lightness components to be dependent on the relative lightness and saturation values of the original image in several resolutions. If the relative saturation contrast for a pixel within an image window is greater than the relative lightness contrast, then the new lightness component is set equal to the relative saturation contrast. Otherwise, the lightness component remains unchanged. Through this, strong changes in the saturation values are carried over into the lightness component. Subsequently, the saturation component is changed in a similar way depending on the relative lightness contrast of the original image. This procedure appears very heuristic and does not correspond to knowledge of human color perception.

An entirely different technique for treating relative intensity contrast and relative saturation contrast in color images is based on a modification of the *HSI* color space [Kim et al. 92]. The modification consists of trying to linearize the relation between color saturation and lightness. In the modified color space the lightness values are first spread to their maximum possible dynamic domain. The modified saturation component, depending on the maximum and minimum saturation value in the (not modified) *HSI* image, is subsequently stretched to the maximum possible domain, and the result is again transformed back into the *RGB* space. However, no statistical investigations yet exist concerning to what extent the use of this modified *HSI* color space leads to an improvement of the results.

5.4.2 Changing the Hue

If the hue value is changed exclusively or in addition to the other color components, then the result is a color image in pseudocolor representation. Such a representation can be attained, such that in addition to a modification of the histogram of the color saturation, a modification of the histogram of the hues is also implemented (see [Fre88]). The hues no longer match with the hues of the original, but modest hue changes in the original become visible as large changes of the hue in the treated image. Through this, the visual evaluation of the image can be likewise supported. In expansion of the contrast definitions in Section 1.2.5, one can discuss the increase of the relative hue contrast. This designation is not usual.

From perception psychology it is known that simultaneous color contrast influences the way in which colors are perceived [Dav91]. In this connection, a red influences the perception of a green object and a yellow influences the perceived blue. This effect is strengthened when a surface is surrounded by others. From the view of perception psychology, the visual detection of colored objects is influenced (and/or the color image is enhanced) when significant image domains

Figure 5.7. *Illustrations of the color image "Shawl" reduced in color saturation by 50% (left), the results of the contrast increase by modification of the hue histogram (center), and by three-dimensional histogram equalization (right).*

are represented in component colors as differently as possible. However, the resulting image is as a rule not suitable for a subsequent treatment by a computer vision technique. Figure 5.7 shows illustrations of the color image "Shawl" (see original image Fig. 5.4a) reduced in color saturation by 50% and the results of the increase of the relative saturation contrast by modification of the hue histogram and by three-dimensional histogram equalization.

For the production of the image reduced in color saturation by 50%, the image data was transformed into the *HSI* space, the saturation values were multiplied by the factor 0.5, and the data was subsequently again transferred back into the *RGB* space. The reinforcement of the relative saturation contrast is only certainly recognizable in the gray-level representations.

5.5 REFERENCES

[Arg et al. 91] F. Argenti, M. Barni, V. Cappellini, A. Mecocci. Vector median deblurring filter for color image restoration. *Electronics Letters* **27** (1991), pp. 1899-1900.

[Bar et al. 00] M. Barni, F. Buti, F. Bartolini, V. Cappellini. A quasi-Euclidean norm to speed up vector median filtering. *IEEE Transactions on Image Processing* **9** (2000), pp. 1704 –1709.

[Boc86] I.M. Bockstein. Color equalization method and its application to color image processing. *J. Optical Society of America A* **3** (1986), pp. 735-737.

[Cas et al. 00] V. Caselles, G. Sapiro, D.H. Chung: Vector median filters, inf-sup operations, and coupled PDE's: theoretical connections. *J. Math. Imaging and Vision* **12** (2000), pp. 109-120.

[Che93] R. Chellappa. *Digital Image Processing*. IEEE Computer Society Press, Los Alamitos, California, 2nd ed., 1993.

[Dav91] J. Davidoff. *Cognition Through Color*. MIT Press, Cambridge, Massachusetts, 1991.

[Den et al. 92] P. Deng-Wong, F. Cheng, A.N. Venetsanopoulos. Adaptive morphological filters for color image enhancement. *Proc. SPIE* **1818**, *Visual Communications and Image Processing*, 1992, pp. 358-365.

[Fre88] H. Frey. *Digitale Bildverarbeitung in Farbräumen*. Ph.D. thesis, University Ulm, Germany, 1988.

[GonWin87] R.C. Gonzalez, P. Wintz. *Digital Image Processing*. 2nd ed., Addison-Wesley, Reading, Massachusetts, 1987.

[GupCha96] A. Gupta, B. Chanda. A hue preserving enhancement scheme for a class of colour images. *Pattern Recognition Letters* **17** (1996), pp. 109-114.

[Har et al. 87] R.M. Haralick, S.R. Sternberg, X. Zhuang. Image analysis using mathematical morphology. *IEEE Transactions on Pattern Analysis and Machine Intelligence* **9** (1987), pp. 532-550.

[HarSha91] R.M. Haralick, L.G. Shapiro. Glossary of computer vision terms. *Pattern Recognition* **24** (1991), pp. 69-93.

[Hsi et al. 92] T. Hsieh, R.D. McGrath, V. Quintana, W. Wiercienski. Color linear image sensor for vision applications. *Proc. SPIE* **1822**, Optics, Illumination, and Image Sensing for Machine Vision *VII*, 1992, pp. 12-20.

[Kim et al. 92] J.-Y. Kim, J.-C. Shim, Y.-H. Ha. Color image enhancement based on modified IHS coordinate system. *Proc. SPIE* **1825**, Intelligent Robots and Computer Vision XI, 1992, pp. 366-377.

[KleZam96] R. Klette, P. Zamperoni. *Handbook of Image Processing Operators*. Wiley, New York, 1996.

[PitTsa91] I. Pitas, P. Tsalides. Multivariate ordering in color image filtering. *IEEE Transactions on Circuits and Systems for Video Technology* **1** (1991), pp. 247-259.

[PlaVen00] K.N. Plataniotis, A.N. Venetsanopoulos. *Color Image Processing and Applications*. Springer, Berlin, Germany, 2000.

[Pra91] W.K. Pratt. *Digital Image Processing*. 2nd ed., Wiley, New York, 1991.

[RegTes97] C.S. Regazzoni, A. Teschioni. A new approach to vector median filtering based on space filling curves. *IEEE Transactions on Image Processing* **6** (1997), pp. 1025-1037.

[RodYan94] J.J. Rodriguez, C.C. Yang. Effects of luminance quantization error on color image processing. *IEEE Transactions on Image Processing* **3** (1994), pp. 850-854.

[RodYan95] J.J. Rodriguez, C.C. Yang: High-resolution histogram modification of color images. *Graphical Models and Image Processing* **57** (1995), pp. 432-440.

[San97] S.J. Sangwine. Fourier transforms of colour images: The quaternion FFT. *Proc. 4th Int. Workshop on Systems, Signals and Image Processing*, M. Domanski, R. Stasinski (eds.), Poznan, Poland, 1997, pp. 207-210.

[SanEll01] S.J. Sangwine, T.A. Ell. Hypercomplex Fourier transforms of color images. *Proc. Int. Conference on Image Processing*, Thessaloniki, Greece, 2001, Vol. I, pp. 137-140.

[SchSte97] B.E. Schmitz, R.L. Stevenson. The enhancement of images containing subsampled chrominance information. *IEEE Transactions on Image Processing* **6** (1997), pp. 1052-1056.

[SohSch78] J.M. Soha, A.A. Schwartz. Multidimensional histogram normalization contrast enhancement. *Proc. 5th Canadian Symposium on Remote Sensing*, 1978, pp. 86-93.

[Str et al. 87] R.N. Strickland, C.-S. Kim, W.F. McDonnell. Digital color image enhancement based on the saturation component. *Optical Engineering* **26** (1987), pp. 609-616.

[Toe92] A. Toet. Multiscale color image enhancement. *Pattern Recognition Letters* **13** (1992), pp. 167-174.

[Tra et al. 96] P.E. Trahanias, D. Karakos, A.N. Venetsanopoulos. Directional processing of color images: theory and experimental results. *IEEE Transactions on Image Processing* **5** (1996), pp. 868-880.

[TraVen92] P.E. Trahanias, A.N. Venetsanopoulos. Color image enhancement through 3-d histogram equalization. *Proc. 11th Int. Conference on Pattern Recognition*, The Hague, Netherlands, 1992, Vol. III, pp. 545-548.

[TraVen93] P.E. Trahanias, A.N. Venetsanopoulos. Vector directional filters - a new class of multichannel image processing filters. *IEEE Transactions on Image Processing* **2** (1993), pp. 528-534.

[Val et al. 91] K.P. Valavanis, J. Zheng, J.M. Gauch. On impulse noise removal in color images. *Proc. Int. Conference on Robotics and Automation*, Sacramento, California, 1991, pp. 144-149.

[Var et al. 01] M.I. Vardavoulia, I. Andreadis, P. Tsalides. A new vector median filter for colour image processing. *Pattern Recognition Letters* 22 (2001), pp. 675-689.

[Vri et al. 92] M. Vriesenga, G. Healey, K. Peleg, J. Sklansky: Controlling illumination color to enhance object discriminability. *Proc. Int. Conference on Computer Vision and Pattern Recognition*, Champaign, Illinois, 1992, pp. 710-712. [Vri et al. 95] M. Vriesenga, G. Healey, J. Sklansky, K. Peleg. Colored illumination for enhancing discriminability in machine vision. *J. Visual Communication and Image Representation* **6** (1995), pp. 244-255.

[Wic et al. 92] R. Wichman, K. Öistämö, Q. Liu, M. Grundström, Y. Neuvo. Weighted vector median operation for filtering multispectral data. *Proc. SPIE* **1818**, Visual Communications and Image Processing, 1992, pp. 376-383.

[YanRod96] C.C. Yang, J.J. Rodriguez. Saturation clipping in the LHS and YIQ color spaces. *Proc SPIE* **2658**, Color Imaging: Device-Independent Color, Color Hardcopy, and Graphic Arts, San Jose, California, 1996, pp. 297-307.

[Zha et al. 96] Q. Zhang, P.A. Mlsna, J.J. Rodriguez. A recursive technique for 3-d histogram enhancement of color images. *Proc. IEEE Southwest Symposium on Image Analysis and Interpretation*, 1996, San Antonio, Texas, pp. 218-223.

[Zhe et al. 93] J. Zheng, K.P. Valavanis, J.M. Gauch. Noise removal from color images. *J. of Intelligent and Robotic Systems* **7** (1993), pp. 257-285.

[Zhu et al. 92] W. Zhu, N.P. Galatsanos, A.K. Katsaggelos. Regularized multichannel restoration of color images using cross-validation. *Proc. SPIE* **1818**, Visual Communications and Image Processing, 1992, pp. 345-356.

6 EDGE DETECTION IN COLOR IMAGES

Digital image functions are generally degraded during the image formation process by distortions of various causes. One goal of early processing steps, apart from the removal of these distortions (e.g., noise), is to detect significant discontinuities (edges) of the image function. The accuracy in detecting these discontinuities (*edge detection*) and the efficiency in implementing these operations are important criteria for using an algorithm in the area of computer vision. This applies equally to the extraction of edges in color images. Inaccuracies in color edge detection directly influence the results of a subsequent color image processing technique such as edge-based color image segmentation (see Section 7.3), edge-based stereo analysis (see Section 9.3.1), or edge-based tracking or recognition of colored objects in image sequences.

While edge detection in gray-level images is a well-established area that is covered in most textbooks on digital image processing, edge detection in color images has not received the same attention. Since color images contain more information than gray-level images, more edge information is expected from color edge detection in general. Most textbooks address this topic only very briefly or not at all. This is based on the fact that the transition from scalar to vector-valued image functions has not yet been generally accepted in this area. In addition, monochromatic-based formulations are still in extensive use for detecting edges in color images.

Several monochromatic-based definitions for image edges and their inadequacies were already discussed in Section 1.2.3. The basic idea of monochromatic-based techniques consists of applying a technique known from gray-level image processing separately on each vector component of the color image and to "suitably" combine the results attained in this manner. A combination can consist, for example, of the union of the edge points determined in the individual vector components, the calculation of the sum of absolute values of the gradients for the three color components, or a regularized fusion of the edge results in every color component [Sal et al. 96a]. However, techniques of this type ignore the connection between the vector components. Since a color image does not represent a scalar, but rather a vector-valued function, a discontinuity of the chromatic information should also be defined as vector-valued.

Up to now, the signal-theoretic fundamentals for color images have not been presented sufficiently. While wavelets and Gabor functions contributed substantially to theoretical examinations of edge detection in gray-level images (see, e.g., [Ayd et al. 96], [Lee96], and [Mal96]), the inclusion of wavelets and Gabor functions in investigations in digital color image processing has not gained much attention outside this area of color image coding. A technique for the detection of colored textures using Gabor functions is found in [JaiHea98] and a technique for a classification of multispectral data is introduced in [Mal et al. 97] by using adaptive wavelets. Fundamental signal-theoretic investigations are urgently needed here for color images. In the following, various vector-valued techniques for detecting discontinuities in color images are described.

6.1 VECTOR-VALUED TECHNIQUES

In some early publications on color edge detection (see [PieHar86], [Shi87], and [Sol85]), vector-valued techniques were suggested that replaced gray-level differences of adjacent pixels in some way by vector differences. However, these simple difference operators do not represent the state of the art in edge detection either in gray-level image processing or in color image processing. Thus, they are not presented in this section.

6.1.1 Color Variants of the Canny Operator

Novak and Shafer suggest an extension of the *Canny operator* [Can86] for color edge detection. Kanade introduced this approach in [Kan87]. The philosophy of the Canny operator consists of first determining the first partial derivatives of the smoothed image function according to x and y, and on the basis of these values, finding the magnitude and direction of the "best" edge. For a color pixel and/or color vector $\mathbf{C}(x,y)=(R,G,B)$ in the *RGB* space, the variation of the image function is described, as mentioned in Section 1.2.3, at location (x,y) by the equation $\Delta\mathbf{C}=\mathbf{J}\Delta(x,y)$. The Jacobian matrix is indicated by $\mathbf{J}$, which contains the first partial derivatives for each component of the color vector. In the *RGB* space $\mathbf{J}$ is given by

$$\mathbf{J}=\begin{pmatrix} R_x & R_y \\ G_x & G_y \\ B_x & B_y \end{pmatrix}=\left(\mathbf{C}_x,\mathbf{C}_y\right).$$

The indexes x and y designate the respective partial derivatives of the functions, for example:

$$R_x = \frac{\partial R}{\partial x} \text{ and } R_y = \frac{\partial R}{\partial y}.$$

That direction in the image, along which the largest change and/or the largest discontinuity in the chromatic image function occurs, is represented by the eigenvector of $\mathbf{J}^T\mathbf{J}$ corresponding to the largest eigenvalue.

This technique can likewise be used for multichannel images and/or image sequences. The direction θ of a color edge defined in such a way is determined in an individual image with any norm by

$$\tan(2\theta) = \frac{2 \cdot \mathbf{C}_x \cdot \mathbf{C}_y}{\|\mathbf{C}_x\|^2 - \|\mathbf{C}_y\|^2},$$

whereby $\mathbf{C}_x$ and $\mathbf{C}_y$ are the partial derivatives of the color components, for example, in the *RGB* space

$$\mathbf{C}_x = (R_x, G_x, B_x).$$

The magnitude m of an edge is indicated by

$$m^2 = \|\mathbf{C}_x\|^2 \cos^2(\theta) + 2 \cdot \mathbf{C}_x \cdot \mathbf{C}_y \cdot \sin(\theta)\cos(\theta) + \|\mathbf{C}_y\|^2 \sin^2(\theta).$$

Finally, after the direction and the magnitude are determined for each edge, *non-maximum suppression* is used based on a threshold value in order to eliminate "broad" edges.

The technique introduced above is theoretically well founded. Its practical application is, however, very computationally costly. Therefore, as an alternative, less time-consuming procedures were tested at Carnegie Mellon University. Here each operator is subdivided into three individual processing steps that are defined below. Then, after k, $k = 0, \cdots, 3$ processing steps for the individual components of the color vectors have been implemented, these results can be combined with a "combination operator." The following processing steps are implemented next for the attained scalar values. By this combination, the vector-valued technique is transferred into a monochromatic-based technique. For the definition of the "combination operators," different mathematical norms can be used as a basis, such as the L_1-norm (sum of the absolute values), the L_2-norm (Euclidian norm) or the L_∞-norm (maximum of the absolute values). A color edge operator can now be described by the k processing steps that were implemented for the individual color channels and the index of the norm used for the combination. The Canny operator can be subdivided into three processing steps:

I. Determine the partial derivatives
II. Calculate edge direction and edge magnitude
III. Implement the nonmaximum suppression.

In accordance with the convention specified above, the I/2 Canny operator consists of determining the partial derivatives for each component of the color vectors (processing step I), combining the results using the Euclidean norm L_2, and executing the remaining processing steps for the combined values. The 0/1 and 0/2 color operators would consist of determining an intensity image of the color vectors and subsequently executing the standard Canny operator. Kanade [Kan87] summarizes the results attained for a selected series of color test images as follows:

1. The color edges describe object geometry in the scene better than the intensity edges, although over 90% of the edges are identical.
2. II/∞ proved to be the best multilevel operator (i.e., computation of the edge magnitude and direction separately for each color channel and subsequent selection of the edge with the strongest magnitude).
3. A similar, but not as good, result can be obtained with the I/∞ operator. This is faster than the II/∞ operator since the combination of the color channels takes place earlier.
4. The II/∞ operator produced almost exactly the same edges as the theoretical operator based on the Jacobian analysis described above.

Note that so far there have been no investigations as to what extent the choice of the smoothing operator and/or the choice of a suitable standard deviation for the Gaussian smoothing influences the detected results. Furthermore, the fourth statement in particular has not yet been sufficiently examined. To our knowledge, the theoretical investigations necessary for this have not yet been performed. In a study conducted at Stanford University in 1997 on color variants of the Canny operator, an efficient computation of the operator applying parallel algorithms was investigated. This study is mentioned here since the color variant of the Canny operator at Stanford University was also applied on our color test images and the results are shown in Section 6.2.

6.1.2 Cumani Operator

For edge detection in color or multispectral images Cumani suggests the extension of procedures based on the second partial derivatives of the image functions (see [Cum91] and [Cum et al. 91]). A three-channel color image $\mathbf{C}$ is regarded as a two-dimensional vector field

$$\mathbf{C}(x,y) = \left(C_1(x,y), C_2(x,y), C_3(x,y)\right)$$

with the three components $C_1(x,y)$, $C_2(x,y)$, and $C_3(x,y)$. In the *RGB* space these vector components correspond to the components $R(x,y)$, $G(x,y)$, and $B(x,y)$ for the red, green, and blue color channels (or the long-, middle-, and short-wave spectral transmission, respectively). The notation $C_i(x,y)$ is given at this point, on the one hand, for a compact representation. On the other hand, it should be made clear that this technique is applicable in general for n-channel (multispectral) images. In this connection it is always assumed that a Euclidian metric exists for the n-dimensional vector space. Therefore, this technique cannot be easily used for edge detection in the *HSI*, *CIELUV*, or *CIELAB* space.

The squared local contrast $S(\mathbf{p};\mathbf{n})$ at $\mathbf{p}=(x,y)$ is defined as a quadratic norm of the directional derivatives of the image function $\mathbf{C}$ toward the unit vector $\mathbf{n}=(n_1,n_2)$ by

$$S(\mathbf{p};\mathbf{n}) = K \cdot n_1^2 + 2 \cdot F n_1 n_2 + H \cdot n_2^2 .$$

The abbreviations are defined as

$$K = \sum_{i=1}^{3} \frac{\partial C_i}{\partial x} \cdot \frac{\partial C_i}{\partial x}, \quad F = \sum_{i=1}^{3} \frac{\partial C_i}{\partial x} \cdot \frac{\partial C_i}{\partial y}, \quad \text{and} \quad H = \sum_{i=1}^{3} \frac{\partial C_i}{\partial y} \cdot \frac{\partial C_i}{\partial y}.$$

The eigenvalues of the matrix

$$A = \begin{pmatrix} K & F \\ F & H \end{pmatrix}$$

correspond to the extreme values of $S(\mathbf{p};\mathbf{n})$ and are obtained if $\mathbf{n}$ is the corresponding eigenvector. The extreme values $\lambda_\pm$ and the corresponding eigenvectors are given by

$$\lambda_\pm = \frac{K+H}{2} \pm \sqrt{\frac{(K+H)^2}{4} + F^2} \tag{6.1}$$

and

$$\mathbf{n}_\pm = (\cos(\theta_\pm), \sin(\theta_\pm)) \tag{6.2}$$

with

$$\theta_- = \theta_+ + \frac{\pi}{2}$$

and

$$\theta_+ = \begin{cases} \frac{\pi}{4} & \text{if } (K-H)=0 \text{ and } F>0 \\ -\frac{\pi}{4} & \text{if } (K-H)=0 \text{ and } F<0 \\ \text{undefined} & \text{if } K=F=H=0 \\ \frac{1}{2}\tan^{-1}\left(\frac{2F}{K-H}\right) & \text{else.} \end{cases}$$

In the one-channel case, $\lambda_\pm$ corresponds to the gradient, and $\mathbf{n}_\pm$ and $\theta_\pm$ give the direction of the strongest and the weakest magnitude, respectively. The two latter terms correspond thus to the gradient direction. Since only the direction of the steepest magnitude is of importance for the extraction of edge points, λ_-, n_-, and θ_- are not further addressed in the following.

The squared local contrast of the vector-valued image function $\mathbf{C}$ dependent on location and direction is defined by $S(\mathbf{p};\mathbf{n}_+)$. The maximum squared local contrast λ_+ was clearly defined as a maximum of $S(\mathbf{p};\mathbf{n}_+)$ over the possible directions $\mathbf{n}_+$, while the direction of the maximum magnitude is determined only up to the orientation. Edge points (i.e., discontinuities of the image function that are characterized by a particularly high contrast) are sought. The maxima of λ_+ must be calculated. This takes place by deriving the function $\lambda_+(\mathbf{p})$, which is represented as a function of the location. Subsequently, the zeros of the derivative, which represent the maxima, are to be determined. In order to find the zeros of $\lambda_+(\mathbf{p})$ defined in Eq. (6.1), the derivatives of this function can also be formed in direction $\mathbf{n}_+$ (see Eq. (6.2)). It can be shown [Cum et al. 91] that

$$\nabla\lambda_+ \cdot \mathbf{n}_+ = \nabla S(\mathbf{p};\mathbf{n}_+)\cdot \mathbf{n}_+$$

holds. Therefore, the derivative of λ_+ is defined by $D_S(\mathbf{p};\mathbf{n})$ with

$$\begin{aligned} D_S(\mathbf{p};\mathbf{n}) &:= \nabla\lambda_+ \cdot \mathbf{n}_+ \\ &= K_x n_1^3 + (K_y + 2F_x)n_1^2 n_2 + (H_x + 2F_y)n_1 n_2^2 + H_y n_2^3 , \end{aligned} \tag{6.3}$$

whereby the indexes x and y denote the corresponding derivatives to x and y, respectively, and the index + sign in the components of $\mathbf{n}_+$ are omitted for simplification. In the one-channel case, λ_+ corresponds to the absolute value of the gradient. $D_S(\mathbf{p};\mathbf{n}_+)$, as the derivative of $\lambda_\pm$ to $\mathbf{n}_+$, corresponds in the one-channel case to the derivative of the absolute value of the gradient in the gradient direction.

Altogether $D_S(\mathbf{p};\mathbf{n}_+)$ is a form that is based on the second directional derivatives of the image function. The edge points, which were defined as the maximum points of the first derivative of the image function, are represented therefore in $D_S(\mathbf{p};\mathbf{n}_+)$ by zeros (or zero-crossings in the digital grid). For the detection of these *zero-crossings* (with regard to, for example, a 4- or 8-neighborhood), neighboring function values with different signs must be sought. The sign of $D_S(\mathbf{p};\mathbf{n}_+)$ is so far not uniquely defined. The definition of $\mathbf{n}_+$ as the eigenvector of a matrix results in the fact that it is uncertain whether $\mathbf{n}_+$ or $(-\mathbf{n}_+)$ is the sought-after vector. Since $\mathbf{n}_+$ cubically arises in $D_S(\mathbf{p};\mathbf{n}_+)$, $D_S(\mathbf{p};\mathbf{n}_+)$ is directly dependent on the sign of $\mathbf{n}_+$. For the solution of this problem Cumani [Cum91] recommends an investigation into the subpixel domain using a bilinear interpolation.

Alshatti and Lambert [AlsLam93] propose a modification of Cumani's technique in order to resolve the ambiguities in the gradient directions. Since λ_+ is an eigenvalue of the matrix $\mathbf{A}$, the associated eigenvector $\mathbf{n}_+$ can be directly determined. Thereby the complex approximation in the subpixel domain, as suggested by Cumani, is avoided (see [AlsLam93] for a detailed representation).

In both techniques specified above, the partial derivatives of K, F, and H to x and y must be determined. These computationally costly calculations can be accomplished more efficiently if these derivatives are determined directly without first calculating and storing K, F, and H. If the quadratic norm is selected for the three-channel RGB space, then the term K is defined by

$$K = \sum_{i=1}^{3} \frac{\partial C_i}{\partial x} \cdot \frac{\partial C_i}{\partial x} = \frac{\partial R}{\partial x} \cdot \frac{\partial R}{\partial x} + \frac{\partial G}{\partial x} \cdot \frac{\partial G}{\partial x} + \frac{\partial B}{\partial x} \cdot \frac{\partial B}{\partial x} .$$

Thus, the partial derivative from K to x (indicated by K_x) is determined by

$$K_x = \frac{\partial K}{\partial x} = \frac{\partial\left(\sum_{i=1}^{3} \frac{\partial C_i}{\partial x} \cdot \frac{\partial C_i}{\partial x}\right)}{\partial x} = \sum_{i=1}^{3}\left(\frac{\partial\left(\frac{\partial C_i}{\partial x} \cdot \frac{\partial C_i}{\partial x}\right)}{\partial x}\right)$$

$$= \sum_{i=1}^{3}\left(\frac{\partial C_i}{\partial x} \cdot \frac{\partial^2 C_i}{\partial x^2} + \frac{\partial^2 C_i}{\partial x^2} \cdot \frac{\partial C_i}{\partial x}\right) = \sum_{i=1}^{3} 2 \cdot \frac{\partial C_i}{\partial x} \cdot \frac{\partial^2 C_i}{\partial x^2} .$$

The other terms K_y, F_x, F_y, H_x, and H_y can be described similarly by

$$F_x = \sum_{i=1}^{3}\left(\frac{\partial C_i}{\partial x} \cdot \frac{\partial^2 C_i}{\partial x \partial y} + \frac{\partial C_i}{\partial y} \cdot \frac{\partial^2 C_i}{\partial x^2}\right), \quad F_y = \sum_{i=1}^{3}\left(\frac{\partial C_i}{\partial x} \cdot \frac{\partial^2 C_i}{\partial y^2} + \frac{\partial C_i}{\partial y} \cdot \frac{\partial^2 C_i}{\partial x \partial y}\right),$$

$$K_y = 2 \cdot \sum_{i=1}^{3} \frac{\partial C_i}{\partial x} \cdot \frac{\partial^2 C_i}{\partial x \partial y}, \quad H_x = 2 \cdot \sum_{i=1}^{3} \frac{\partial C_i}{\partial y} \cdot \frac{\partial^2 C_i}{\partial x \partial y}, \text{ and}$$

$$H_y = 2 \cdot \sum_{i=1}^{3} \frac{\partial C_i}{\partial y} \cdot \frac{\partial^2 C_i}{\partial y^2}.$$

The first directional derivative of the quadratic contrast function can be determined efficiently with the equations specified above.

In implementing the equations above it must still be specified how the partial derivatives of the image functions are to be determined. Alshatti and Lambert [AlsLam93] and Cumani [Cum91] applied several 3×3 – convolution masks for this. From the investigations of Marr and Hildreth [MarHil80] it is well known that the use of convolution masks of a fixed size of 3×3 pixels is not always suitable for the complex problem of determining discontinuities in image functions. For better accuracy in the results of the color edge detection it is therefore recommended to also include convolution masks of various sizes in the calculation process. Therefore, for the determination of the partial derivatives, masks that are based on the two-dimensional Gaussian function and their partial derivatives are suggested here. These masks are called *Gaussian masks* in the following and can be parameterized by the standard deviation σ.

The partial derivatives of the two-dimensional Gaussian function can be calculated simply. The size of the Gaussian masks can be specified by those function values that are, for example, larger than 0.1 percent of the maximum function value of the Gaussian function for a standard deviation σ. Thus, the choice of a standard deviation of, for example, $\sigma = 0.5$ corresponds to a mask the size of 3×3 pixels. A larger value for the standard deviation σ produces larger convolution masks and a stronger smoothing of the image function. This also simultaneously increases the necessary computing time.

Here it is expressly pointed out that the Cumani operator can be parameterized over the standard deviation σ if Gaussian masks are included in the calculations of the partial derivatives. Thereby an application of this operator is also possible in different resolutions [Kos95]. The use of Gaussian masks is, however, not entirely necessary here for the scalability of the operator. Other functions, such as Gabor functions, can also be used.

The results of the edge detection with the Cumani operator using Gaussian masks with the standard derivatives $\sigma = 0.5$ and $\sigma = 1.0$ are presented for a selected color image in Fig. 6.2. There it can be recognized that better results are obtained for the selected color image using the larger convolution mask.

6.1.3 Operators Based on Vector Order Statistics

Following the use of morphological operators for edge detection in gray-level images (see [Lee et al. 86], [Har et al. 87], and [KleZam96]), Trahanias and

Venetsanopoulos suggest vector-valued ranking operators for edge detection in color images [TraVen92], [TraVen93]. The scheme of the detection and combination of local minima and maxima of the image function, as it is used for a morphological edge extraction in gray-level images, cannot be extended directly to vector-valued color images. No exact equivalent of the scalar "Min-Max" operator exists for vector-valued variables. In [TraVen93] the application of vector order statistics is therefore suggested for this.

Ordering of vector-valued data cannot be uniquely defined. Therefore, a set of techniques for the arrangement of an ordering scheme for vector-valued data was introduced, which generally can be classified into marginal (*M-ordering*), reduced (*R-ordering*), conditional (*C-ordering*), and partial ordering (*P-ordering*) [TraVen93]. Trahanias and Venetsanopoulos propose the use of reduced ordering for edge detection in vector-valued color images. This is because this scheme contains a natural definition of the vector median (see Section 5.3.2) for the first sample in the arranged vector sequence and "vector outliers" occupy the upper ranks in this vector sequence. Furthermore, the other ordering schemes appear less suitable for color image processing. Marginal ordering (*M-ordering*) corresponds to a componentwise monochromatic-based processing and partial ordering (*P-ordering*) implies the construction of a convex hull, which is difficult in the three-dimensional. Conditional ordering (*C-ordering*) represents simply an ordering according to a specific selected component and thus does not use the information content from the other signal components.

$\mathbf{x} = (x_1, x_2, \cdots, x_p)^T$ represents a p-dimensional (multivariate) term provided with several characteristic variables, whereby the x_l, $l = 1,2,\cdots, p$, are random variables and $\mathbf{x}_i$, $i = 1,2,\cdots,n$, is an observation of $\mathbf{x}$. Each $\mathbf{x}_i$ represents a p-dimensional vector. In a reduced ordering (*R-ordering*), each multivariate observation is reduced as a function of a distance criterion to a scalar value d_i. If the sum of the distances of the vector $\mathbf{x}_i$ to each vector from the set $\mathbf{x}_1, \mathbf{x}_2, \cdots, \mathbf{x}_n$ is selected as a distance metric, then d_i is represented by

$$d_i = \sum_{k=1}^{n} \left\| \mathbf{x}_i - \mathbf{x}_k \right\| ,$$

where $\|\cdot\|$ represents a suitable vector norm. An arrangement of the $d_i s$ in ascending order, $d_1 \le d_2 \le \cdots \le d_n$, associates the same ordering to the multivariate $\mathbf{x}_i s$, $\mathbf{x}_1 \le \mathbf{x}_2 \le \cdots \le \mathbf{x}_n$. In this arranged sequence $\mathbf{x}_1$ is the vector median (see Eq. (5.10)) of the data samples. It is defined as that vector contained in the given set whose distance to all other vectors is a minimum. Furthermore, vectors that have a higher rank in this arranged sequence are those vectors that diverge the most from the other data (outliers).

A color image is now regarded as a vector field, represented by a discrete vector-valued function $\mathbf{C}: Z^2 \rightarrow Z^m$, where $m = 3$ for three-channel color images. **F** indicates a window over the image function that contains n pixels (color vectors). If reduced ordering is specified for all color vectors lying within the window, then $\mathbf{x}_i$ indicates the ith vector in this ordering. Based on the prefaces to vector order statistics, Trahanias and Venetsanopoulos [TraVen93] present a simple color edge operator *VR*, which they call *vector rank operator*. *VR* is defined by

$$VR = \left\| \mathbf{x}_n - \mathbf{x}_1 \right\| . \tag{6.4}$$

VR describes the deviation of "vector outliers" in a quantitative way with the highest rank of the vector median within the window **F**. Thus, VR delivers a small value in a uniform image domain in which the vectors differ only slightly. In contrast to this, the operator supplies a large value to an edge, since $\mathbf{x}_n$ would be selected among the vectors along an edge (in the smaller domain), while $\mathbf{x}_1$ would be selected among the vectors on the other side of the edge (the larger domain). Edges in color images can be determined in such a way by indicating a threshold value for *VR*. *VR* would be very sensitive, however, in relation to impulse noise, since the vectors lying in the upper order of rank can correspond with the noisy data.

In order to eliminate this disadvantage, Trahanias and Venetsanopoulos [TraVen93] consider *dispersion measures*. A general class of *vector dispersion edge detectors (VDED)* can be defined by using a linear combination of the arranged vectors

$$\begin{aligned} VDED &= OSO\left(\left\| \sum_{i=1}^{n} \alpha_{i1}\mathbf{x}_i \right\|, \left\| \sum_{i=1}^{n} \alpha_{i2}\mathbf{x}_i \right\|, \ldots, \left\| \sum_{i=1}^{n} \alpha_{ik}\mathbf{x}_i \right\| \right) \\ &= OSO_j\left(\left\| \sum_{i=1}^{n} \alpha_{ij}\mathbf{x}_i \right\| \right), \quad j = 1,2,\cdots,k , \end{aligned}$$

where *OSO* indicates an operator based on ordering statistics. In principle, edge operators can be derived from the above equation by a suitable choice of an OSO and a set of coefficients α_{ij}. In order to limit this difficult task, some requirements on an edge operator are observed in [TraVen93]. First, the edge operator should not be sensitive to impulse and Gaussian noise, and second, the edge operator should supply a reliable answer for *ramp edges*.

Since the vectors afflicted by impulse noises appear in the higher ranks in the set of ordered vectors, the detector can be made insensitive to this kind of noise. Instead of a single difference, as in Eq. (6.4), k sets are determined by differences and a minimization is implemented

$$\min_j \left(\left\| x_{n-j+1} - x_1 \right\| \right) \quad j = 1,2,\ldots,k;\ k < n.$$

Impulses (up to $k-1$) caused by isolated noisy pixels are not detected by the implementation of minimization. The choice of a suitable value for k in the above equation depends on n, the size of the observed work window **F**. However, no general formula for the definition of k can be indicated. In [TraVen93] it is proposed to interpret k as the number of the pixels belonging to the "smaller side of the edge" if **F** is centered on an edge pixel.

In order to make the operator insensitive to Gaussian noise, Trahanias and Venetsanopoulos replace the vector median $\mathbf{x}_1$ by a so-called vector-valued "α-trimmed" mean value (*vector α-trimmed mean VαTM*, $\sum_{i=1}^{l}(\mathbf{x}_i / l)$). The resulting edge operator *MSD*, based on *minimum vector dispersion (MVD)*, is defined by

$$MVD = \min_j \left(\left\| \mathbf{x}_{n-j+1} - \sum_{i=1}^{l} \frac{\mathbf{x}_i}{l} \right\| \right), \quad j = 1,2,\cdots,k; l < n.$$

The value for the parameter l in the above equation cannot be formally determined. Trahanias and Venetsanopoulos [TraVen93] argue that a duality exists between l and k in that l describes the number of pixels that are on the "larger side of an edge" if **F** is centered on an edge pixel. They argue further that suitable values can thus be subjectively determined for the parameters. The choice of the parameters k and l in the above equation is, however, subjective and heuristic.

6.2 RESULTS OF COLOR EDGE OPERATORS

Various techniques for edge detection in color images were presented in the previous section. This section will cover how significant the differences are in the results when differing techniques for edge detection are applied. A discussion of several criteria for the evaluation of edge operators (in gray-level images) can be found in [Sal et al. 96b] and general overviews of evaluation criteria for computer vision techniques are presented in [CouTha01] and [YitPel03]. The topic of edge detection in color images is, however, not covered there.

According to a definition from the IAPR (International Association for Pattern Recognition), *benchmarking* in computer vision is "an objective measure of the performance of a vision technique obtained by evaluating its performance on test data" [IAPR01]. One problem of benchmarking color edge operators lies in the fact that so far neither do generally recognized and available color test patterns exist, nor is it clear how an objective measurement for the evaluation of a "color edge" is to be specified. Here the intention is not to present "benchmarking" of color edge operators. Rather, several results determined by different color edge operators are presented for a selected color image. These results are not representative. They are meant to give a rough impression of the results that are attained with different color edge operators.

The results of an investigation [Kan87] of different color variants of the Canny operator were already described in Section 6.1.1. Here resulting images for the vector-valued variants of the Canny operator are presented. The results for a selected color test pattern, obtained with the Cumani operator, including Gaussian masks (compare Section 6.1.2), are compared directly to these results. In addition, a resulting image with a monochromatic technique is visualized. For a monochromatic-based technique, the classic *Mexican Hat operator* (and/or LoG operator) was selected as an example. The Mexican Hat operator, which is based on physiological realizations of the human visual system, is defined by the negative Laplacian derivative of a two-dimensional Gaussian distribution $-\nabla^2 GAUSS(x,y)$ [MarHil80]. It applies

$$-\nabla^2 GAUSS(x,y) = \frac{x^2+y^2-2\sigma^2}{2\pi\sigma^2}\exp\left(-\frac{x^2+y^2}{2\sigma^2}\right). \tag{6.5}$$

The operator can be parameterized over the standard deviation σ. The size of the convolution masks were fixed by those function values that are greater than 0.1 percent of the maximum function value of the Gaussian function for a standard deviation σ. It should be taken into consideration that the definition of the mask size has an influence on the processing result [Kos89]. The convolution mask produced for a selected σ is applied to all three spectral transmissions of the color image. A pixel in a color image is declared as part of a color edge if a zero-crossing was detected in at least one of the resulting images achieved in this manner.

Figure 6.1. *The color image "Block".*

One result of edge detection by means of this monochromatic-based color variant of the Mexican Hat operator is presented in the following. Figure 6.1 shows the color image "Block". In Fig. 6.2, some results of color edge detection are presented. The results can by interpreted as follows. Many pixels in the image background are determined as edge points by applying the monochromatic-based color variant of the Mexican Hat operator (see Fig. 6.2a).

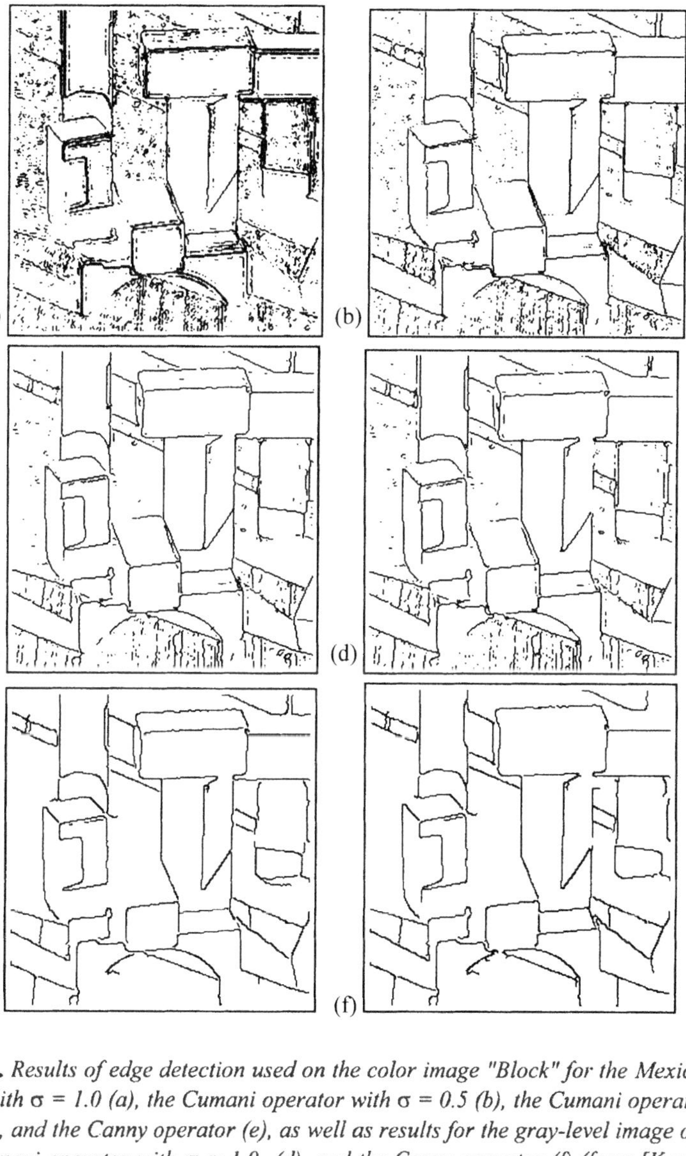

(a) (b) (c) (d) (e) (f)

Figure 6.2. *Results of edge detection used on the color image "Block" for the Mexican Hat operator with σ = 1.0 (a), the Cumani operator with σ = 0.5 (b), the Cumani operator with σ = 1.0 (c), and the Canny operator (e), as well as results for the gray-level image of Block for the Cumani operator with σ = 1.0 (d), and the Canny operator (f) (from [KosAbi05], (e) and (f) with friendly permission of John Owens, Stanford University).*

In addition, many gaps develop at the same time in the detected edges. The results of the Mexican Hat operator can be improved by defining a larger standard deviation. In Fig. 6.2, the result for the standard deviation $\sigma = 1.0$ was selected in order to show such a comparison to the Cumani operator, which was parameterized over the same value for the standard deviation. Better results are achieved with the Cumani operator (see Fig. 6.2b and c). Here the quality of the results is continually improved if Gaussian masks with a greater standard deviation are used instead of a 3×3 convolution mask with $\sigma = 0.5$. A comparison of the results, which the Cumani operator supplies for the color image "Block" (see Fig. 6.2c) and for the corresponding gray-level image (see Fig. 6.2d), is interesting. It is to be noted that several edges that had not been determined in the gray-level image were detected in the color image. Further investigations have shown that edge detection in color images is more robust in relation to noise than an appropriate edge detection in the associated gray-level image. This applies especially to weakly contrasted images.

From the results obtained with the Canny operator it can also be recognized that some edges could be detected in the color image (see Fig. 6.2e) that were not determined in the gray-level image (see Fig. 6.2f). This statement applies likewise to the color image "Lena," for which a color representation and the results obtained with the Canny operator are indicated in Fig. 6.3. With a comparison of the results for the Cumani operator, indicated in Fig. 6.2, and the Canny operator for the color image "Block", it is to be recognized that more edges were detected with the Cumani operator than with the Canny operator. This statement cannot be generalized, however, and applies only to the results presented in Fig. 6.2. The inclusion of the results that can be obtained with vector-valued ranking operators, as they were described in Section 6.1.3, remains the subject of future work. It can be said that the results that are determined in color images are at least as good as or better than the results that are determined in gray-level images. Additional

Figure 6.3. *Results of edge detection applied to the color image "Lena" (left) for a color variant of the Canny operator (center) and the gray-level algorithm of the Canny operator (right) (with friendly permission from John Owens, Stanford University).*

discussions and comparisons of color edge detection operators can be found in [KriBha98], [PlaVen00], [RuzTom01], and [Wes et al. 00].

Apart from a qualitative evaluation of the results of color edge detection, a quantitative evaluation is also of interest. As a function of the processed image, about 90% of all detected edges is identical in the color image and intensity image [Kan87]. Also of concern during the detection of edges in color images is the detection of the remaining 10% of the edges. It depends on the respective application whether the expenditure is justified for the detection of this additional 10%. This question cannot be answered in general. We consider the following example.

In a factory building there are two objects bordering each other whose surfaces exhibit different colors of the same brightness. These objects can be work pieces, containers, cardboard, or other articles. The border between the two objects cannot be determined in the gray-level image. Color edge detection is necessary for this. If the technique for edge detection is part of a program for obstacle detection and collision avoidance for a mobile robot, then detecting the border edge between the two objects is not necessary. It is not important whether the object is recognized as a large obstacle or as two small obstacles that border one another. In order to save time, the more time-efficient technique of edge detection in the gray-level image is recommended here. However, if the robot's task consists of grasping the objects, then it is of crucial importance to know whether it faces one large object or two small objects.

The detection of edges in the image is not a setting of tasks locked in itself, but rather is always a component of a processing chain with different aims. In an edge-based stereo analysis (see Section 9.3.1), only those edges can be assigned that were also detected in both images. A missing edge that was not detected can lead to a complete misinterpretation within shape reconstruction (see [Kle et al. 98] for the topic of shape reconstruction). Furthermore, not detecting an edge also has a decided influence on the result of an edge-based segmentation process. This chapter it not concerned with deciding whether the additional edges are needed. Rather, vector-valued techniques were introduced that make it possible to at least partly detect the remaining edges. In the following section, it is shown that color information can be used, under certain conditions, for classifying edges.

6.3 CLASSIFICATION OF EDGES

In addition to quantitative and qualitative advantages of color edge detection, color information allows for classification of the edges. Edges in images are characterized by discontinuities in the image function. They can have completely different causes due to the geometrical and photometric conditions within a scene. Different types of edges are outlined in Fig. 6.4. Edges can be distinguished into the following five classes:

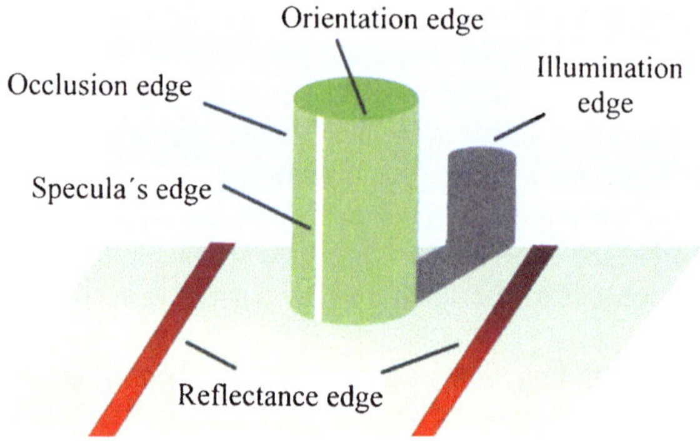

Figure 6.4. *Different types of edges in a representation of a scene.*

1. *Object edges,* or *orientation edges,* arise from a discontinuity of the vector normal of continuous surfaces
2. *Reflectance edges* arise from a discontinuity of the reflectance of object surfaces, for example, by a change of surface material
3. *Illumination edges*, or *shadow edges*, arise from a discontinuity of the intensity of the incident lighting
4. *Specula's edges*, or *highlight edges*, arise from a special orientation between the light source, the object surface, and the observer, and are due to material properties
5. *Occlusion edges* are boundaries between an object and the background as seen by the observer. Occlusion edges do not represent a physical discontinuity in the scene. They exist due to a special viewing position.

In many areas of digital image processing a classification of edges is necessary and/or advantageous. For example, only orientation edges, reflectance edges, and illumination edges should be matched in stereo vision. Specula's edges and occlusion edges should not be matched because their occurrence in the images depends on the viewing position of both cameras, and they do not represent the identical physical locus in the scene. Illumination edges should not be matched if motion analysis is applied. The classification of edges by their physical origin is difficult or even impossible in gray-level images. Color image processing can at least partly supply a way out of this misery.

6.3.1 Physics-Based Classification

If it is, for example, known that the objects in the scene consist of inhomogeneous dielectric materials, then a physics-based classification of the color edges is possible. The dichromatic reflection model (DRM) describes the reflection on inhomogeneous, dielectric materials, such as plastic or paint (see Section 7.4.1). It indicates general, hybrid reflections, without specifically modeling the specular reflection component. Several vector-valued techniques for edge classification in color images are based on the DRM. They can be summarized as follows:

Reflectance edges and/or material changes can at least be partly recognized by a rejection scheme using spectral crosspoints (see [RubRic82]). A minimum of two spectral samples (e.g., the values in the red and green color channels) is needed for the application of this technology. The limits of this classification technique are discussed in [Ger et al. 92].

Illumination edges can be classified by active shadow recognition (see [FunBaj93]) or by analyzing the structure of the shadows in color images [Ger et al. 92]. Furthermore, the retinex theory of color constancy has been applied to color images to detect shadow edges and to remove shadows [Fin et al. 02]. Note that the retinex algorithm also provides some dynamic range compression that yields to a change of the colors in the image. Furthermore, several recent approaches have been published on shadow detection in gray-level images (see, e.g., [Hsi et al. 03], [Mat et al. 02] and [Pra et al. 01]).

Highlight edges can be classified on the basis of methods for highlight detection (see [Baj et al. 96], [Kli et al. 90], [SchTes95], and [TsaTsa97]). These methods separate the specular and the diffuse reflection component applying the dichromatic reflection model (see Section 7.4.1). The analysis of highlights in color images is the subject of Section 8.1. In general, highlight analysis techniques can be subdivided into global and local techniques. While global techniques have to consider the entire image for finding color clusters (see [Kli et al. 90] and [SchTes95]), local techniques perform local analysis on pixels [SchKos00]. However, the latter technique requires that more than one image is available of an object to be analyzed.

Orientation edges and *occlusion edges* can be classified by using gradient estimation techniques. These two latter classes of edges can be determined without evaluation of color information.

The techniques mentioned above do not offer a complete solution to the problem of edge classification. They do represent, however, a first step toward edge classification. This applies particularly if no complete classification need be accomplished, but to decide only in individual cases whether, for example, an edge is caused by a highlight. One disadvantage of the procedures specified above using the dichromatic reflection model is that certain knowledge of the material properties of the objects in the scene must be present.

A goal for the future is to achieve an edge classification solely on the basis of the vector signals in the color space and without further knowledge. This will not be easy to ensure. Techniques in this direction are presented, for example, in [MaxSha97].

6.3.2 Classification Applying Photometric Invariant Gradients

While the techniques in the previous section where mainly designed to detect one specific class of edges, Gevers and Stokman [GevSto03] proposed a technique for an automatic classification of color edges into the three classes:

1. shadow-geometry (orientation and occlusion)
2. highlight
3. material transitions

Although their technique can be applied to hyperspectral data, we focus here on the case of a three-channel color image. In addition to the *RGB* space, Gevers and Stokman investigate normalized colors $c_1 c_2$ defined by

$$c_1(R,G,B) = \arctan\left(\frac{R}{B}\right), \quad c_2(R,G,B) = \arctan\left(\frac{G}{B}\right)$$

and the two-dimensional opponent color space defined by

$$o_1(R,G,B) = \frac{(R-G)}{2}, \quad o_2(R,G,B) = \frac{B}{2} - \frac{(R+G)}{4}.$$

The gradients in the three considered color spaces are denoted by ∇C_{RGB}, $\nabla C_{c_1c_2}$ and $\nabla C_{o_1o_2}$. From several investigations one may conclude that ∇C_{RGB} measures the presence of (1) shadow-geometry, (2) highlight, and (3) material edges. Further, $\nabla C_{c_1c_2}$ measures the presence of (2) highlight and (3) material edges, while $\nabla C_{o_1o_2}$ measures the presence of (1) shadow-geometry and (3) material edges. As a result, a taxonomy of color edges can be specified; see Table 6.1. By applying automatic threshold setting to the three gradients, an automatic physics-based edge classification into the three classes (1) shadow-geometry, (2) highlight, and (3) material edges can be obtained.

	Orientation edge	Shadow edge	Highlight edge	Material edge
∇C_{RGB}	+	+	+	+
$\nabla C_{c_1c_2}$	–	–	+	+
$\nabla C_{o_1o_2}$	+	+	–	+

Table 6.1. *Classification of color edges based upon the sensitivity of the different color edge models with respect to the imaging conditions. – denotes invariant and + denotes sensitivity of the color edge model to the imaging condition (after [GevSto03]).*

This classification is computationally very costly and so far no extensive investigations exist on false-positive and false-negative classification results. Note that although this taxonomy allows identifying highlight, material, and shadow-geometry edges, it is not capable of distinguishing among orientation, occlusion, and shadow edges. Here additional classification techniques need to be applied.

6.4 COLOR HARRIS OPERATOR

It is not always neccessary to extract a whole list of edges. Only the matching of significant image features, such as corners, is required in several applications. These significant features are called *interest points* and are frequently used in image registration, motion detection, tracking, 3D modeling, and object recognition. Among the proposed *interest point detectors* the most commonly used one is the Plessey operator [HarSte88] (often referred to as the *Harris corner detector* or *Harris operator*). The algorithm for the Harris operator to detect corners in grayscale images can be outlined as follows:

Input: Grayscale image $E(x,y)$, Gaussian smoothing window (window typically has a radius of 3 times the standard deviation σ), k value, threshold T.
Output: Map indicating position of each detected corner.

1. For each pixel in the grayscale image $\mathbf{E}(x,y)$, calculate the autocorrelation matrix $\mathbf{M}$:

$$\mathbf{M} = \begin{bmatrix} M_{11} & M_{12} \\ M_{21} & M_{22} \end{bmatrix}$$

   with

$$M_{11} = S_\sigma(E_x^2),\ M_{22} = S_\sigma(E_y^2), \text{ and}$$
$$M_{12} = M_{21} = S_\sigma(E_x E_y) \quad ,$$

   where $S_\sigma()$ stands for a Gaussian smoothing that can be obtained by a convolution with a Gaussian window (e.g. $\sigma = 0.7$).
2. Construct the "cornerness map" by calculating the cornerness measure $\mathbf{C}_M(x,y)$ for each pixel (x, y):

$$\mathbf{C}_M(x,y) = Det(\mathbf{M}) - k(Trace\,(\mathbf{M}))^2 ,$$

   where $Det\,(\mathbf{M}) = M_{11}M_{22} - M_{21}M_{12}$, $Trace\,(\mathbf{M}) = M_{11} + M_{22}$, and $k = \text{constant}$ (e.g. $k = 0.04$).

3. Threshold the interest map by setting all $\mathbf{C}_M(x, y)$ below a threshold T to zero.
4. Perform nonmaximal suppression to find local maxima.
5. All nonzero points remaining in the cornerness map are corners.

Example results obtained when applying this algorithm to image "South College" are depicted in Fig. 6.5. The corner detector was extended to the Color Harris Operator in 1998 by Montesinos, Gouet and Deriche [Mon et al. 98]. They extended the calculation of the autocorrelation matrix to color images by replacing step 1 in the detection algorithm with step $1'$.

$1'$. For each pixel in the color image **C**(x,y), calculate the autocorrelation matrix $\mathbf{M}'$:

$$\mathbf{M}' = \begin{bmatrix} M'_{11} & M'_{12} \\ M'_{21} & M'_{22} \end{bmatrix}$$

with

$$M'_{11} = S_\sigma(R_x^2 + G_x^2 + B_x^2),$$

$$M'_{22} = S_\sigma(R_y^2 + G_y^2 + B_y^2), \text{ and}$$

$$M'_{12} = M_{21} = S_\sigma(R_x R_y + G_x G_y + B_x B_y),$$

where $S_\sigma()$ stands for a Gaussian smoothing that can be obtained by a convolution with a Gaussian window (e.g. $\sigma = 0.7$).

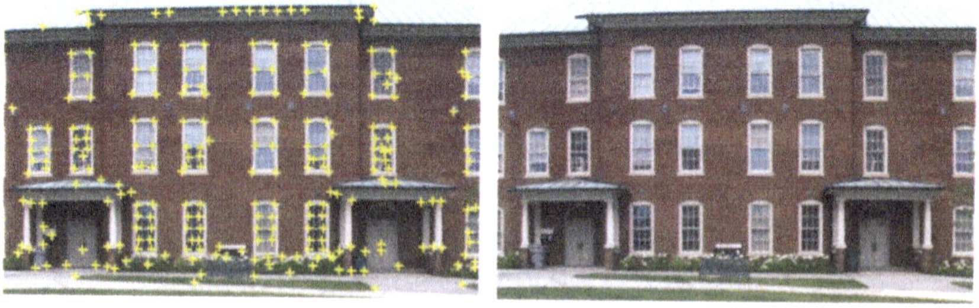

Figure 6.5. *Results of Harris operator (left) applied to image "South College" (right).*

The rest of the modified algorithm is identical to the graysale version. For additional information on the color version of the Harris operator see also [Gou et al. 98], [GouBou02], and [Mon et al. 00].

6.5 REFERENCES

[AlsLam93] W. Alshatti, P. Lambert. Using eigenvectors of a vector field for deriving a second directional derivative operator for color images. *Proc. 5th Int. Conference on Computer Analysis of Images and Patterns*, D. Chetverikov, W.G. Kropatsch (eds.), Budapest, Hungary, 1993, pp. 149-156.

[Ayd et al. 96] T. Aydin, Y. Yemez, E. Anarim, B. Sankur. Multidirectional and multiscale edge detection via m-band wavelet transform. *IEEE Transaction on Image Processing* **5** (1996), pp. 1370-1377.

[Baj et al. 96] R. Bajcsy, S.W. Lee, A. Leonardis. Detection of diffuse and specular interface reflections and inter-reflections by color image segmentation. *Int. J. of Computer Vision* **17** (1996), pp. 241-272.

[Can86] J. Canny. A computational approach to edge detection. *IEEE Transactions on Pattern Analysis and Machine Intelligence* **8** (1986), pp. 679-698.

[CouTha01] P. Courtney, N. A. Thacker. Performance characterization in computer vision: The role of statistics in testing and design. In *Imaging and Vision Systems: Theory, Assessment and Applications*, J. Blanc-Talon, D. Popescu (eds.), NOVA Science Books, 2001.

[Cum91] A. Cumani. Edge detection in multispectral images. Computer Vision, Graphics, and Image Processing: Graphical Models and Image Processing **53** (1991), pp. 40-51.

[Cum et al. 91] A. Cumani, P. Grattoni, A. Guiducci. An edge-based description of color images. *Computer Vision, Graphics, and Image Processing: Graphical Models and Image Processing* **53** (1991), pp. 313-323.

[Fin et al. 02] G. Finlayson, S. Hordley, M. Drew. Removing shadows from images using Retinex. *Proc. Color Imaging Conference*, Scottsdale, Arizona, 2002, pp. 73-79.

[FunBaj93] G. Funka-Lea, R. Bajcsy. Active color image analysis for recognizing shadows. *Proc. 13th Int. Joint Conference on Artificial Intelligence*, Chambéry, France, 1993, Vol. 2, pp. 1573-1578.

[Ger et al. 92] R. Gershon, A.D. Jepson, J.K. Tsotsos. Ambient illumination and the determination of material changes. In: *Physics-Based Vision Principles and Practice: Color*, G.E. Healey, S.A. Shafer, L.B. Wolff (eds.), Jones and Bartlett, Boston, 1992, pp. 101-108.

[GevSto03] T. Gevers, H. Stokman. Classifying color edges in video into shadow-geometry, highlight or material edges. *IEEE Transactions on Multimedia* **5** (2003), pp. 237-243.

[Gou et al. 98] V. Gouet, P. Montesinos, D. Pelé. Stereo matching of color images using differential invariants. *Proc IEEE Int. Conference on Image Processing*, Chicago, Illinois, 1998, pp. 152-156.

[GouBou02] V. Gouet, N. Boujemaa, On the robustness of color points of interest for image retrieval. *Proc. IEEE Int. Conference on Image Processing*, Rochester, New York, 2002, pp. 377-380.

[Har et al. 87] R.M. Haralick, S.R. Sternberg, X. Zhuang. Image analysis using mathematical morphology. *IEEE Transaction on Pattern Analysis and Machine Intelligence* **9** (1987), pp. 532-550.

[HarSte88] C. Harris, M. Stephens. A combined corner and edge detector. *Proc. 4th Alvey Vision Conference*, 1988, pp. 147-151.

[Hsi et al. 03] J.-W. Hsieh, W.-F. Hu, C.-J. Chang, Y.-S. Chen. Shadow elimination for effective moving object detection by Gaussian shadow modeling. *Image and Vision Computing* **21** (2003), pp. 505-516.

[IAPR01] International Association for Patter Recognition IAPR Benchmarking, http://peipa.essex.ac.uk/benchmark.

[JaiHea98] A. Jain, G. Healey. A multiscale representation including opponent color features for texture recognition. *IEEE Transaction on Image Processing* **7** (1998), pp. 124-128.

[Kan87] T. Kanade. Image understanding research at CMU. *Proc. Image Understanding Workshop*, Los Angeles, California, 1987, Vol. II, pp. 32-40.

[KleZam96] R. Klette, P. Zamperoni. *Handbook of Image Processing Operators*. Wiley, New York, 1996.

[Kle et al. 98] R. Klette, K. Schlüns, A. Koschan. *Computer Vision: Three-Dimensional Data from Images*. Springer, Singapore, 1998.

[Kli et al. 90] G.J. Klinker, S.A. Shafer, T. Kanade. A physical approach to color image understanding. *Int. J. of Computer Vision* **4** (1990), pp. 7-38.

[Kos89] A. Koschan. How to minimize numerical inaccuracies arising with the implementation of parametric 2D filter functions. *Proc. Int. Symposium on Computer Architecture and Digital Signal Processing*, Hong Kong, 1989, pp. 395-398.

[Kos95] A. Koschan. A comparative study on color edge detection. *Proc. 2nd Asian Conference on Computer Vision ACCV'95*, Singapore, 1995, Vol. III, pp. 574-578.

[KosAbi05] A. Koschan, M. Abidi. Detection and classification of edges in color images. *IEEE Signal Processing Magazine* **22** (2005), pp. 64-73.

[KriBha98] R. Krishnamoorthi, P. Bhattacharya. Color edge extraction using orthogonal polynomials based zero crossing scheme. *Information Sciences* **112** (1998), pp. 51-65.

[Lee et al. 86] J.S. Lee, R.M. Haralick, L.G. Shapiro: Morphologic edge detection. *Proc. 8th Int. Conference on Pattern Recognition*, Paris, France, 1986, pp. 369-373.

[Lee96] T.S. Lee. Image representation using 2d Gabor wavelets. *IEEE Transaction on Pattern Analysis and Machine Intelligence* **18** (1996), pp. 959-971.

[Mal et al. 97] Y. Mallet, D. Coomans, J. Kautsky, O. de Vel. Classification using adaptive wavelets for feature extraction. *IEEE Transaction on Pattern Analysis and Machine Intelligence* **19** (1997), pp. 1058-1066.

[Mal96] S. Mallat. Wavelets for a vision. *Proc. of the IEEE* **84** (1996), pp. 604-614.

[MarHil80] D. Marr, E. Hildreth. Theory of edge detection. *Proc. of the Royal Society of London* **B207** (1980), pp. 187-217.

[Mat et al. 02] Y. Matsushita, K. Nishino, K. Ikeuchi, M. Sakauchi. Shadow elimination for robust video surveillance. *Proc IEEE Workshop on Motion and Video Computing*, Orlando, Florida, 2002, pp. 15-21.

[MaxSha97] B.A. Maxwell, S.A. Shafer. Physics-based segmentation of complex objects using multiple hypotheses of image formation. *Computer Vision and Image Understanding* **65** (1997), pp. 269-295.

[Mon et al. 98] P. Montesinos, V. Gouet, R. Deriche. Differential invariants for color images. *Proc. International Conference on Pattern Recognition*, Brisbane, Australia, 1998, Vol. 1, pp. 838-840.

[Mon et al. 00] P. Montesinos, V. Gouet, R. Deriche, D. Pelé. Matching color uncalibrated images using differential invariants. *Image and Vision Computing* **18** (2000), pp. 659-671.

[PieHar86] M. Pietikäinen, D. Harwood. Edge information in color images based on histograms of differences. *Proc. Int. Conference on Pattern Recognition*, Paris, France, 1986, pp. 594-596.

[PlaVen00] K.N. Plataniotis, A.N. Venetsanopoulos. *Color Image Processing and Applications*. Springer, Berlin, Germany, 2000.

[Pra91] W.K. Pratt. *Digital Image Processing*. 2nd ed., Wiley, New York, 1991, pp. 548-553.

[Pra et al. 01] A. Prati, R. Cucchiara, I. Mikic, M.M. Trivedi. Analysis and detection of shadows in video streams: A comparative evaluation. *Proc. Int. Conference on Computer Vision and Pattern Recognition,* 2001, Kauai , Hawaii, Vol. II, pp. 571-576.

[RubRic82] J.M. Rubin, W.A. Richards. Color vision and image intensities: when are changes in material? *Biological Cybernetics* **45** (1982), pp. 215-226.

[RuzTom01] M.A. Ruzon, C. Tomasi. Edge, junction, and corner detection using color distributions. *IEEE Transactions on Pattern Analysis and Machine Intelligence* **23** (2001), pp. 1281-1295.

[Sal et al. 96a] R.A. Salinas, C. Richardson, M. Abidi, R. Gonzalez. Data fusion: Color edge detection and surface reconstruction through regularization. *IEEE Transactions on Industrial Electronics* **43** (1996), pp. 355-363.

[Sal et al. 96b] M. Salotti, F. Bellet, C. Garbay. Evaluation of edge detectors: critics and proposal. *Proc. Workshop on Performance Characteristics of Vision Algorithms*, H.I. Christensen, W. Förstner, C.B. Madson (eds.), Cambridge, England, 1996, http://www.vision.auc.dk/~hic/perf-proc.html.

[SchKos00] K. Schlüns, A. Koschan. Global and local highlight analysis in color images. *Proc. 1st Int. Conf. on Color in Graphics and Image Processing*, Saint-Etienne, France, 2000, pp. 300-304.

[SchTes95] K. Schlüns, M. Teschner. Analysis of 2d color spaces for highlight elimination in 3d shape reconstruction. *Proc. Asian Conference on Computer Vision*, Vol. II, Singapore, 1995, pp. 801-805.

[Shi87] Y. Shirai. *Three-Dimensional Computer Vision*. Springer, Berlin, 1987.

[Sol85] J.C. Solinsky. The use of color in machine edge detection. *Proc. VISION'85*, Detroit, Michigan, 1985, pp. 4-34 - 4-52.

[TraVen92] P.W. Trahanias, A.N. Venetsanopoulos. Color edge detectors based on multivariate ordering. *Proc. SPIE* **1818** Visual Communications and Image Processing, 1992, pp. 1396-1407.

[TraVen93] P.W. Trahanias, A.N. Venetsanopoulos. Color edge detection using vector order statistics. *IEEE Transactions on Image Processing* **2** (1993), pp. 259-264.

[TsaTsa97] W.H. Tsang, P.W.M. Tsang. Suppression of false edge detection due to specular reflection in color images. *Pattern Recognition Letters* **18** (1997), pp. 165-171.

[Wes et al. 00] S. Wesolkowski, M.E. Jernigan, R.D. Dony. Comparison of color image edge detectors in multiple color spaces. *Proc. Int. Conference on Image Processing*, 2000, pp. 796-799.

[YitPel03] Y. Yitzhaki, E. Peli. A method for objective edge detection evaluation and detector parameter selection. *IEEE Transactions on Pattern Analysis and Machine Intelligence* **25** (2003), pp. 1027-1033.

7 COLOR IMAGE SEGMENTATION

Image segmentation describes the process of dividing an image into nonoverlapping, connected image areas, called *regions*, on the basis of criteria governing similarity and homogeneity. Analogous to that, *color image segmentation* describes the process of extracting from the image domain one or more connected regions satisfying uniformity (homogeneity) criteria that are based on features derived from spectral image components. These components are defined in a chosen color space (see Chapter 3). The segmentation process can be supported by some additional knowledge about the objects in the scene such as geometric and optical properties.

An introduction to the fundamental principles of segmenting gray-level images can be found in numerous textbooks on digital image processing (see, e.g., [GonWoo02], [Pit93], [Pra91], [Son et al. 93]) and an overview in an article [PalPal93]. This chapter focuses on techniques using color information for image segmentation. Four classes of segmentation techniques are introduced and discussed.

Perhaps one of the most important features of a segmentation process is the region definition. Roughly four types of region definitions can be differentiated:

1. Region is a connected component of a pixel set specified by a class membership function defined in a color space. The grouping of the color signals is carried out in the color space. One condition for grouping can be that the color of the pixel lies within a plane or given polyhedra in the color space.
2. Region is a (maximally) connected set of pixels in the image plane for which the uniformity condition is satisfied. In contrast to type 1, the grouping of the color signals takes place in the image plane instead of the color space. A uniform region is obtained, for example, when larger, nonuniform regions are split or when a region is determined by merging other pixels (or blocks of pixels) in the neighborhood of a starting pixel.
3. Region is a connected set of pixels bounded by edge pixels creating a color contour. The color contour is determined by applying an operator for edge

detection on the color image (see Chapter 6) and possibly by an ensuing filling of the gaps in the contour. In a certain sense the regions are also uniform for they represent the complementary set of a nonuniform set created by edge pixels.

4. Region is a connected component of a pixel set whose grouping results from a physical modeling of the color signal in the color space. The objective of the segmentation is to extract regions in the color image that correspond to the surfaces of objects in the scene, each consisting of one homogeneous material. Shading, shadow, and highlight should have no influence on the result of this image segmentation, although the color values in the image are changing.

The region definitions of the first two types use a *uniformity predicate*, which for type 1 is pixelbased and for type 2 is areabased. The definition of type 3 uses a *nonuniformity predicate*. Thus, this classification is consistent and complete. The region definition of type 4 represents a supplementary subclass of regions that can also be viewed as a special case of regions of type 1.

The additional region definition of type 4 is introduced because the objective of image segmentation and the assumptions concerning the material characteristics of objects in the scene differ from those of the other three types. For example, light red and dark red color vectors are always assigned to the same region of type 4 for it is assumed that the brightness differences were caused by shading or shadows and the pixels correspond to color vectors that represent the same surface area. None of the segmentation techniques using region type 4 is an extension of an intensity-based approach. These techniques can be used exclusively for color images.

Furthermore, these techniques that employ a type 4 region definition belong to a new class of image processing techniques. These techniques have become known in the last few years as *physics-based vision techniques*. The additional type 4 region definition serves to better distinguish a technique of this class from the other techniques.

Segmentation processes are subdivided into four classes according to the use of a type of region definition. For type 1 they are called *pixel-based techniques*, for type 2 they are *area-based techniques*, for type 3 they are *edge-based techniques* and for type 4 they are *physics-based techniques*. Several approaches to color image segmentation will be discussed for all four classes in the following sections. Furthermore, a watershed transformation for color image segmentation will be introduced as an example. This technique is selected since investigations have shown good segmentation results for color images.

7.1 PIXEL-BASED SEGMENTATION

Segmentation techniques that employ a type 1 region definition in a color space are explained in this section. These techniques can be subdivided into two groups:

1. Histogram-based techniques that use the surrounding intervals for pixel classification starting from one or more maxima in the histogram
2. Techniques that implement cluster analysis in the color space

Both groups of techniques are explained in the following.

7.1.1 Histogram Techniques

The use of histograms and threshold values in the segmentation of an image can be seen in the example of a gray-level image. In the first processing step, the frequency of the appearance of each gray level is calculated in an intensity histogram over the entire image (see Fig. 7.1). Next, the maxima and minima (peaks and valleys) are determined in the function given by frequencies in the histogram. The minima (e.g., T_1 and T_2 in Fig. 7.1) form the interval boundaries for the segmentation. In this example, the image is subdivided into three intensity classes, the intensity values between 0 and T_1, T_1 and T_2, and T_2 and, for example, 255.

This technique can be expanded to the segmentation of color images. In the simplest case the color signal can be combined to a one-dimensional function F. Function F can be defined, for example, (see [BonCoy91]) for the RGB values r, g, and b, standardized by Eq. (3.1), by

$$F(r,g,b) = \frac{(0.25r + 4g - b)}{5}.$$

Better results are generally achieved when one-, two-, or three-dimensional histograms are calculated for the components of the color signal. For this procedure a color space must first be selected. Which color space is best suited depends in turn on the application. When one dimensional histograms are used, a histogram is calculated for each color channel and the maxima and minima are determined in each histogram. A priority list can be subsequently drawn up for the maxima that are found. These can be arranged according to the number of determined frequencies, or a weighting can be established (e.g., hue is more of a deciding factor than brightness). Color image segmentation can be implemented with this technique using the following processing steps:

0. Create an empty region list for the entire image.
1. Observe the next region to be segmented. If there are no more, the segmentation is finished.
2. Choose the maximum with the highest priority. If there are no more, mark the region as uniform and go to step 1.
3. Determine the threshold values.
4. Select connected regions and enter them into the region list.
5. Go to step 1.

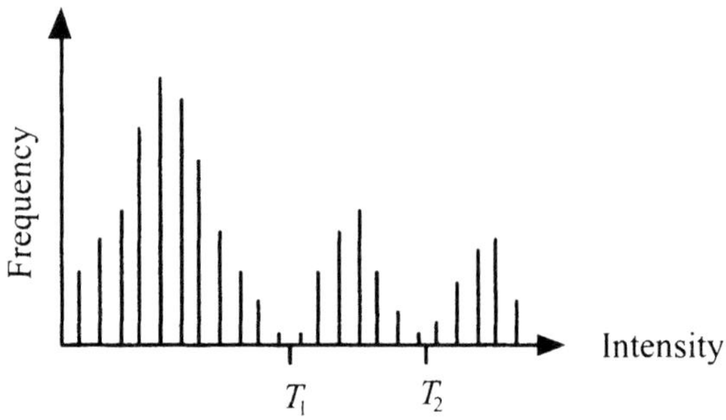

Figure 7.1. *Example of an intensity histogram that can be subdivided into three areas by two threshold values.*

Additionally, neighborhood relationships (e.g., in an 8-neighborhood) can be included in the segmentation in order to fill small holes in the regions.

Newer approaches for segmenting color images determine the maxima and minima in two-dimensional histograms (e.g., in the u*v*-histogram in the CIELUV color space) or in three-dimensional histograms in the color space. This is an alternative to determining the maxima and minima in a one-dimensional histogram by a monochromatic-based technique. Figure 7.2 shows an example of a three-dimensional color histogram. Three-dimensional color histograms are employed as well for a model-based description of colored objects (see [SwaBal91]). Model-based object recognition in digital color images in video real time can result by applying histogram-based *color indexing techniques*. A histogram-based color indexing is also used for image retrieval in image data-bases.

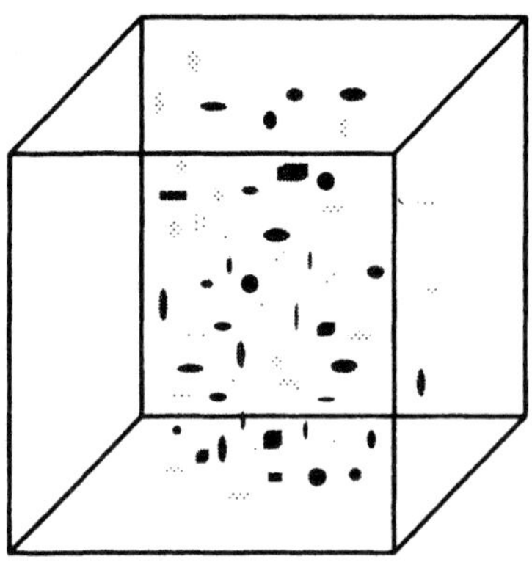

Figure 7.2. *Example of a three-dimensional color histogram where the gray tone in the cube represents the frequency of a color in the image.*

7.1.2 Cluster Analysis in the Color Space

One reliable technique for segmenting gray-level images is *cluster analysis*, which is described in many textbooks (e.g., [Pra91]). This technique is expanded to include vector-valued color images. Starting from a number of given vector-valued cluster centers $\mathbf{Z}_1,\ldots,\mathbf{Z}_m$ in the color space, a m-dimensional probability vector $(p_{i1},\ldots,p_{im})$ is determined for each pixel $\mathbf{q}_i$ in color image $\mathbf{C}$. Here a component p_{ik} assumes the probability that the pixel $\mathbf{q}_i$ belongs to class $\mathbf{Z}_k$, $k=1,\ldots,m$. The cluster centers define the midpoints of each of the classes to which the pixels should be assigned. The probability p_{ik} can be defined by

$$p_{ik} = \frac{\| \mathbf{C}(\mathbf{q}_i) - \mathbf{Z}_k \|^{-1}}{\sum_{l=1}^{m} \| \mathbf{C}(\mathbf{q}_i) - \mathbf{Z}_l \|^{-1}} .$$

For cluster analysis in color images, a color space and an affiliated color distance measurement that meets the requirements of a norm are selected.

The cluster centers must already be known for the practical application of cluster analysis. For some applications the cluster centers can be determined beforehand; for example, relevant features are extracted by means of principal component analysis in a certain number of images with similar content. These relevant features establish the cluster centers in advance. This technique has been successfully employed in the identification of skin tumors [Umb et al. 93] or in the classification of the degree of ripeness of citrus fruits [FerVid92].

However, for the general purpose of color image segmentation, generally no information exists about the colors appearing in the image. Here the cluster centers must be individually determined for each image. This can result, in analogy to the histogram techniques, by determining the maxima in the one-, two- or three-dimensional color histograms. Campadelli, Medici, and Schettini [Cam et al. 95] employ a neuronal Hopfield net for determining relevant cluster centers.

7.2 AREA-BASED SEGMENTATION

Area-based segmentation techniques employ a condition of uniformity in the image area. They can be subdivided into *region-growing techniques* and those that segment an image through *split-and-merge techniques*. The region-growing techniques begin from given start values (*seeds*) in order to connect adjoining pixels (e.g., in a 4- or an 8-neighborhood) to regions with differing strategies. In contrast to this, techniques that segment an image by split and merge begin with nonuniform image areas. They split up the area as long as needed until a uniformity criterion is reached. The regions obtained in this manner are subsequently merged again in order to obtain uniform regions of maximal size. Both kinds of techniques are used in the segmentation of color images.

7.2.1 Region-Growing Techniques

The region-growing techniques view the pixel values, for example, in an 8-neighborhood around the given start values. If a neighboring pixel meets the condition of uniformity, it will belong to the same region as the start pixel. Figure 7.3 illustrates an example of a region-growing technique for gray-level images. The underlined values 2 and 8 represented in Fig. 7.3a are the start pixels. The criterion used for including a pixel in the same region is that the absolute difference between the gray value of a pixel and the gray value of the start pixel is smaller than a threshold T. $T = 3$ results in the outcome shown in Fig. 7.3b. The sample image is segmented into two regions marked "a" and "b."

The condition of uniformity must be modified when using a region-growing technique in color images. A vector-valued segmentation technique can be executed in various color spaces. Instead of the difference of the gray-levels, a corresponding color distance measurement (see Chapter 3) is used for determining the condition of uniformity for the color vectors in the selected color space. The ensuing course of the procedure corresponds to the technique for gray-level images.

7.2.2 Split-and-Merge Techniques

In the techniques that segment an image by split and merge, the entire image is viewed at the beginning of the segmentation process. A region is split, for example, into four subregions until a given uniformity condition is achieved for the (sub)regions. For gray-level images, this condition can be that the variance of

---

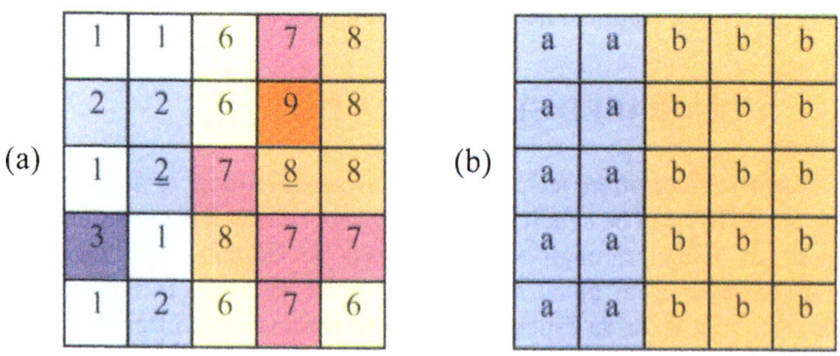

Figure 7.3. *Example of a region-growing technique using known seed points.*

the gray-levels within a region is smaller than a given threshold value *T*. When this condition for a region is not met, this region is further split up. Figure 7.4 illustrates an example. There *R* indicates the entire image. Each node corresponds to a (sub)region, whereby in this example only region R_4 was further devided up. If the image is divided up only into regions, adjoining regions will be similar in the final division. These will be merged together in a following step according to the given condition of uniformity in order to attain uniform regions of maximal size.

The principle of this segmentation, as in the region-growing technique, can be easily expanded to vector-valued color images in which a color distance measurement for the color vectors is included in the uniformity condition. However, the division of the image into regions can be carried out differently from that of the above-named patterns. In contrast to the segmentation of gray-level images, the use of color information also enables the inclusion of perceptual attributes in the *HSI* color space for image segmentation.

In addition to this, a formulation consists of evaluating hue and saturation for the segmentation only there, where the values are perceptually sensible. Otherwise, the segmentation is carried out only from the intensity values. There are three problems to consider when using color attributes in the *HSI* color space for segmenting color images. First, the hue value is meaningless if the intensity is very high or very low. Second, the hue value is not stable if the color saturation is very poor. Third, the saturation value is also meaningless if the intensity is very high or very low.

Tseng and Chang [TseCha92] therefore suggest subdividing the color image into chromatic and achromatic areas. Figure 7.5 illustrates the definition. A pixel lies in the achromatic area if case 1 or case 2, presented in Fig. 7.5, is present. Based on this definition, it must be determined for an image area whether this area is chromatic or achromatic. Moreover, it can be agreed that, for example, an area is called "achromatic" if at least 60% of the pixels lying within the area fulfill the definition in Fig. 7.5.

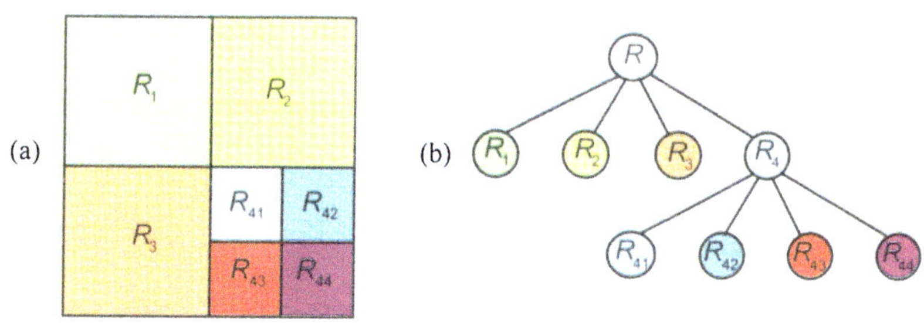

Figure 7.4. *(a) The partitioned image; (b) The accompanying quadtree.*

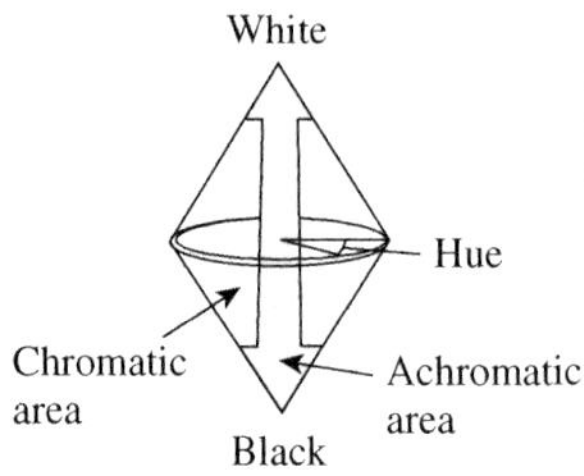

A Pixel is termed "achromatic" if case 1 or case 2 applies.
Case 1: (intensity > 95) or (intensity ≤ 25).
Case 2: (80 < intensity ≤ 95) and (saturation < 18),
(60 < intensity ≤ 80) and (saturation < 20),
(50 < intensity ≤ 60) and (saturation < 30),
(40 < intensity ≤ 50) and (saturation < 40),
(25 < intensity ≤ 40) and (saturation < 60),

Figure 7.5. *Definition of chromatic and achromatic areas in the HSI color space (according to [TseCha92]).*

After the division of the color image into chromatic and achromatic areas, these areas can be further divided with a hue histogram in the chromatic areas and with an intensity histogram in achromatic areas. The regions attained in this manner are subsequently merged together again by means of a region-growing technique. Tseng and Chang [TseCha92] obtained good segmentation results with this heuristic technique.

7.3 EDGE-BASED SEGMENTATION

Edge-based segmentation techniques use nonuniform measurements or discontinuities in the image function for the division of an image into regions. Local and global techniques can be distinguished from one another in principle. Local techniques use only the information in a pixel's local neighborhood for the detection of an edge pixel. In contrast, global techniques implement a type of global optimization for the entire image and thus identify an edge pixel only after several optimization steps and changes in large areas of the image. Most previously known global techniques for color image segmentation ([Dai89], [Hua et al. 92], [LiuYan94], [PanHea93], [PanHea95], [PerKoc94]) use differing types of Markov random fields. Common to them are computationally costly optimization techniques and long processing times. Here a limitation on local techniques results from reasons of practicability. An overview of several selected global techniques was presented in [SkaKos94].

7.3.1 Local Techniques

In addition to vector-valued formulas, monochromatic-based formulas are also common for the detection of edges in color images in edge-based segmentation of color images (see Chapter 6). In the following, a monochromatic-based formula is discussed.

Lanser [Lan93] proposed a color image segmentation that initially detects edges separately in the vector components in the CIELAB color space and subsequently unites the group of the resulting edge pixels. For segmentation it is crucial that closed contours are detected, or that only small gaps in the contours are to be closed. For this, Lanser uses the intersection of the edge pixels, which he attains by single and twice-repeated application of a modification of the Deriche operators [LanEck92]. Each of the edges is extended by one pixel on the edge ends.

The transition from detected contours to regions results from complement formation. By means of a morphological opening, Lanser opens small dividers between two regions in order to merge similar regions. The remaining holes in the regions are expanded in conclusion by a controlled region-growing technique. The order of events of this technique is identical to the order for the segmentation of gray-level images except for the unification of the results determined in the individual components in the CIELAB color space. A vector-valued formula for edge-based segmentation was proposed by Chapron [Cha92]. For the detection of edges in the vector-valued color images he uses a modification of the approach of DiZenzo [DiZ86], which is similar to the Cumani Operator (see Section 6.1.2). After closing the small gaps in the contours, the transition, likewise by complement formation, ensues to the regions.

7.3.2 Segmentation by Watershed Transformation

Segmentation by watershed transformation can be seen as a region-growing technique and thus could have also been described in Section 7.2. Watershed transformation forms the basis of a morphological segmentation of gray-level images. It was developed by Meyer and Beucher [MeyBeu90] and converted by Vincent and Soille [VinSoi91] into a digital algorithm for gray-level images. The technique can be applied to the original image data or, as described in the following, to gradient images. In the latter case it is based on the discontinuities of image function and for this reason it is indirectly a type of edge-based technique.

Good segmentation results have been achieved for medical gray-level images [Weg et al. 96] as well as color images [Saa94] with the watershed transformation. For this reason the watershed transformation is presented here. For an efficient parallel implementation of the technique (for gray-level images) one can refer to [MeiRoe95]. Before its use on color images is explained, the principle should be first clarified by an example of a gray-level image segmentation.

If a gray-level image is viewed as a topographical relief, then the image value $E(\mathbf{p})$ denotes the height of the surface area at position $\mathbf{p}$. A *path* P of length l between two pixels $\mathbf{p}$ and $\mathbf{q}$ is a l+1-tupel $(\mathbf{p}_0, \mathbf{p}_1, \ldots, \mathbf{p}_{l-1}, \mathbf{p}_l)$ with $\mathbf{p}_0 = \mathbf{p}$, $\mathbf{p}_l = \mathbf{q}$, and $(\mathbf{p}_i, \mathbf{p}_{i+1}) \in G$ for all $i \in [0, l)$. G denotes here the basis grid. A set M of pixels is called *connected* if for each pair of pixels $\mathbf{p}, \mathbf{q} \in M$ a path exists between $\mathbf{p}$ and $\mathbf{q}$, which only passes through pixels from M. A *connected component* is a nonempty maximal connected set of pixels. A *regional minimum* of E in the height of h is a connected component of pixels $\mathbf{p}$ with $E(\mathbf{p}) = h$. According to this it is impossible to reach a point of lesser height without overcoming a point of greater height.

Let us assume that small holes are bored into each minimum of the topographical surface area and the relief is slowly immersed into a water basin. The valleys of the relief are filled and small reservoirs emerge by the gradual ascent of the water level. A dam is constructed at the positions (pixels) where the water from two or more reservoirs would flow together (see Fig. 7.6). The number of dams that have emerged after completing the immersion process is called the *watershed transform* of image E.

A is a pixel set and $\mathbf{a}$, $\mathbf{b}$ are two pixels in A. The geodesic distance $d_A(\mathbf{a}, \mathbf{b})$ between two pixels $\mathbf{a}$ and $\mathbf{b}$ is defined as the infinum of the lengths of all paths from $\mathbf{a}$ to $\mathbf{b}$ in A. $B \subseteq A$ is partitioned into k connected components B_i, that is it holds that

$$B = \bigcup_{i=1}^{k} B_i \ .$$

The *geodesic influence zone* $iz_A(B_i)$ of a component B_i within A is defined as the geometric locus of all pixels $\mathbf{p}$ in A, whose geodesic distance to B_i is less than from B to every other component Bj. It holds that

$$iz_A(B_i) = \left\{ p \in A \,\middle|\, \forall j \in \{1,\ldots,k\} \setminus \{i\} : d_A(\mathbf{p}, B_i) < d_A(\mathbf{p}, B_j) \right\} \ .$$

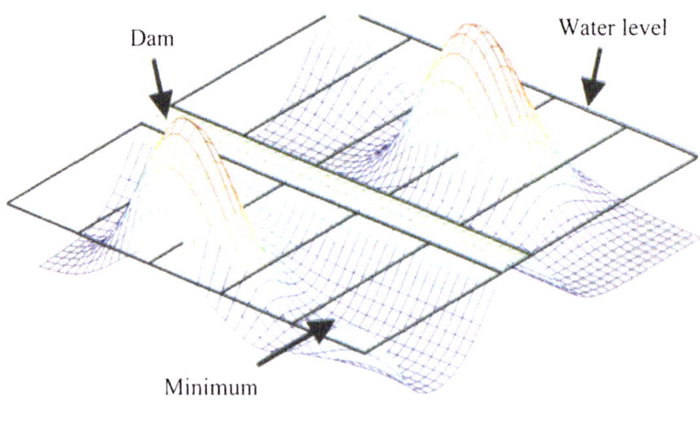

Figure 7.6. *Example of building a dam in places where water from two reservoirs would flow together.*

The set $iz_A(B)$ is defined as the union of the influence zones of the connecting components of B:

$$IZ_A(B) = \bigcup_{i=1}^{k} iz_A(B_i) .$$

The complement $SKIZ_A(B)$ of the set $iz_A(B)$ within A, that is $SKIZ_A(B) = A \setminus IZ_A(B)$, is called *skeleton by influence zones*. The set $T_h(E) = \{p \in D \,|\, E(p) \le h\}$ is denoted as the threshold value set of E on level h; $h_{\min}$ and $h_{\max}$ denote the minimum and the maximum of the gray-levels in the digital image, respectively. Min_h indicates the union of all regional minima at height h. The principle of a watershed transformation is presented in Fig. 7.7.

The algorithmic implementation of the watershed transformation can be subdivided into three processing steps:

1. Calculation of the gradient image
2. Sorting of the gray-levels of the gradient image
3. Calculation of the watersheds and reservoirs

The gradient image represents the absolute values of the gradient of the image function E and is given by Eq. (1.4). For reasons of efficiency it is advantageous when all pixels in the gradient image having the same gray-level can be directly accessed. In addition, the gray-levels of the gradient image are sorted in ascending order and stored in a table. The watersheds and the reservoirs can be determined in the image by this table and a neighborhood relation $N_G(\mathbf{p})$ that represents the set of the (4-, 6- or 8-) neighbors of a pixel **p** with regard to a grid G.

Apart from the table, two matrixes of the gradient image size are the most important data structures in this algorithm. The positions of the reservoirs and the watersheds are stored in one of the matrixes, here named work image E_{seg}. The

---

Define the following recursion:

$$X_{h_{min}}(E) = \{\mathbf{p} \in D \,|\, E(\mathbf{p}) = h_{min}\},$$

$$X_{h+1}(E) = Min_{h+1} \cup IZ_{T_{h+1}(E)}(X_h(E)),\ h = h_{min}, \ldots, h_{max} - 1 .$$

Intuitively, $X_h(E)$ can be interpreted as the set of those pixels p that lie in the same reservoir and for which $E(\mathbf{p}) \le h$ is fulfilled. The watershed transform of the image E is the complement of $X_{h_{max}}(E)$ in D. It holds that

$$Wshed(E) = D \setminus X_{h_{max}}(E) .$$

Figure 7.7. *Principal algorithm of a watershed transformation.*

other matrix is indicated as an additional image E_{aux} and is designed for the storage of the geodesic distances of each pixel to the closest image area with a lower threshold value.

Assuming the relief was immersed into a water basin to height h. In the work image E_{seg} all pixels of the already flooded reservoir are clearly marked as being assigned to a minimum. The threshold value $h+1$ is now viewed and $IZ_{T_{h+1}(E)}(X_h(E))$ is calculated. The geodesic distances of all pixels of the gray level $h+1$ to the pixels of the set X_h are stored in the additional image E_{aux}. In the work image, all pixels that are marked show the gray level $h+1$. In connection with this the geodesic influence zone of already registered reservoirs is calculated within this gray-level. All pixels that lie in the influence zone of a minimum are marked as belonging to this minimum. Thus, existing reservoirs are expanded. The minima newly occurring in a gray-level are registered while a clear mark is assigned to them. In this way the existence of reservoirs is first recognized. After completion of the simulated immersion $(h = h_{\max} - 1)$, the work image E_{seg} contains the (preliminary) segmentation result.

One difficulty in segmentation by means of watershed transformation exists in the oversegmentation of an image. By this the division of homogeneous regions into many small regions (e.g., on the basis of noise) is understood (see Fig. 7.9, lower left). In watershed transformation, reservoirs result directly from the existence of watersheds without consideration of the relative brightness contrast. Therefore, watersheds that are traced back solely to image noise result in the gradient image. The sought-after contours from a gradient image segmented in this manner are properly localized. They decline, however, on account of oversegmentation into a multitude of nonrelevant contours.

One possibility of preventing this oversegmentation exists in applying a threshold value operation on the calculated gradient image, as this is implemented by Saarinen [Saa94]. Thus, watersheds can originate only there, where the amount of the gradient has a certain minimum size. The problem (of each threshold value operation) consists in determining the "best" threshold value for the respective image. Another possibility presented here for the treatment of oversegmentation without using a threshold value consists of a hierarchical growth process. All watersheds and reservoirs of the original image are first calculated. Each reservoir is then assigned a representative value that is calculated, for example, from the gray-levels of the corresponding region in the original image.

The result is a simplified original image that is assembled from small regions with uniform gray-levels. Such an image is called the *first mosaic image*. The hierarchical growth process lies in the fact that watershed transformation is performed once again on this mosaic image. Thereby the regions of the mosaic image of the first order merge into larger regions. The second mosaic image can be determined from this, on which watershed transformation is used once again. Instead of using watershed transformation directly on the mosaic image, it can be used on a modeling of the mosaic image by graphs [Weg et al. 96]. This procedure is described in the following.

7.3.3 Use of Watershed Transformation in Graphs

For the implementation of watershed transformation in the presegmented regions (in the mosaic image), the regions and their neighboring relationships are represented by a graph. The calculation of watersheds on graphs is carried out analogous to the calculation of watersheds on the basis of pixels. Instead of the original image pixels, the previously determined regions form the basis for the calculation of watersheds.

$G = (V, K)$ is a graph whereby $V = \{v_1, v_2, \ldots, v_n\}$ indicates the set of nodes and $K = \{k_1, k_2, \ldots, k_n\}$ denotes the set of edges. For $v, w \in V$ and $k \in K$, $(v, w) = k$ if the edge k binds nodes v and w together. The vicinity of a node v is given by $N_K(v) = \{w \in V | (v, w) \in K\}$. The geodesic distance $d_K(v_1, v_2)$ between two nodes $v_1, v_2 \in V$ is equal to the length of the shortest path from v_1 to v_1. If no connection between v_1 and v_1 exists, then the geodesic distance is defined as infinite.

Since a graph defines a grid as well as a digital image, the use of mathematical morphology in graphs represents a generalization of its use in gray-level images. From the definition of geodesic distance for graphs, the set *SKIZ*, determined by the skeletalization through the influence zones and reservoirs, can be formed. Deciding differences between the gray-level images and the graphs derived from the regions lie, for example, in the nonhomogeneous neighborhood structure of the graphs. In this sense a node in the graph no longer represents just one pixel, but rather under certain conditions a large region. Finally, it is important whether the result of a graph transformation is to be interpreted again as a gray level image.

The regions of a mosaic image are represented by a *region graph* constructed by analysis of the watersheds. The watersheds separate adjacent regions from each other and thus the neighborhood relation can be easily read from them. Each node of the graph represents a region in the mosaic image. If the regions neighbor two nodes, then this relation is modeled by an edge between both nodes. Figure 7.8 shows an example of a region graph for a presegmented mosaic image. The value of a node $G_R(v_i)$ in the region graph can be determined by one or more characteristics of the region (e.g., region size, mean, and variance of the gray-levels within a region, etc.).

One problem in using watershed transformation in a region graph is that the graph does not necessarily contain information about the gray-level differences of neighboring regions. Saarinen [Saa94] proposes moreover to weight the edges in the region graph based on the size of the regions and the Euclidean distance between the image values of the regions. Another procedure, similar to determining the gradient image from the original image, is to calculate a *gradient graph* from the region graph.

The determination of such a gradient graph can result from each region of the mean of differences being assigned to the adjacent regions. Watershed transformation can then be used in this gradient graph whose structure is similar to that of the region graph. Since, however, the mean represents only an approximate convergence with the factual gray-level differences between the regions, it can occur that the contours of the segmented regions only roughly agree with those of the original image. One way to overcome this problem is to model the gradient at the graph level by a contour graph. This technique is further pursued in the following.

Each edge of the region graph corresponds to a node in the contour graph. The value of a node $G_K(v_{i,j})$ in the contour graph is defined by the absolute value of the difference of two adjacent regions i and j according to

$$G_K(v_{i,j}) = \left| G_R(v_i) - G_R(v_j) \right| .$$

All nodes whose contours surround a reservoir are connected to each other. In contrast to the region graph, which is planar, this does not have to apply to the contour graph (see Fig. 7.8).

In a contour graph defined like this, the immersion can again be simulated in a water basin. With this, the nodes of the contour graph are entered in ascending order into a table. Finally, the watershed algorithm for this table and the neighboring relation $N_K(v)$, defined in the contour graph, are implemented. Those nodes of the contour graph that belong to the same newly calculated reservoir get the same label in the contour graph (see example in Fig. 7.8). Nodes that separate the reservoirs from one another are marked as watersheds (shaded area in Fig. 7.8). From the label of the contour graph it is decided which regions merge together and which remain separate.

For the formation of a mosaic image of the next-highest order the contour graph is first transferred into a region graph of the next-highest order. If a contour node is marked as watershed, then both regions (-nodes) from which the contour nodes came may not be combined. If, on the other hand, a label was assigned to a contour node (e.g., x or y in Fig. 7.8), then both regions (-nodes) from which the contour nodes came will be combined under this label. The mosaic image of the next highest order can be calculated from the region graph of the next-highest order (see example in Fig. 7.8). This procedure can be implemented several times (e.g., six times) in order to merge the oversegmented image areas. The number six is in this connection a heuristic size that has resulted from experience with the examined test images.

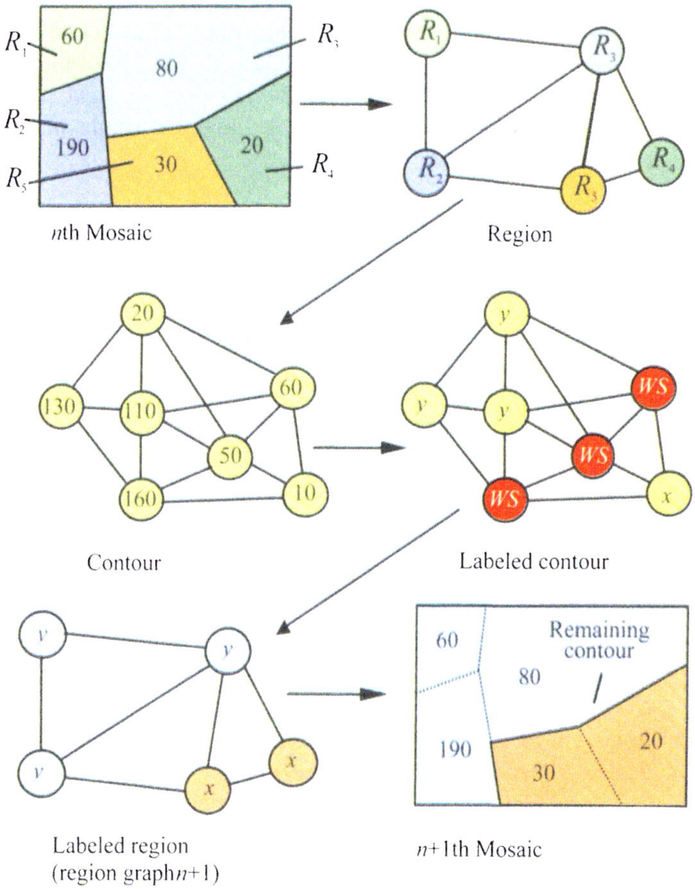

Figure 7.8. *Example of the development of the next mosaic image with corresponding graphs (adapted from [Weg et al. 96]).*

7.3.4 Expansion of the Watershed Transformation for Color Images

So far, the watershed transformation has been viewed exclusively for gray-level images. Now an expansion of this technique is introduced for the segmentation of color images. A three-channel color image $\mathbf{C}(x, y) = (R(x, y), G(x, y), B(x, y))$ is viewed in the *RGB* color space. If the watershed transformation were carried out separately on each of the three vector components of the color signal, then three results with differing regions and watersheds would be produced at the end of the segmentation process. The information from the vector components of the color signal can be combined for the segmentation of color images.

In a first processing step the original image is filtered by means of a vector median (see Section 5.3.2) in order to suppress noise in the image function. Afterward a gradient image is calculated for each vector component. The compound gradient image G_C is determined from the three resulting gradient images G_R, G_G, and G_B for the red, green, and blue channel. In addition, vectors are formed from the three gradient images and the lengths of the vectors are stored in an image G_C where

$$G_C(x,y) = \sqrt{G_R(x,y)^2 + G_G(x,y)^2 + G_B(x,y)^2} \tag{7.1}$$

(see [Saa94]). The regions and watersheds can now be calculated with the above-named processing step for the compound gradient image G_C. The color image determined by means of a vector median is listed for each region in the mosaic image of the first order. For continuation of the segmentation process, the mosaic image is transferred into a region graph. Each region in the region graph is characterized by the representative values for each vector component $(v_r, v_g, v_b)^T$. Finally, the region graph is transferred into a contour graph. The value of a node $G_K(v_{k,l})$ in the contour graph is defined by means of the maximum norm in the *RGB* color space for two adjacent regions *k* and *l* with

$$G_K(v_{k,l}) = \max\left\{ \left|v_r^k - v_r^l\right|, \left|v_g^k - v_g^l\right|, \left|v_b^k - v_b^l\right| \right\}.$$

The next mosaic image can be calculated from this contour graph. This growing process is iteratively carried out (e.g., six times).

In Figs. 7.9 and 7.10, the watershed transformation and the growth process is illustrated for a selected sample image "Pepper." In addition to the original image and the compound gradient image, the first to the sixth mosaic images are represented. The sixth mosaic image represents the segmentation result. Apart from small disturbances in the lower portion of the pepper in front, good segmentation was achieved by the expansion of the watershed transformation to color images.

7.4 PHYSICS-BASED SEGMENTATION

Segmentation techniques in which physical models in a color space are used for the division of an image into regions that correspond to surfaces and/or objects in the scene are discussed in this section. The goal of these techniques is to segment a color image at object borders and not at shadows or specular highlights in the image. This is a difficult task since the image presenting a surface area is influenced by various effects such as highlights, shadows, shading, sensor noises, uneven lighting, and surface textures.

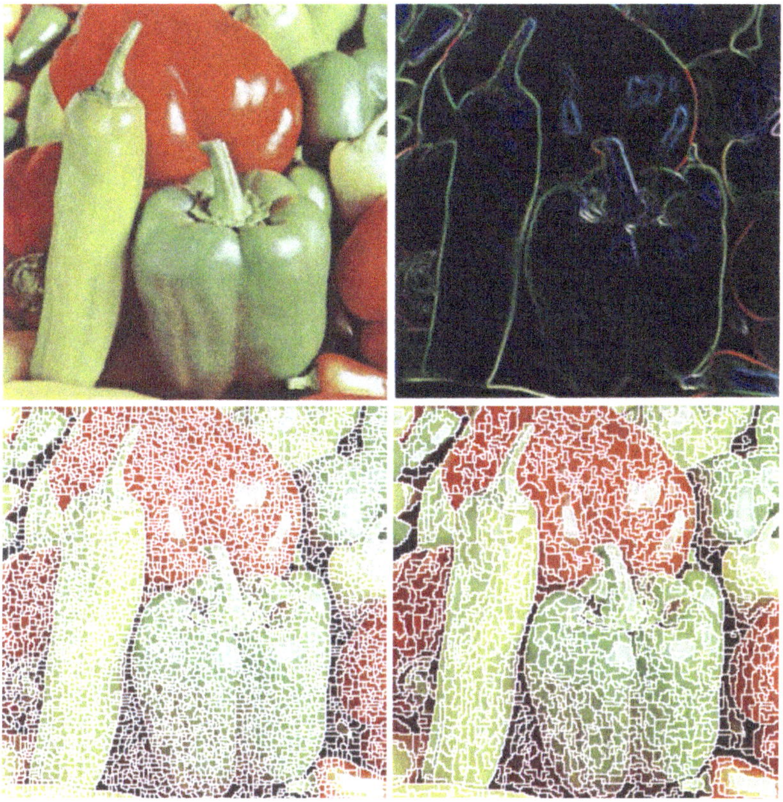

Figure 7.9. *The original color image "Pepper" (upper left), the compound gradient image of the original data filtered by means of vector median (upper right), the first mosaic image (lower left), and the second mosaic image (lower right).*

Most of the techniques described in the previous sections implement uniformity criteria for the definition of regions based on a uniform or a minimal color distance of color signals distorted by noise. These algorithms frequently produce results for realistic images in which a surface area of an object is segmented into many regions. A highlight can be segmented as a separate region in the image or the surface area of a bent object can be subdivided into many regions on the basis of intensity changes caused by shading. This would further impede object recognition. Physics-based segmentation techniques facilitate the segmentation of real images on the basis of physical models for image formation. The fundamental mathematical methods that are used in the physics-based techniques are frequently similar to those techniques already discussed in the previous sections.

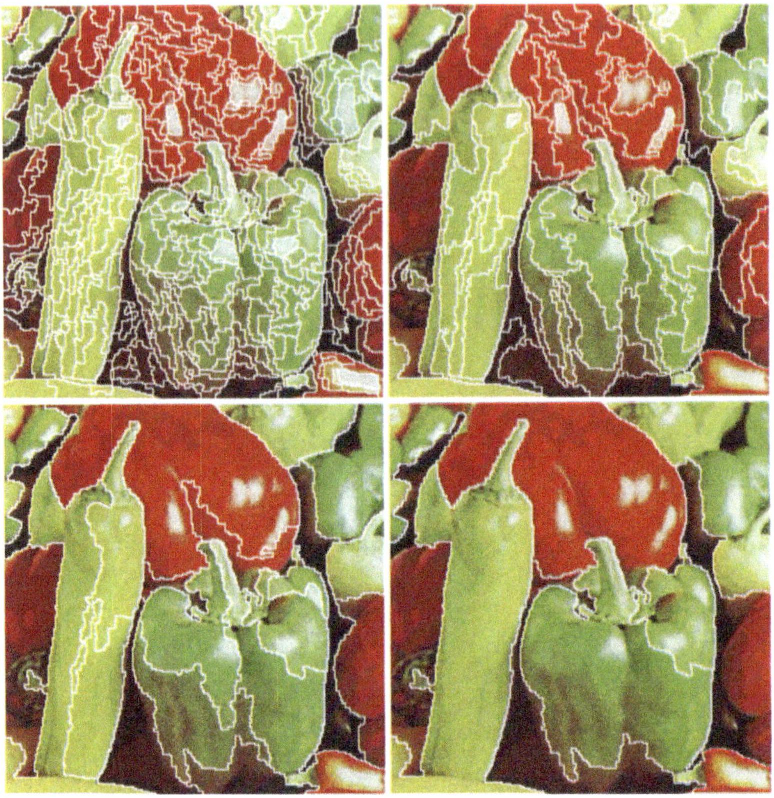

Figure 7.10. *The third mosaic image (upper left), fourth mosaic image (upper right), fifth mosaic image (lower left) and sixth mosaic image (lower right) for the color image "Pepper."*

For example, Healey [Hea90] used a region-splitting technique controlled by a previously explained edge-detection technique for the classification of regions. The formulations for physics-based segmentation and the techniques already described in the previous sections do not differ in the basic mathematical methods. They differ, rather, in the reflection models that are employed in color image segmentation. Therefore, physics-based segmentation has so far been limited to the determination of changes in reflection on materials whose reflection characteristics are known and can be sufficiently well modeled.

Several methods can therefore be seen as preprocessing for the actual segmentation process. For example, material changes of shadows or highlights need to be distinguished. This classification is necessary in order to get segmentation results that represent surface areas or objects in the scene and not changes in irradiance. Maxwell and Shafer [MaxSha96] proposed for this 36 hypotheses on whose basis they implement grouping color signals in the color space to corresponding regions in the image (see [MaxSha96] and [MaxSha97]). Before some techniques for physics-based segmentation are discussed, the dichromatic reflection model is first introduced. General descriptions of surface area reflections that also include the dichromatic reflection model are given in [Kle et al. 98].

7.4.1 Dichromatic Reflection Model

The *dichromatic reflection model* (*DRM*) describes the reflection on optically inhomogeneous, dielectric materials, such as plastic or paint (see [Sha85]). It indicates general, hybrid reflections, without specially modeling the specular reflection component. The surface area construction of these materials consists of an interface and an optically neutral medium in which color pigments are found. The construction of such a material is illustrated in Fig. 7.11.

The interface separates the surface area from the medium surrounding it, which is generally air. A portion of the radiation that appears on the surface area does not penetrate into the material, but rather is reflected on the interface. This reflection is called *Fresnel reflection* (*interface reflection* or *surface reflection*) and has approximately the same spectral distribution as the light of the illumination source.

The light that is not reflected from the interface penetrates into the material. There it is scattered onto the color particles, partially absorbed, partially passed on, and finally a portion of the penetrated light is reflected through the interface into the surrounding medium. This final process is indicated as *body reflection*. When the division of the color pigments is homogeneous and the pigments demonstrate the same optical behavior, it can be assumed that the light penetrating into the material does not show any particular direction when exiting the surface area.

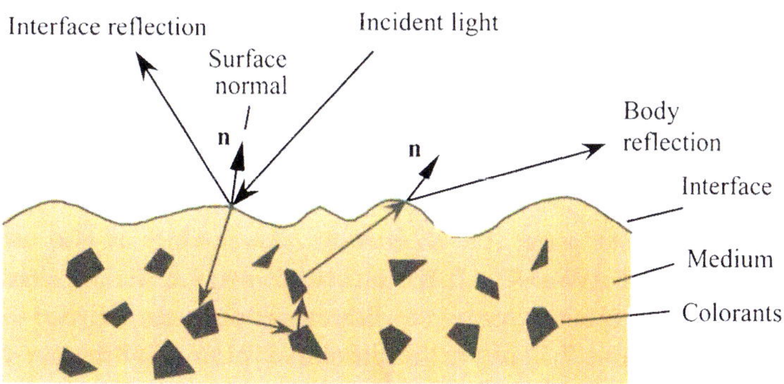

Figure 7.11. *Reflection on an inhomogeneous dielectric material.*

The light energy L that falls on the sensor depends on the wavelength λ, the surface normal $\mathbf{n}$, the lighting direction $\mathbf{s}$, and the viewer direction $\mathbf{v}$. L is the sum of the radiance of the interface reflection L_s and the body reflection L_b. The interface reflection describes the specular reflection component, while the diffuse reflection component is described by the body reflection. The dichromatic model is formulated mathematically by

$$\begin{aligned} L(\lambda,\mathbf{n},\mathbf{s},\mathbf{v}) &= L_s(\lambda,\mathbf{n},\mathbf{s},\mathbf{v}) + L_b(\lambda,\mathbf{n},\mathbf{s},\mathbf{v}) \\ &= m_s(\mathbf{n},\mathbf{s},\mathbf{v})\cdot c_s(\lambda) + m_b(\mathbf{n},\mathbf{s},\mathbf{v})\cdot c_b(\lambda). \end{aligned} \tag{7.2}$$

By a dichromatic reflection model it is assumed that the geometric components m_s and m_b can be separated from the spectral components $\mathbf{c}_s$ and $\mathbf{c}_b$. $\mathbf{c}_s$ is denoted as *interface reflection color* and $\mathbf{c}_b$ as *body reflection color*. In addition, if the existence of a neutral reflecting interface is assumed, then $\mathbf{c}_s$ describes the spectral distribution of the lighting. This special case of the dichromatic reflection model is called the *neutral interface reflection model* (*NIRM*; see [Lee90]).

Three-channel color images represent in general three spectral transmissions of visible light in the red, green, and blue areas. The scene radiance can therefore be represented as a three-dimensional color vector and the model of the scene radiance can be described by (see [Kle et al. 98])

$$\mathbf{L} = \begin{pmatrix} L(red,\mathbf{n},\mathbf{s},\mathbf{v}) \\ L(green,\mathbf{n},\mathbf{s},\mathbf{v}) \\ L(blue,\mathbf{n},\mathbf{s},\mathbf{v}) \end{pmatrix} = \mathbf{L}_s + \mathbf{L}_b \tag{7.3}$$

$$= m_s(\mathbf{n},\mathbf{s},\mathbf{v})\cdot \mathbf{c}_s + m_b(\mathbf{n},\mathbf{s},\mathbf{v})\cdot \mathbf{c}_b \tag{7.4}$$

$$= m_s(\mathbf{n},\mathbf{s},\mathbf{v})\cdot \begin{pmatrix} c_{s,R} \\ c_{s,G} \\ c_{s,B} \end{pmatrix} + m_b(\mathbf{n},\mathbf{s},\mathbf{v})\cdot \begin{pmatrix} c_{b,R} \\ c_{b,G} \\ c_{b,B} \end{pmatrix}. \tag{7.5}$$

Since m_s and m_b can be any scaling factors, vectors $\mathbf{c}_s$ and $\mathbf{c}_b$ form a plane in the *RGB* color space, the *dichromatic plane*, which is also called the *color-signal plane* [TomWan89]. If the object has many differing surface area orientations, then the color vectors in the dichromatic plane are assigned to *T*- and *L*-shaped clusters. Figure 7.12 shows the dichromatic plane of a bent, one-colored, hybrid-reflecting object. If the object consists of many hybrid-reflecting materials, then a special cluster develops for each material.

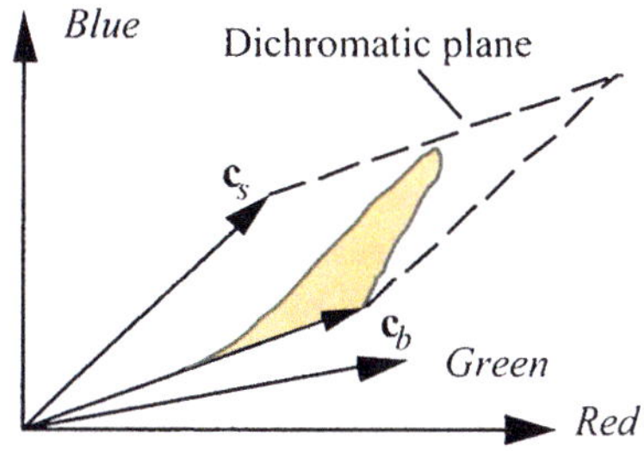

Figure 7.12. *Dichromatic plane of a bent, one-colored, hybrid-reflecting object (reprinted from [Kle et al. 96] with permission from Vieweg).*

The color spaces represented in the figure are likewise called *color histograms,* although the entry in this color histogram is only binary. It is noted only whether a color appears in the scene. In color image processing the DRM as well as the NIRM are utilized. By analyzing the cluster it is possible, among other things, to separate reflection components in order to eliminate highlights from the images (see Section 8.1) and to carry out color image segmentation. A good approximation of the DRM and the NIRM was shown with the help of spectral measurements (spectral radiometer measurements) for various materials (see [Hea89], [TomWan89], and [Lee90]).

7.4.2 Classification Techniques

The goal of physics-based techniques is, as already mentioned, the segmentation of a color image at the object boundaries and not at shadow or highlight boundaries in the image. Gershon, Jepson, and Tsotsos [Ger et al. 86] proposed, for example, a formula that makes it possible to distinguish between shadow boundaries and material changes in the scene. By application of a spectral reflection model they analyze the cases of ideal shadows by which the ambient and the direct lighting have the same spectral composition, and cases of nonideal shadows. For the quantization of the deviation of a shadow boundary from the ideal case, they define a pull factor P_f in the RGB color space by

$$P_f = \frac{|gR - rG|}{R^2 + G^2}.$$

This work is used to estimate the probability that a shadow and an illuminated region belong to the same material. The stronger the effect of the additional ambient lighting, the more the shadow region is distinguished from the illuminated region. Therefore, it is more improbable that both regions belong to the same material.

For distinguishing between shadow boundaries and material changes, biologically inspired color operators are used that are based on the organization of receptive fields in the cortex of primates. The operators have an antagonistic organization for the center and the surround and their responses depend on the pull factor for the shadow boundary. The response of a monochromatic On/Off unit is defined for (R+) center / ($R-$) surround by

$$RESP(\mathbf{p};\sigma) = G(\mathbf{p};\sigma_c) * L_R(\mathbf{p}) - G(\mathbf{p};\sigma_s) * L_R(\mathbf{p}).$$

On this occasion * describes the convolution, $L_R(\mathbf{p})$ is the logarithm of the red component in the color image, and $G(\mathbf{p};\sigma_i)$ is a Gauss function of the form

$$G(\mathbf{p};\sigma_i) = \frac{1}{2\pi\sigma_i^2}\exp\left(\frac{-|\mathbf{p}|^2}{2\sigma_i^2}\right).$$

The response of an antagonistically organized double-opponent color operator is calculated, for example, for ($R+G-$) center / ($R-G+$) surround by

$$\begin{aligned} RESP(\mathbf{p};\sigma) &= G(\mathbf{p};\sigma_c) * (L_R(\mathbf{p}) - L_G(\mathbf{p})) - G(\mathbf{p};\sigma_s) * (L_R(\mathbf{p}) - L_G(\mathbf{p})) \\ &= (L_R(\mathbf{p}) - L_G(\mathbf{p}))DOG(\mathbf{p}) \end{aligned}.$$

On this occasion, $L_G(\mathbf{p})$ is the logarithm of the green vector component in the color image and *DOG* (*difference of Gaussians*) indicates a *DOG* filter with $\sigma_s / \sigma_c = 2.5$ in the experiments from [Ger et al. 86]. The responses of the operators were combined to distinguish material changes from shadow boundaries. For the red-green case, the *relative amplitude response* (*RAA*) is determined by

$$RAA = \frac{|\text{peak response of } R+G-/R-G+|}{[(\text{peak response of } R+/R-)^2 + (\text{peak response of } G+/G-)^2]^{1/2}}.$$

If the *RAA* value is larger than the expected pull factor, then it is assumed that the discontinuity in the area was not caused by a material change, but rather by a shadow.

A somewhat more extensive formulation for determining areas representing the scene material in a color image was suggested by Healey ([Hea89], [Hea90], [Hea92]). The main idea is based on the splitting of regions using a standardized color space. It is assumed that no interreflections exist in the scene (see Section 8.2 for the minimization of interreflections in color images). In view of the spectral reflection characteristics of metal surface areas and dielectric materials, the measured sensor response depends on the sensor, the lighting, and the reflecting material, but not on the scene geometry [Hea90]. For the standardization of sensor responses Healey [Hea92] proposes the application of the L_2 norm.

On the basis of normalization, dark-brown and light-brown pixels are assigned to the same material class. These pixels have spectral reflection functions that roughly represent a scalar multitude of each other. Therefore, it is assumed that they belong to the same material and are differentiated solely on the basis of object geometry (shading effects). Furthermore, Healey assumes that the affiliation of a pixel to a material class occurring in the image is equally probable for all material classes. The division of the measured sensor responses is modeled by a multidimensional standard division function for each material class. By using Bayesian theory, a set of distinguishing functions is defined from which the affiliation of a pixel is fixed to an already-noted material class or to a new material class. Through this the number of materials represented in the color image is determined at the same time.

Furthermore, possible candidates for highlight regions are subjected to an additional examination. If the sensor values within a candidate region lie on a straight line in the dichromatic plane that intersects the origin, then the pixels do not belong to a highlight region but rather to a material class. This formulation was developed for *N* spectral transmissions (i.e., it is not limited to three-channel *RGB* images). It is rather easily conceivable that the results improve in determining the differing material classes when more than three spectral transmissions are used. These spectral transmissions are easily produced with a black-and-white camera and various color filters (see Section 4.1.2).

Klinker, Shafer, and Kanade [Kli et al. 88], [Kli et al. 90] presented an extensive examination of the influence of highlights, shadows, and camera characteristics (as attenuation, clipping, and blooming of the sensor signal, as well as the chromatic aberration of the lens; see Section 4.3) on the results of color image segmentation. By using the dichromatic reflection model for inhomogeneous, dielectric materials, they classify the physical occurrences from the measured color variation in the color space. They have shown that a color cluster whose pixel contains a matte and a highlight region looks like a "skewed T" in the color space.

A hypothesis-based segmentation algorithm was developed using this classification. The algorithm searches in a bottom-up process for color clusters in image areas that show a characteristic form. When a promising cluster in the color space is found for an image area, a hypothesis is made that describes the object color or highlight color and determines the shadowing and highlight components for each pixel in this area. By using a region-growing technique the algorithm determines the exact size of the region in the image for which the new hypothesis applies. The applicability of the hypothesis is verified by this step.

The analysis consists therefore of many small interpretation cycles that link the bottom-up process with the results of the top-down processing. This formulation can be used to determine the number of inhomogeneous dielectric materials in a scene acquired under unknown lighting conditions. A detailed discussion of highlights and interreflections in color images follows later in Chapter 8.

An algorithmically simpler formulation for distinguishing between material changes and changes in the image function on the basis of highlights, shadows, shading, or interreflections was proposed by Bajcsy, Lee, and Leonardis [Baj et al. 90ab]. The assumption is made that the image consists of *patches* of object surface areas that have a uniform chromaticity type (not intensity). The image can be then subdivided into regions with uniform hue and saturation without considering the surface area structure.

For this, an algorithm in the *HSI* color space is used for inhomogeneous dielectric materials under consideration of the dichromatic reflection model. It is assumed that the highlights have the same spectral composition as the lighting. For the lighting, a white balance is carried out by means of a reference map with known reflection factors (see Section 4.4.3). Furthermore, Bajcsy, Lee, and Leonardis describe the resulting structures of the color cluster in the *HSI*-color space for phenomena such as shadowing, shading, highlights, and interreflections. Toward the end of this analysis, a histogram technique is applied to the hue component to segment the individual surface areas. A further segmentation is applied to the saturation component. The following observations are based on this:

1. Shadows, highlights, shading, and interreflections change the intensity.
2. Shadows and shading change neither the hue nor the color saturation.
3. A highlight reduces the saturation value.
4. Interreflections generally cause a change of the hue and the saturation.

Apart from these observations and by using threshold values on the color saturation values, the color clusters for highlights and for most small interreflections can be separated from those of the body reflection. This algorithm can be used for the segmentation of color images that were taken under controlled illumination and represent inhomogeneous dielectric materials. In practice, it is to be noticed that the above-named observations of Bajcsy, Lee, and Leonardis hold for the ideal case and do not consider effects such as sensor noise and the dynamic range of the camera.

7.5 COMPARISON OF SEGMENTATION PROCESSES

Allen and Huntsberger [AllHun89] compared the results of two edge-based techniques ([Nev77] and [You86]) with the results of a pixel-based histogram technique [Oht et al. 80] and a fuzzy-cluster analysis [Hun et al. 85]. The criteria for the assessment of the techniques were accuracy, numerical stability, and robustness toward noise. The four techniques were applied to noisy synthetic color images as well as to natural color images. For the synthetic images the deviations of the detected region borders of the actual region borders were used for assessment. For the natural images the visual impressions of the results were used as a measurement for the quality of the algorithms. In this investigation the best results for the selected test images were always achieved with the histogram technique.

In another investigation of color image segmentation, Gauch and Hsia [GauHsi92] evaluated the results of a region-growing technique, an edge-based technique, and a recursive split-and-merge technique in four color spaces (*RGB*, *HSI*, *YIQ*, and CIELAB; see Chapter 3). All three techniques were used on a natural image of a street scene and a synthetic image with six different surface areas. As a result of this investigation, none of the four color spaces proved to be the best for all three techniques or for the synthetic and the natural images. In the comparison of the techniques, the best results in this investigation were achieved by the split-and-merge technique.

Other application-based assessment criteria, such as the real-time capability of a segmentation technique, have not yet been examined. For example, Priese and Rehrmann [PriReh93] proposed a hybrid segmentation process that is composed of a bottom-up region-growing and a top-down separation technique. They use a hexagonal structure in the examination of local neighborhoods and a heuristic color uniformity function in the *HSV* color space. With this technique traffic signs can be segmented in video real-time using color information.

More recent segmentation techniques, such as watershed transformation or physics-based techniques, have not yet been examined in comparative studies. With the physics-based techniques, comparison with the other techniques is made difficult because they are as a rule applicable in only certain materials available in the scene. Here other models must be included for additional materials in the segmentation process. For the future, it is important here (just as for edge classification) that a classification of the materials based solely on the order of the vector signals in the color space be possible without any previous knowledge. First steps in this direction can be found in [MaxSha97]. The physics-based techniques are so far more suitable for a preprocessing, in the sense of a detection of highlights, shadows, and so on, than for the actual segmentation. Good segmentation results were always achieved in color images in my own investigations of watershed transformation. A comparison of the results of this technique with the other techniques would be very informative here.

In conclusion, it is stressed that in many applications the assessment of the hue is very successful for the segmentation of color images. However, caution is necessary if the images are very dark. Neither a universal algorithm nor a "best" color space exists for color image segmentation. Many differing color spaces were already used in segmentation. Nevertheless, no general advantage of one color space over the other color spaces has yet been established. Furthermore, increased vector-valued formulations also should be pursued for color image segmentation. Perez and Koch compiled a general overview of techniques for the segmentation of color images in [PerKoc94]. An interesting selection of publications on physics-based segmentation of color images can be found in [Hea et al. 92] and a more current overview in [MaxSha97].

7.6 REFERENCES

[AllHun89] J.T. Allen, T. Huntsberger. Comparing color edge detection and segmentation methods. *Proc. IEEE 1989 Southeastcon*, 1989, pp. 722-728.

[Baj et al. 90a] R. Bajcsy, S.W. Lee, A. Leonardis. Color image segmentation with detection of highlights and local illumination induced by inter-reflections. *Proc. 10th Int. Conference on Pattern Recognition*, Atlantic City, New Jersey, 1990, pp. 785-790.

[Baj et al. 90b] R. Bajcsy, S.W. Lee, A. Leonardis. Color image segmentation and color constancy. *Proc. SPIE* **1250**, Perceiving, Measuring and Using Color, 1990, pp. 245-255.

[BonCoy91] L. Bonsiepen, W. Coy. Stable segmentation using color information. *Proc. 4th Int. Conference on Computer Analysis of Images and Patterns*, R. Klette (ed.), Dresden, Germany, 1991, pp. 77-84.

[Cam et al. 95] P. Campadelli, D. Medici, R. Schettini. Using Hopfield networks to segment color images. *Proc. 8th Int. Conference on Image Analysis and Processing*, San Remo, Italy, 1995, pp. 25-29.

[Cha92] M. Chapron. A new chromatic edge detector used for color image segmentation. *Proc. 11th Int. Conference on Pattern Recognition*, The Hague, Netherlands, 1992, Vol. III, pp. 311-314.

[Dai89] M.J. Daily. Color image segmentation using Markov random fields. *Proc. Int. Conference on Computer Vision and Pattern Recognition*, San Diego, 1989, pp. 304-312.

[DiZ86] S. DiZenzo. A note on the gradient of a multi-image. *Computer Vision, Graphics, and Image Processing* **33** (1986), pp. 116-125.

[FerVid92] F. Ferri, E. Vidal. Colour image segmentation and labeling through multiedit-condensing. *Pattern Recognition Letters* **13** (1992), pp. 561-568.

[GauHsi92] J. Gauch, C.-W. Hsia. A comparison of three color image segmentation algorithms in four color spaces. *Proc. SPIE* **1818**, Visual Communications and Image Processing, 1992, pp. 1168-1181.

[Ger et al. 86] R. Gershon, A.D. Jepson, J.K. Tsotsos. Ambient illumination and the determination of material changes. *J. Optical Society of America A* **3** (1986), pp. 1700-1707.

[GonWoo02] R.C. Gonzalez, R.E. Woods. Digital Image Processing. 2nd ed., Prentice-Hall, Upper Saddle River, New Jersey, 2002.

[Hea89] G. Healey. Color discrimination by computer. *IEEE Transactions on Systems, Man, and Cybernetics* **19** (1989), pp. 1613-1617.

[Hea90] G.E. Healey. Using physical color models in 3-d machine vision. *Proc. SPIE* **1250**, Perceiving, Measuring and Using Color, 1990, pp. 264-275.

[Hea92] G.E. Healey. Segmenting images using normalized color. In: G. Healey, S.A. Shafer, L.B. Wolff (eds.). *Physics-Based Vision: Principles and Practice Color*. Jones and Bartlett, Boston, Massachusetts, 1992, pp. 166-198.

[Hea et al. 92] G. Healey, S.A. Shafer, L.B. Wolff (eds.). *Physics-Based Vision: Principles and Practice Color*. Jones and Bartlett, Boston, Massachusetts, 1992.

[Hua et al. 92] C.-L. Huang, T.-Y. Cheng, C.-C. Chen. Color images′ segmentation using scale space filter and Markov random fields. *Pattern Recognition* **25** (1992), pp. 1217-1229.

[Hun et al. 85] T.L. Huntsberger, C.L. Jacobs, R.L. Cannon. Iterative fuzzy image segmentation. *Pattern Recognition* **18** (1985), pp. 131-138.

[Kle et al. 96] R. Klette, A. Koschan, K. Schlüns. *Computer Vision: Räumliche Information aus digitalen Bildern.* Vieweg, Braunschweig/Wiesbaden, Germany, 1996.

[Kle et al. 98] R. Klette, K. Schlüns, A. Koschan. *Computer Vision: Three-Dimensional Data from Images.* Springer, Singapore, 1998.

[Kli et al. 88] G.J. Klinker, S.A. Shafer, T. Kanade. Image segmentation and reflection analysis through color. *Proc. Image Understanding Workshop*, Vol. II, Cambridge, Massachusetts, 1988, pp. 838-853.

[Kli et al. 90] G.J. Klinker, S.A. Shafer, T. Kanade. A physical approach to color image understanding. *Int. J. of Computer Vision* **4** (1990), pp. 7-38.

[Lan93] S. Lanser. Kantenorientierte Farbsegmentation im CIE-Lab Raum. *Proc. 15th DAGM-Symposium Mustererkennung*, S.J. Pöppl, H. Handels (Hrsg.), Lübeck, 1993, pp. 639-646.

[LanEck92] S. Lanser, W. Eckstein. A modification of Deriche's approach to edge detection. *Proc. 11th Int. Conference on Pattern Recognition*, The Hague, Netherlands, 1992, Vol. III, pp. 633-637.

[Lee90] H.-C. Lee. Illuminant color from shading. *Proc. SPIE* **1250**, Perceiving, Measuring and Using Color, 1990, pp.236-244.

[LiuYan94] J. Liu, Y.-H. Yang. Multiresolution color image segmentation. *IEEE Transactions on Pattern Analysis and Machine Intelligence* **16** (1994), pp. 689-700.

[MaxSha96] B.A. Maxwell, S.A. Shafer. Physics-based segmentation: Looking beyond color. *Proc. Image Understanding Workshop*, Palm Springs, 1996, pp. 867-878.

[MaxSha97] B.A. Maxwell, S.A. Shafer. Physics-based segmentation of complex objects using multiple hypotheses of image formation. *Computer Vision and Image Understanding* **65** (1997), pp. 269-295.

[MeiRoe95] A. Meijster, J.B.T.M. Roerdink. A proposal for the implementation of a parallel watershed algorithm. *Proc. 6th Int. Conference on Computer Analysis of Images and Patterns*, S. Hlavác (ed.), Prague, Czech Republic, 1995, pp. 790-795.

[MeyBeu90] F. Meyer, S. Beucher. Morphological segmentation. *J. Visual Communications and Image Representation* **1** (1990), pp. 21-45.

[Nev77] R. Nevatia. A color edge detector and its use in scene segmentation. *IEEE Transactions on Systems, Man, and Cybernetics* **7** (1977), pp. 820-826.

[Oht et al. 80] Y.-I. Ohta, T. Kanade, T. Sakai. Color information for region segmentation. *Computer Graphics and Image Processing* **13** (1980), pp. 222-241.

[PalPal93] N.R. Pal, S. Pal. A review on image segmentation techniques. *Pattern Recognition* **26** (1993), pp. 1277-1294.

[PanHea93] D.K. Panjwani, G. Healey. Unsupervised segmentation of textured color images using Markov random field models. *Proc. Int. Conference on Computer Vision and Pattern Recognition*, New York, 1993, pp. 776-777.

[PanHea95] D.K. Panjwani, G. Healey. Markov random field models for unsupervised segmentation of textured color images. *IEEE Transactions on Pattern Analysis and Machine Intelligence* **17** (1995), pp. 939-954.

[PerKoc94] F. Perez, C. Koch. Toward color image segmentation in analog VLSI: Algorithm and hardware. *Int. J. of Computer Vision* **12** (1994), pp. 17-42.

[Pit93] I. Pitas. *Digital Image Processing Algorithms*. Prentice Hall, Hertfordshire, England, 1993.

[Pra91] W.K. Pratt. *Digital Image Processing*. 2nd ed., Wiley, New York, 1991.

[PriReh93] L. Priese, V. Rehrmann. On hierarchical color segmentation and applications. *Proc. Int. Conference on Computer Vision and Pattern Recognition*, New York, 1993, pp. 633-634.

[Saa94] K. Saarinen. Color image segmentation by a watershed algorithm and region adjacency graph processing. *Proc. Int. Conference on Image Processing*, 1994, Vol. III, pp. 1021-1025.

[Sha85] S.A. Shafer. Using color to separate reflection components. *Color Research Applications* **10** (1985), pp. 210-218.

[SkaKos94] W. Skarbek, A. Koschan. *Colour image segmentation: A survey*. Technical Report 94-32, Technical University Berlin, Dept. of Computer Science, October 1994.

[Son et al. 93] M. Sonka, V. Hlavac, R. Boyle. *Image Processing, Analysis and Machine Vision*. Chapman & Hall, London, 1993.

[SwaBal91] M.J. Swain, D.H. Ballard. Color indexing. *Int. J. of Computer Vision* **7** (1991), pp. 11-32.

[TomWan89] S. Tominaga, B.A. Wandell. Standard surface-reflectance model and illuminant estimation. *J. Optical Society of America A* **6** (1989), pp. 576-584.

[TseCha92] D.-C. Tseng, C.-H. Chang. Color segmentation using perceptual attributes. *Proc. 11th Int. Conference on Pattern Recognition*, The Hague, Netherlands, 1992, Vol. III, pp. 228-231.

[Umb et al. 93] S.E. Umbaugh, R.H. Moss, W.V. Stoecker, G.A. Hance. Automatic color segmentation algorithms with application to skin tumor feature identification. *IEEE Engineering in Medicine and Biology* **12** (1993), pp. 75-82.

[Vin89] L. Vincent. Graphs and mathematical morphology. *Signal Processing* **16** (1989), pp. 365-388.

[VinSoi91] L. Vincent, P. Soille. Watersheds in digital spaces: an efficient algorithm based on immersion simulations. *IEEE Transactions on Pattern Analysis and Machine Intelligence* **13** (1991), pp. 583-598.

[Weg et al. 96] S. Wegner, T. Harms, H. Oswald, E. Fleck. Medical image segmentation using the watershed transformation on graphs. *Proc. Int. Conference on Image Processing*, Lausanne, Switzerland, 1996, Vol. III, pp. 37-44.

[You86] R. Young. Simulation of human retinal function with the Gaussian derivative model. *Proc. Int. Conference on Computer Vision and Pattern Recognition*, Miami Beach, Florida, 1986, pp. 564-569.

8 HIGHLIGHTS, INTERREFLECTIONS, AND COLOR CONSTANCY

An image function is influenced during its generation by a number of physical factors in the scene. In image processing it is frequently assumed that a scene consists only of matte objects. This leads to a strong restriction of the applicableness of the procedures, since our environment usually also contains objects with specular surfaces. These surfaces cause highlights in the image.

If several objects are present in the scene or if an object is not convex, then each surface receives not only the light of a single source of light, but also light reflected from other surfaces, depending on the position of the objects or surfaces relative to each other and to the source of light. These possible interreflections affect both the brightness and the color of the surfaces in the image.

The spectral composition of the light reflected by the surfaces in the scene also changes if the lighting conditions change. A color classification of surfaces from color images, invariant in relation to lighting changes, called *color constancy*, cannot be easily achieved. It is, nonetheless, of importance for color object recognition or color object tracking in an image sequence.

All these phenomena are well known. Yet they are usually neglected in image processing although they can have a crucial influence on the result of a procedure. In this chapter, different vector-valued techniques for the analysis of these phenomena are discussed. An interesting collection of articles covering these groups of topics was compiled in [Hea et al. 92].

8.1 HIGHLIGHT ANALYSIS IN COLOR IMAGES

The detection and treatment of highlights in color images presents an interesting challenge and at the same time a new possibility for digital color image processing [Hea et al. 92]. Highlights are often misinterpreted as light objects or light patterns on surfaces. Therefore, the results of image segmentation or correspondence analysis in stereo images may be significantly falsified. In contrast to this, highlights in the color image can be located and, depending on existing requirements, eliminated if suitable reflection models are included in the sensor model of the color image generation.

Reflection that can be modeled consists of a diffuse and a specular component. The *diffuse* or *Lambertian reflection* can be described by the Lambertian cosine law. This model or law is generally used also in digital image processing, although it is only conditionally suitable for the description of reflection on very rough surfaces. The underlying model and its limitations were pointed out in detail in [Kle et al. 98].

For the description of the specular reflection component there exists a set of models that are based on physical and geometrical optics. Two well-known models are the *Beckmann-Spizzichino model* from physical optics and the *Torrance-Sparrow model* from geometrical optics. A comparison of these two models is represented in [Nay et al. 91a]. Simplifications of these two models are used in computer graphics and image processing for the description of the specular component. These simplifying models are to be found in many books on computer graphics (in [Fol et al. 95]). The dichromatic reflection model described in Section 7.4.1 has attained special importance in color image processing. It is used both during the analysis of highlights and for the analysis of interreflections.

The analysis of highlights in color images can serve different goals. For example, Lee [Lee86] shows how the color of the light source can be determined from the highlights on two or more surfaces. This is of importance for the treatment of the color constancy problem (see Section 8.3). In a somewhat more general technique, Lee [Lee90] shows the determination of the color of the light source from the image of an individual material. With a Lambertian reflector it cannot be differentiated whether the image shows a white surface lit with yellow light or a yellow surface lit with white light. If the surface consists of an inhomogeneous dielectric material and the dichromatic reflection model can be used as a basis, then the color signal can be split up into the body reflection component and the interface reflection component. On the assumption that the highlight possesses the same spectral composition as the lighting, a clear distinction is possible. Then the color of the light source can be determined. For example, a yellow highlight indicates a white surface lit with yellow light while a white highlight indicates a yellow surface lit with white light. Due to the special importance that highlight analysis has attained in digital color image processing, several procedures for this are discussed in the following.

8.1.1 Klinker-Shafer-Kanade Technique

As mentioned previously in Section 7.4.2, Klinker, Shafer, and Kanade (see [Kli93], [Kli et al. 88], and [Kli et al. 90]) have presented an extensive examination of the influence of highlights, shading, and camera characteristics on the results of color image segmentation. By using the dichromatic reflection model for inhomogeneous dielectric materials (see Section 7.4.1), they classify the physical events on the basis of the measured color variation in the image. In the *RGB* space, the color signal of a pixel can be split up in accordance with Eq. (7.4) into

$$\mathbf{C}(x,y) = m_s(x,y)\cdot \mathbf{c}_s + m_b(x,y)\cdot \mathbf{c}_b$$

according to the dichromatic reflection model. Thus, $\mathbf{c}_s$ indicates the interface reflection color and $\mathbf{c}_b$ the body reflection color (see Section 7.4.1). $m_s(x,y)$ and $m_b(x,y)$ are two scaling factors that depend on the surface normal $\mathbf{n}$, the lighting direction $\mathbf{s}$, and the viewer direction $\mathbf{v}$.

Within the dichromatic planes spanned by both vectors $\mathbf{c}_b$ and $\mathbf{c}_s$ (see Section 7.4.1) are the pixels that represent a common surface arranged according to the physical event in a certain location. Klinker, Shafer, and Kanade showed that a color cluster that contains pixels of a matte and a highlight region looks like a "skewed T" or a "skewed L" in the color space (see Fig. 8.1a). By determining the location of a pixel in the color cluster, an allocation of the pixels to the matte or to the highlight region is thus in principle possible.

For the production of a matte image the color values must be projected onto a pixel belonging to a highlight region along vector $\mathbf{c}_s$ to vector $\mathbf{c}_b$. For this, the location of vectors $\mathbf{c}_s$ and $\mathbf{c}_b$ must be determined in the *RGB* space. In real images taken with a camera, the color cluster usually does not possess the idealized form represented in Fig. 8.1a. In real images the color cluster consists rather of a point cloud, as it is represented in an example of a real, orange watering can in Fig. 8.1b.

With the practical application of this technique for highlight elimination the exact determination of the intersection between the highlight cluster and the body

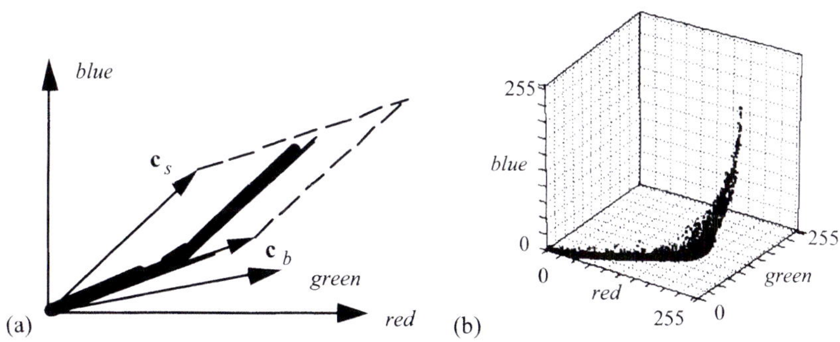

Figure 8.1. *(a) Dichromatic plane of a curved, hybridly reflecting object showing one body color. (b) An L-shaped color cluster of a real, orange watering can represented as a color histogram (reprinted from [Kle et al. 98] with permission from Vieweg).*

reflection cluster presents a problem. Formulated differently, the question arises as to where a highlight cluster begins and where it ends. The L-shaped color cluster in Fig. 8.1 shows that this question is not always easily answered for real color images. Here linear approximations for determining vectors $\mathbf{c}_s$ and $\mathbf{c}_b$ in the point cloud must be executed.

A further difficulty develops if the color signal is additionally affected by interreflections. The form of the pixels in the color area affected by interreflections is similar to the form of a highlight cluster. For the solution of this problem, Klinker [Kli93] uses the heuristic that the highlight cluster begins somewhere in the "upper 50%" of the body reflection cluster. A more exact determination of the location of the highlight cluster and the interreflection cluster can take place via evaluation of the color histogram (see [Nov92] and [NovSha92]). However, this will not be covered here in greater detail.

For highlight analysis it is important that in each case only individual surfaces are included in the analysis. If several different or equivalent objects in the image are viewed at the same time, then the color clusters in the *RGB* space overlay each other. A separation of highlight clusters, interreflection clusters, and body reflection clusters is then not possible in this case. Thus for highlight analysis, physics-based segmentation of the color image must be accomplished (see Section 7.4). The segmentation results are affected on their part by the appearance of highlights and interreflections.

Therefore, for the practical execution of the Klinker-Shafer-Kanade technique, the results of the segmentation should be coupled with the results of the highlight analysis. As already mentioned in Section 7.4.2, the entire technique consists of many small interpretation cycles that link the respective intermediate results with one another. In Fig. 8.2, an algorithm for highlight analysis is presented according to Klinker, Shafer, and Kanade [Kli et al. 88].

---

1. Identify a region in the color image that represents an individual surface.
2. Assume that this surface consists of an inhomogeneous dielectric material.
3. Calculate a histogram of the color values for the pixels within this region.
4. Specify vectors $\mathbf{c}_s$ and $\mathbf{c}_b$ for the interface reflection color and/or the body reflection color in the color histogram.
5. Project for each pixel in the region its color value $\mathbf{C}(x,y)$ on both reflection color vectors in order to determine both scaling factors $m_s(x,y)$ and $m_b(x,y)$ for the border layer and the body reflection.

Figure 8.2. *Principal algorithm of highlight analysis according to Klinker, Shafer, and Kanade.*

8.1.2 Tong-Funt Technique

One of the main difficulties in highlight analysis, according to Klinker, Shafer, and Kanade, exists in physics-based segmentation of the color image. Errors or inaccuracies in the segmentation results directly affect highlight analysis. Tong and Funt [TonFun88] suggest a modification of this technique. Instead of determining vectors $\mathbf{c}_s$ and $\mathbf{c}_b$ and the scaling functions $m_s(x, y)$ and $m_b(x, y)$ for each individual segmented region, as in the algorithm described earlier, they first combine the information on the regions in order to calculate vector $\mathbf{c}_s$.

According to the dichromatic reflection model, the interface reflection color $\mathbf{c}_s$ is identical to the color of the illumination and is thus equal for all regions. The associated dichromatic planes for all pixels are determined after a rough segmentation of the image. Vector $\mathbf{c}_s$ is given as the intersection of all dichromatic planes attained in this manner. Instead of intersecting all planes directly, the line that is most parallel to all planes is sought. To find this line, the normals for each plane are calculated and then a least-squares fit is used to find the line closest to the perpendicular to all plane normals.

The basic idea of determining vector $\mathbf{c}_s$ only once for the entire image is rather efficient. However, the single dichromatic planes must be separated and the corresponding normals must be calculated when this technique is applied to a color image. In the test images used by Tong and Funt, exclusively red, blue, yellow, and green objects that exhibit sufficiently different colors are contained. But if a color image were to contain light red, dark red, and violet objects, then the dichromatic planes could not be so easily separated. Thus, no meaningful calculation of the normals is ensured.

8.1.3 Gershon-Jepson-Tsotsos Technique

Gershon, Jepson, and Tsotsos introduced a similar technique for highlight analysis at about the same time as Klinker, Shafer, and Kanade (see [Ger87], [Ger et al. 87a], and [Ger et al. 87b]). Here the assumption is also made that a chromatic distortion of the pixels in the direction of the color of the light source is an indication of a highlight. The analysis is based likewise on the observation that the color values of a hybridly reflecting surface form a cluster within the chromatic level in the color space that exhibits the form of a "skewed L" or, as Gershon states, the form of a dog leg.

In place of extensive physics-based segmentation in the Klinker-Shafer-Kanade technique, in the Gershon-Jepson-Tsotsos technique only a rough segmentation is accomplished first by means of splitting and merging (see Section 7.2.2). The *RGB* values are subsequently transformed into a color $C_1C_2C_3$ space (see Section 8.3). Here, for color constancy, a linear technique is used with which both the spectral reflectance factor of the surface material and the spectral power distribution of the illumination are approximated in each case by a linear combination of three basis vectors (see Section 8.3.1).

This procedure is quite complex to implement. In addition, spectral sensitivities of the subsensors must be known, which is what Gershon, Jepson, and Tsotsos ensure by the use of a black-and-white camera with three Kodak Wratten filters (see Section 4.2.1). With commercial CCD color cameras these sensitivities are usually not known. Extensive physics-based segmentation in the Klinker-Shafer-Kanade technique is replaced in the Gershon-Jepson-Tsotsos technique by an extensive determination of color constancy.

If the image exists transformed in the $C_1C_2C_3$ space, then it is independent of the lighting. Hence it follows that the color transition between matte regions and highlights includes a change in the $C_1C_2C_3$ space independent of the lighting. The additional highlight analysis takes place similarly to the Klinker-Shafer-Kanade technique. Lines in the region are approximated in each case for the matte and the specular reflection component of a material. The intersection of the two lines determines the transition from the matte to the highlight portion. The procedure of highlight analysis according to Gershon, Jepson, and Tsotsos is outlined in Fig. 8.3. The generation of the matte image is not explicitly described in [Ger87], [Ger et al. 87a], and [Ger et al. 87b]. However, it can take place similarly to the Klinker-Shafer-Kanade technique in accordance with processing step 5 in Fig. 8.2.

The procedure can be well implemented under natural lighting if many materials are present in the scene. The large number of materials is needed so that color constancy can be computed representatively. With the Gershon-Jepson-

---

1. Segment the image in the *RGB* space (e.g., by split-and-merge).
2. Transform the *RGB* values into color constant values in the $C_1C_2C_3$ space.
3. Approximate in each region the values in the $C_1C_2C_3$ space from the smallest square to a line in the three-dimensional space.

For each pair of bordering regions:

4. Verify whether the lines are parallel, if yes, examine a new pair of regions.
5. Determine the intersection between the two lines and examine whether the intersection lies in the $C_1C_2C_3$ space.
6. Determine the region in which at least 75% of the C_1 values lie over the intersection and mark them as highlight candidates. If both regions pass the test, examine a new pair of regions.
7. Verify whether the lines that represent the candidate regions lie in the direction of an ideal reflector.

Figure 8.3. ***Principal algorithm of highlight analysis according to Gershon, Jepson, and Tsotsos.***

Tsotsos technique, the more materials present in the scene, the more precise the result. In contrast, with the Klinker-Shafer-Kanade technique, the best results are obtained if as few as possible materials are present in the scene. The decision for one of these two procedures will depend additionally on which other procedures are needed within the image processing chain. For example, if color constancy is also to be achieved for color object recognition, then the $C_1C_2C_3$ space, which can be determined for the Gershon-Jepson-Tsotsos technique, must be calculated under certain conditions. Yet in contrast to this, if no color constancy is necessary for the application, good segmentation of the color image must be ensured. In this case one should consider the Klinker-Shafer-Kanade technique.

8.1.4 Schlüns-Teschner Technique

A complex analysis in a three-dimensional color space is necessary with all procedures specified so far for highlight treatment and analysis. Here the color values that represent a material should be distributed as closely as possible within the dichromatic level. Yet this is not the case with objects having nonconstantly differentiable surfaces, particularly with polyhedric objects. Here in each case gaps are present within the cluster, which makes a robust analysis more difficult in the three-dimensional area. For overcoming this problem, Schlüns and Teschner [SchTes9ab] suggest a procedure for highlight elimination that is limited to two two-dimensional analyses of the color values. They use for this the standardized u and v values in a $(YUV)'$ space (see Section 3.2.4), which is defined by

$$(Y,U,V)' = (R,G,B)\begin{pmatrix} \frac{1}{3} & \frac{1}{2} & \frac{-1}{2\sqrt{3}} \\ \frac{1}{3} & 0 & \frac{1}{\sqrt{3}} \\ \frac{1}{3} & \frac{-1}{2} & \frac{-1}{2\sqrt{3}} \end{pmatrix}.$$

Surfaces of an ideal matte material correspond to exactly one point in the uv space. Thus, a dichromatic matte cluster (with or without gaps) also corresponds to exactly one point. In contrast to this, dichromatic highlight clusters (without gaps) form line segments in the uv space. These line segments begin in the point that represents the matte cluster and are aligned in the direction of the color of the light source. Yet they do not reach the place that represents the color of the light source due to the additive composition of the dichromatic line segments. Gaps in the highlight cluster also cause gaps in the line segments, which entail no impairment of highlight analysis. In Fig. 8.4, the structure of a dichromatic cluster is indicated in the uv space for a matte surface and three hybridly reflecting surfaces.

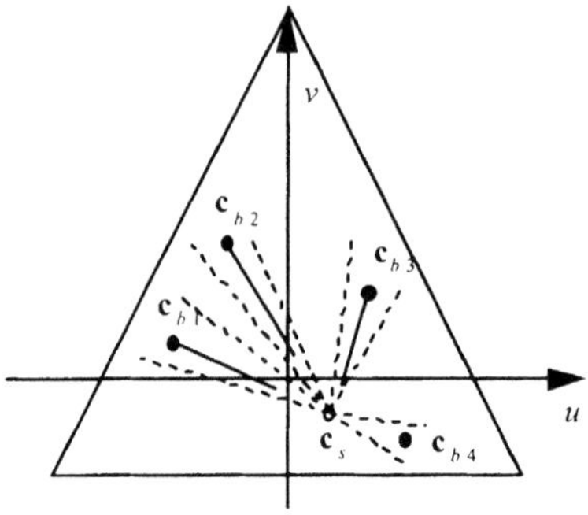

Figure 8.4. *Representation of three hybridly reflecting surfaces and one matte surface in the uv space (according to [SchTes95b]).*

In addition, a one-dimensional hue space (h space) that contains the number of pixels per angle α is used for highlight analysis. In this connection, α is an angle in the uv space between two lines. One line passes through the points of the lighting color and the pixel color in the uv space. The other line passes through the point of the lighting color and a defined fixed point in the uv space. Except for the point of the lighting color, any point in the uv space can be defined as the fixed point. It must, nevertheless, be constant for the entire image.

For the elimination of highlights in a color image, the interface reflection color $\mathbf{c}_s$ and the body reflection color $\mathbf{c}_b$ are to be calculated for each pixel. The elimination then occurs in which the scaling factor m_s is set to zero. The color of the light source either must be known for this technique or it is determined by reference images with the Macbeth ColorChecker (see Section 3.6.2).

For determining the body reflection color $\mathbf{c}_b$, the values in the uv and h space are evaluated. Each maximum in the h space corresponds to a hue component of a sought after body reflection color. For each local maximum $m_h = h(\alpha)$ in the h space, a local maximum $m_{uv} = uv(h(\alpha))$ is looked for along the line l_α determined by the current α. Each value m_{uv} corresponds to a matte color (body reflection color) and the number of maxima indicates the number of different materials in the scene.

If all materials in the scene possess different hues, that uv value is sought along line l_α, which exhibits the greatest distance to the color of the light source. If several local maxima occur along line l_α, then there are materials in the scene with identical hue values containing differing saturation values. However, this does not influence the determination of the matte color. With this technique it is presupposed that the matte colors are visible in the image (i.e., for each hybrid material at least one pixel must exist without a specular reflection component).

In an individual color image, the reflection components cannot be determined independently for each pixel. Therefore, a matte color determined with the above named procedure must be assigned to each pixel. For a matte color lying on line

l_α, all color values lying within an angle segment in the neighborhood of l_α are identified by this matte color in the *uv* space (see Fig. 8.4). The angle segments are defined in the *h* space.

If line l_α contains more than one matte color, then all *RGB* values lying within this segment in the level, which the three-dimensional counterpart of line l_α and the origin (black point) stretches, must be rotated. Thereby the analysis can be accomplished in a two-dimensional color area, since the colors depend only on intensity and color saturation with reference to the color of the light source.

The segmentation of colors in the neighborhood of each line l_α is executed as follows: First, analog to the *h* space, a one-dimensional *i* space is determined that contains the number of pixels with the corresponding intensity and hue values. The greatest occurring intensity value in the *i* space is a measurement for the length of the matte color cluster. Subsequently, each color value that possesses a specular reflection component is projected along vector $\mathbf{c}_s$ onto the respective physically possible matte color vector belonging to l_α. If the intensity of the projected color is greater than the maximal intensity value associated with matte color $\mathbf{c}_b$, then vector $\mathbf{c}_b$ is excluded as a candidate for a body reflection color. This special case can occur if the saturation value of a color of a specular component is larger than the saturation value of a color belonging to another material.

The algorithm for highlight analysis by Schlüns and Teschner is given in Fig. 8.5. In Fig. 8.6a, a real, orange watering can with a large highlight is represented. Fig. 8.6b shows an image of the same watering can after the specular reflection components were eliminated using the Schlüns-Teschner technique.

---

1. Determine the color of the light source $\mathbf{c}_s$ (e.g., with the Macbeth ColorChecker).
2. Transform the *RGB* values into *uv* values and determine the values in the *h* and *i* spaces.
3. Search for each local maximum $m_h = h(\alpha)$ in the *h* space along line l_α determined by the current α a local maximum $m_{uv} = uv(h(\alpha))$ in the *uv* space.
4. If line l_α contains more than one maximum m_{uv}, then rotate all *RGB* values lying within this segment in the level, which is stretched by the three-dimensional counterpart of line l_α and the origin. Otherwise, identify all color values lying within an angle segment in the vicinity of l_α in the *uv* space with m_{uv}.
5. Segment the colors in the vicinity of each one of the lines l_α on the basis of the values in the *i* space and project onto the matte vector.

Figure 8.5. *Principal algorithm of highlight analysis according to Schlüns and Teschner.*

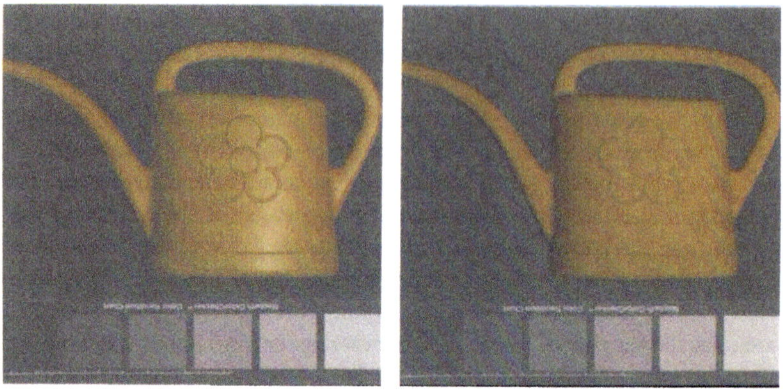

(a) (b)

Figure 8.6. *(a) A real watering can with a large highlight. (b) An image of the same watering can after the highlight was eliminated. In the lower portion of the images the gray surfaces of the Macbeth ColorChecker are represented (reprinted from [Kle et al. 96] with permission from Vieweg).*

In the lower portions of the images, the gray surfaces of the Macbeth ColorChecker (see Section 3.6.2) are represented, which serve for the linearization of the camera characteristic.

8.1.5 Spectral Differencing Using Several Images

The detection of highlights in color images can be simplified when several images are included in the analysis. Lee and Bajcsy [LeeBaj92] propose an algorithm for this by using three different images of the scene, which they call *spectral differencing*. Spectral differencing in this sense does not indicate the determination of partial derivatives of the color image function, but rather the calculation of spectral differences between the pixels in the color images. The technique uses three color images for highlight detection (similar to the form of a trinocular stereo analysis), which were acquired from three different viewing directions with identical lighting direction.

For two color images acquired from different viewing directions in the same lighting direction, the spectral differencing designates an algorithm for the identification of color pixels in an image that do not overlap with any other color pixel of the other image in a three-dimensional color space (e.g., the *RGB* space). In order to find the view-inconsistent color pixels, the algorithm of the spectral differentiation computes the images of the *minimum spectral distances* (*MSD*). It is presupposed here that the color of the lighting and the colors of the objects differ in the scene. Thus, with white light, no white objects may be present in the scene since no separation of the highlight is possible when the highlight color is identical to the color of the object.

The calculation of an *MSD* image can be described according to an example represented in Fig. 8.7. $\mathbf{C}_\alpha$ and $\mathbf{C}_\beta$ are two color images taken from two different views. The notation

$$MSD(\mathbf{C}_\alpha \leftarrow \mathbf{C}_\beta)$$

indicates the *MSD* image of $\mathbf{C}_\alpha$ from $\mathbf{C}_\beta$. The value of a pixel in the *MSD* image $MSD(\mathbf{C}_\alpha \leftarrow \mathbf{C}_\beta)$ is the minimum value of all spectral distances between the pixel in image $\mathbf{C}_\alpha$ and all pixels in image $\mathbf{C}_\beta$. The spectral distance and/or the color distance between the color pixels are defined as the Euclidian distance in the three-dimensional color space. Each *MSD* value that lies above a given threshold value points to a mirroring reflection. The threshold value for the *MSD* image depends only on the sensor noise. Its value can be set independent of the viewer direction.

Figure 8.7a illustrates two different views of an object with specular surface components. The corresponding color clusters in the *RGB* space are represented in Fig. 8.7b. Pixel **p** in image $\mathbf{C}_\alpha$ lies in the *RGB* space removed from all color values of the specular and Lambertian image pixels in image $\mathbf{C}_\beta$. This indicates a specular reflection in pixel **p**. In contrast to this, the Lambertian reflections in views α and β have the same linear cluster in the *RGB* space. Since pixel **r** from the region with Lambertian reflection in image $\mathbf{C}_\alpha$ lies very near to the Lambertian pixels in image $\mathbf{C}_\beta$ within the *RGB* space, its *MSD* value is very small. Therefore, pixel **r** is not detected by spectral differencing. It represents a Lambertian surface pixel.

However, not all highlight pixels can be determined by spectral differencing. In Fig. 8.7b the color value of the image pixel **q** lies within the overlapping region,

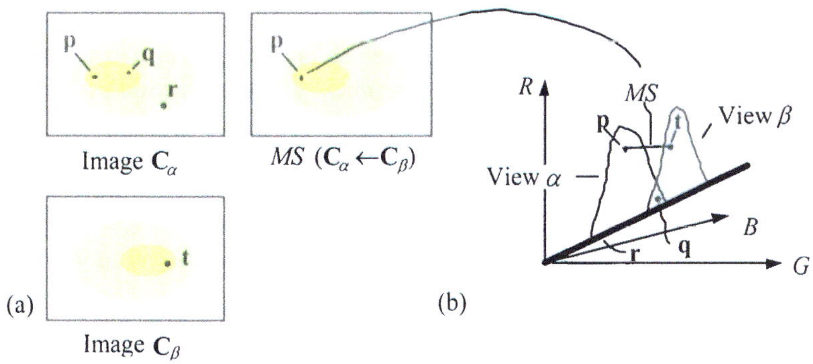

Figure 8.7. *Spectral differencing: (a) Images from different views; (b) associated color clusters in the* RGB *space (according to [LeeBaj92]).*

between the planar clusters for views α and β. Since **q** lies within the planar cluster determined by the specular reflection in image $\mathbf{C}_\beta$, it is difficult to recognize **q** as a highlight pixel if the color values are ordered closely within the planar cluster in view β. The highlight in pixel **q** can be recognized with this algorithm only if very few pixels within the planar cluster in view β occur in the neighborhood of **q**. Here it should be noted that the clusters in the *RGB* space already overlap with several sources of light and several materials in the scene in an image.

Furthermore, pixels that do not represent a visible surface pixel in all images are likewise classified as highlights if the *MSD* value is greater than the given threshold value. Determining a suitable threshold value generally represents a problem. The advantage of the procedure is that no segmentation of the color image is necessary. The algorithm is simple to implement and can be easily paralleled. No conditions are made concerning number and characteristics of the sources of light and camera geometry. Lee and Bajcsy [LeeBaj92] obtained good results for metallic and inhomogeneous dielectric materials with this technique. It should be noted that with this procedure only highlight detection is possible, not the production of matte images.

8.1.6 Photometric Multi-Image Technique

Robust highlight elimination also can be accomplished using several images. For this, a procedure is applied [SchKos00] in which once again three color images are used. These are generated analogous to a photometric stereo analysis technique (i.e., they are taken from an identical viewer direction with three different lighting directions). The procedure is called the *photometric multi-image* technique in the following. The technique of the procedure (similar to the photometric stereo analysis) is based on the fact that for one object point in the scene, three measured color values exist in the three color images. For each pixel at position (x,y), three color vectors, $\mathbf{c}_1$ from image $\mathbf{C}_1$, $\mathbf{c}_2$ from image $\mathbf{C}_2$, and $\mathbf{c}_3$ from image $\mathbf{C}_3$, are represented (see Fig. 8.8a). It is now assumed that the light source color $\mathbf{c}_s$ is identical and known in all three images. According to the dichromatic reflection model for inhomogeneous dielectric materials, three vectors $\mathbf{c}_1, \mathbf{c}_2$, and $\mathbf{c}_3$ always lie in the same dichromatic plane since they represent the same material in the scene and were produced with the same light source color. Figure 8.8b shows the principal position of the three color vectors in the dichromatic plane for any object point.

For matte image generation (as in Klinker [Kli93]), $\mathbf{c}_b$ and $\mathbf{c}_s$ must be known. It was presupposed that the vector of the light source color $\mathbf{c}_s$ is known. For the photometric multi-image technique, the special case is now assumed that the lighting directions are sufficiently different so that (at least) one of the three color vectors does not contain a highlight portion. This vector is then identical with $\mathbf{c}_b$. Since all three color vectors lie in the dichromatic plane, that vector $\mathbf{c}_i$

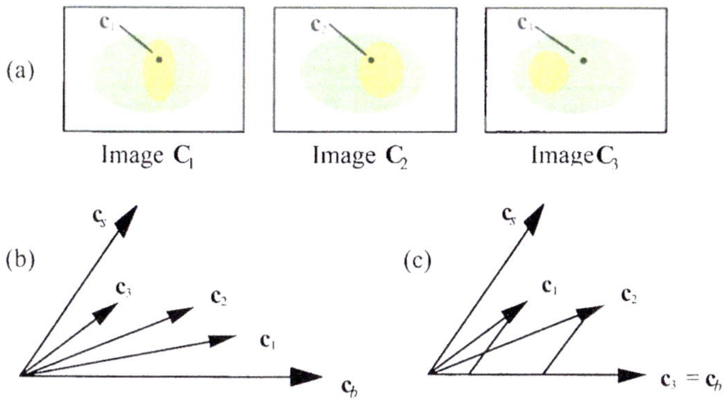

Figure 8.8. *Principle of a photometric multi-image technique. (a) Three color images taken with differing lighting directions; (b) Principal position of three color vectors in the dichromatic plane for any object point; (c) Position of three color vectors in the dichromatic plane, if the color vector in image* $\mathbf{C}_3$ *does not contain a highlight.*

that has the largest angle to $\mathbf{c}_s$ is the sought-after vector identical to $\mathbf{c}_b$. The three angles between $\mathbf{c}_s$ and $\mathbf{c}_i$, $i = 1,\ldots,3$ must be determined and that vector must be selected that exhibits the largest angle. In the example in Fig. 8.8c, $\mathbf{c}_3$ is the sought-after vector.

For the generation of a matte image (or matte images), vectors containing highlights ($\mathbf{c}_1$ and $\mathbf{c}_2$ in Fig. 8.8c) are projected onto the vector that does not contain highlights ($\mathbf{c}_3 = \mathbf{c}_b$ in Fig. 8.8c). Thereby robust highlight elimination is made possible. The photometric multi-image technique works locally in each case on three pixels from three color images and can be implemented very well parallel. In Table 8.1, a comparison of the two techniques for highlight analysis is represented using several examples.

8.1.7 Polarization Technique

Another possibility for detecting highlights is the implementation of polarization filters in image acquisition. Here several images of the object to be examined are provided by using different rotation angles of the polarization filter (see [WolBol91]). The idea of the technique is based on the assumption that the diffuse reflection on an inhomogeneous dielectric material is not polarized, while in contrast the specular reflection component is polarized. The image domains having highlights can be determined by comparing the images acquired from different rotation angles of the polarization filter.

Spectral Differencing	Photometric Multi-Image Technique
Highlights can be only detected, not eliminated.	Highlights can be detected and eliminated.
The color of the light source does not need to be known.	The color of the light source must be known and identical in all images.
Object colors and light source color cannot be identical (i.e., no white objects may be present in white light).	Object colors and light source color cannot be identical (i.e., no white objects may be present in white light).
The technique can be applied to moving or nonrigid objects if it is guaranteed that all three color images are generated at the same time.	The color images must always be produced successively. Thus, the objects must be rigid and sturdy.
Pixels that are not visible in all images can be falsely classified as highlight pixels. Contrarely, highlight pixels that lie within the overlapping region between two planar color clusters are usually not recognized as highlight pixels.	No highlight elimination can be implemented in the image domains that are not illuminated in all three images.
The procedure can be accomplished locally for an individual pixel. However, the color distances of the single pixel to all pixels in the other image must be determined.	The procedure can be implemented locally for a single pixel. Only the three color values from the three color images are needed.

Table 8.1. *Comparison of spectral differencing and the photometric multi-image technique.*

The number of images that are needed for highlight identification using polarization can be limited if color information is also evaluated additionally for polarization. To do this, Lin and Lee [LinLee96], [LinLee97] combine a polarization technique with the color technique of spectral differencing from [LeeBaj92] introduced in Section 8.1.5. Altogether six gray-level images are acquired with a black-and-white camera from two different observer locations by using each of the three differing polarization angles. By a firm allocation between the rotation angles of the polarization filters and the color filters, each color excerpt is polarized differently in the resulting color image. The highlights can be detected analog to the procedure described initially in this section by difference formation of the histograms for the differently polarized color channels.

Similar to the technique of spectral differencing, no matte images can be produced with this technique. Only a detection of highlights is possible. One advantage of this technique (combining spectral differencing with a polarization technique) over the color technique from [LeeBaj92] is that, in addition, highlights whose surface colors are identical to the light source color can be detected on objects. One disadvantage of the combination technique relative to the color technique from [LeeBaj92] results from the need to change the polarization filters between images. Thus, the single images are to be taken successively and the objects must be rigid. Thereby the applicableness of the combination technique is limited in relation to the technique of spectral differentiation.

Another technique for the combined evaluation of polarization and color information was suggested by Nayar, Fang, and Boult [Nay et al. 93], [Nay et al. 97], with which matte images also can be produced. Here it is assumed that the objects consist of inhomogeneous dielectric materials and their surface colors are not identical to the light source color. This procedure of highlight elimination for the treatment of an individual pixel is described in the following. The same procedure is implemented independently for all pixels. The color vector measured using a polarization filter for pixel $\mathbf{p} = (x, y)$ is defined by

$$\mathbf{L} = \mathbf{L}_b + \mathbf{L}_s$$

using the dichromatic reflection model in accordance with Eq. 7.3, whereby $\mathbf{L}_b$ indicates the diffuse and $\mathbf{L}_s$ the specular component. According to the condition the diffuse reflection is unpolarized for a reflection on an inhomogeneous dichromatic material and the specular component is polarized. A rotation of the polarization filter thus affects only the measured values for the specular component. The measured values for the diffuse component remain unchanged under this condition. The specular component thus can be described as the sum of a constant vector $\mathbf{L}_{sc}$ and a cosine function term with the amplitude $\mathbf{L}_{sv}$. For a rotation angle θ_i of the polarization filter, color vector

$$\mathbf{L}_i = \mathbf{L}_b + \mathbf{L}_{sc} + \mathbf{L}_{sv} \cos 2(\theta_i - \alpha)$$

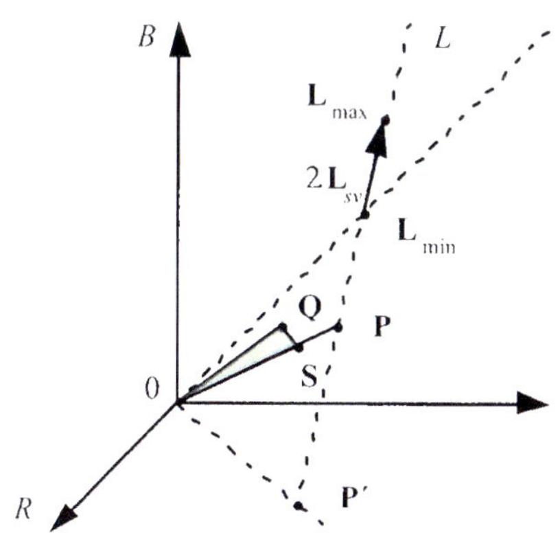

Figure 8.9. *Use of points from the neighborhood of* **p** *for determining the diffuse component (after [Nay et al. 97]).*

results, whereby α describes the phase angle that is determined by projecting the surface normals of the represented object pixels into the plane of the polarization filter (see [Nay et al. 97]). Since a rotation of the polarization filter affects only a change of the specular components, all measured $\mathbf{L}_i$ lie on a line L (a subspace) in the *RGB* space (see Fig. 8.9). By using the above equation and six images with different rotation angles of the polarization filter, an over-determined equation can be set for the calculation of $\mathbf{L}_{sv}$, $\mathbf{L}_c = (\mathbf{L}_b + \mathbf{L}_{sc})$ and α. The two vectors corresponding with the maximum and minimum polarization

$$\mathbf{L}_{\max} = \mathbf{L}_c + \mathbf{L}_{sv} \text{ and } \mathbf{L}_{\min} = \mathbf{L}_c - \mathbf{L}_{sv}$$

are subsequently calculated. From the components of the vectors

$$\mathbf{L}_{\max} = \begin{pmatrix} \mathbf{L}_{\max,R} \\ \mathbf{L}_{\max,G} \\ \mathbf{L}_{\max,B} \end{pmatrix} \text{ and } \mathbf{L}_{\min} = \begin{pmatrix} \mathbf{L}_{\min,R} \\ \mathbf{L}_{\min,G} \\ \mathbf{L}_{\min,B} \end{pmatrix}$$

the degree of polarization can be expressed by the vector

$$\boldsymbol{\omega} = \begin{pmatrix} \omega R \\ \omega G \\ \omega B \end{pmatrix} = \begin{pmatrix} (\mathbf{L}_{\max,R} - \mathbf{L}_{\min,R})/(\mathbf{L}_{\max,R} + \mathbf{L}_{\min,R}) \\ (\mathbf{L}_{\max,G} - \mathbf{L}_{\min,G})/(\mathbf{L}_{\max,G} + \mathbf{L}_{\min,G}) \\ (\mathbf{L}_{\max,B} - \mathbf{L}_{\min,B})/(\mathbf{L}_{\max,B} + \mathbf{L}_{\min,B}) \end{pmatrix}.$$

An image pixel $\mathbf{p}$ is assumed as purely diffuse if the following two conditions are filled: First, all components of vector $\boldsymbol{\omega}$ must be smaller than a given threshold value. Otherwise, the vector signal is not unpolarized. Second, the angle β between vectors $\mathbf{L}_{\max}$ and $\mathbf{L}_{\min}$ must be greater than a second threshold value. Otherwise the color of the specular component is very similar to that of the diffused component and thus the dichromatic model can no longer be used reliably. In the last case no robust estimation of the specular and diffuse component is possible.

If image pixel $\mathbf{p}$ is polarized sufficiently and angle β is large enough, then the determination of the diffuse component $\mathbf{L}_b$ is continued. $\mathbf{L}_b$ cannot be locally computed from the estimation for $\mathbf{L}_c = \mathbf{L}_b + \mathbf{L}_{sc}$ and $\mathbf{L}_{sv}$. Nayar, Fang, and Boult propose for this the observation of image pixels in the environment of $\mathbf{p}$. It is assumed that $\mathbf{L}_b$ corresponds to pixel $\mathbf{P}$ on line L in the *RGB* space (see Fig. 8.9). The position of $\mathbf{P}$ on L can be described by

$$\mathbf{P} = \mathbf{L}_{\min} - d\frac{\mathbf{L}_{sv}}{\|\mathbf{L}_{sv}\|}$$

whereby d is the distance from $\mathbf{P}$ to $\mathbf{L}_{min}$. The determination of $\mathbf{L}_b$ is thus identical to the determination of d. Value d is limited by $0 \le d \le d'$, whereby d' corresponds to pixel $\mathbf{P}'$, which is given by the intersection of L with one of the three planes of the color coordinate system. In the example in Fig. 8.9, L intersects the R-G plane. Generally, L can also intersect the G-B plane or the R-B plane instead. The corresponding plane can be easily found from $\mathbf{L}_{min}$ and $\mathbf{L}_{max}$.

Both vectors $\mathbf{L}_{min}$ and $\mathbf{L}_{max}$ stretch a plane in the RGB space that is indicated by Ω. The basic idea of determining the diffuse component for image pixel **p** consists of finding an image pixel **q** in a local environment of image pixel **p** to which the following two conditions apply (see [Nay et al. 97]):

1. Point **q** must possess a small degree of polarization and thus can be accepted only as diffusely reflecting. Alternatively, a point whose diffuse component has already been computed also can be selected. In both cases, the diffuse component of **q** is designated by **Q**.
2. **Q** must lie close to plane Ω in the RGB space. Furthermore, **Q** must lie between vectors $\mathbf{L}_{min}$ and $\mathbf{P}'$, since the diffuse vector $\mathbf{L}_b$ of **p** lies on the line between $\mathbf{L}_{min}$ and $\mathbf{P}'$. The size of **Q** can differ here from the size of $\mathbf{L}_b$.

If these conditions are filled, then it is assumed that **p** and **q** possess the same diffuse color. The intersection **P** of line L with the line running through the zero point and **Q** results in an estimated value for the diffuse vector $\mathbf{L}_b$.

Due to errors in the measured color and polarization values, it cannot be assumed that the diffuse component **Q** accurately fulfills the conditions specified above. In order to intercept this discrepancy, the angle between the line running through the zero point and **Q** and Ω is calculated. If this angle is greater than a third threshold value, then **Q** is not close to Ω and is not further observed. If the angle is smaller than the third threshold value, then **Q** is projected onto point **S** in plane Ω, and the intersection of the lines through the zero point and **S** with L determines the estimated value for the diffuse vector $\mathbf{L}_b$ (see Fig. 8.9).

The difficulty of physics-based segmentation, which is needed for the Klinker-Shafer-Kanade technique (see Section 8.1.1), is replaced in the Nayar-Fang-Boult technique by the difficulty of identifying suitable threshold values. The selection of threshold values has a direct influence on determining the diffuse component. This influence is particularly great in areas within which the matte color varies, and in the environment of edge pixels (see [Nay et al. 97]. In both the Schlüns-Teschner technique (see Section 8.1.4) and the photometric multi-image procedure (see Section 8.1.6), the light source color must be known. However, with the Nayar-Fang-Boult technique, the light source color need not be known. It can be determined by the procedure.

Nayar, Fang, and Boult used a one-chip CCD color camera and a black-and-white camera with color filters for testing their combination technique [Nay et al. 97]. The chromatic aberration of the lens (see Section 4.3.2) has a negative influence on image quality in the images taken with the black-and-white camera. In contrast to this, difficulties arose with the one-chip CCD color camera due to the bad signal-to-noise ratio of the camera, which the authors tried to adjust by computing the average values for each color channel over all 32 images. The problem of blooming in bright areas (see Section 4.3.1) is common to both acquisition technologies with CCD cameras, especially in image areas with strong highlights. Thereby the values measured for polarization are falsified. Furthermore, the movement of the polarization filter can lead to a lateral shift (of approximately one pixel [Nay et al. 97]) between two consecutive images. In summary, it may be noted that the practical use of this theoretically well-founded technique is strongly impaired by a set of technical boundary conditions.

8.2 INTERREFLECTION ANALYSIS IN COLOR IMAGES

In simplified lighting models, which form the basis of image formation in computer vision, it is frequently assumed that the surfaces in the scene are illuminated by only one identifiable light source. These models do not agree, however, with the actual conditions. If several objects are present in the scene or the object is not convex, then each surface also receives the light reflected from another surface. This phenomenon is called *interreflection* or *mutual illumination*. For image analysis to be as accurate as possible, interreflections in the scene must also be considered since they can unfavorably influence the results of segmentation (see Chapter 7) or photometric stereo analysis (see Chapter 10).

The analysis of interreflections in color images can serve different goals. One of these goals is the recognition and at least partial "elimination" of interreflections in order to aid a simpler or better treatment of the resulting data. Another goal is the use of interreflection analysis in determining surface geometry or the color of surfaces (see [Sha92]). For color image segmentation it is often sufficient to recognize where interreflections are in the image. For color object recognition, however, the "original" color of the surfaces (without overlay from interreflections) should be reconstructed.

Interreflections between objects arise also in gray-level images and influence the result of an intensity-based surface reconstruction (see [Nay et al. 91b]). The errors arising in the surface reconstructions can be partially reduced in gray-level images by suitable procedures (see [Nay et al. 91b]). By analyzing color images instead of gray-level images these errors can be further minimized [NayGon92]. If the task at hand of interreflection analysis is the computation of the original color of the object (e.g., for object recognition), then color images must be analyzed in each case.

The color interreflections appearing in the scene on inhomogeneous dielectric materials can be modeled by an expansion of the dichromatic reflection model (see Section 7.4.1 and [Fun et al. 91], [FunDre92], [Jan91], and [Sha et al. 90]). According to the dichromatic model, each surface reflection consists of an interface reflection component $\mathbf{c}_s$ and a body reflection component $\mathbf{c}_b$. In interreflections from a surface A onto a surface B, the lighting of surface B consists both of the direct lighting by the respective source of light and of the lighting that is determined by the two reflection components of surface A. With the simplified one-bounce model for interreflections described in the following, it is assumed that the total reflection at surface B can be described by a linear combination of these two lighting sources. On the basis of this model, a procedure is described in Section 8.2.2 with which interreflections in material color images can be minimized while simultaneously considering shade (see [Kos97]).

8.2.1 One-Bounce Model for Interreflections

For the investigation of interreflections between two surfaces, two differently colored semi-infinite planes A and B, connected by an opening angle γ (see Fig. 8.10), are considered. Interreflections can also occur if surfaces are not connected. For better illustration, a constraint takes place here on the constellation specified above (see [FunDre92]). According to the notation in Section 4.3, the spectral power distribution of the illumination is designated with $E(\lambda)$. The spectral reflectance factor of surface A is designated by $R^A(\lambda)$ and that of surface B by $R^B(\lambda)$. Furthermore, it is presupposed that only Lambertian surfaces are present in the scene. The RGB values measured with the camera are indicated by ρ_k, $k = 1,\ldots,3$. They result from the direct irradiation of the two surfaces plus the interreflections between them.

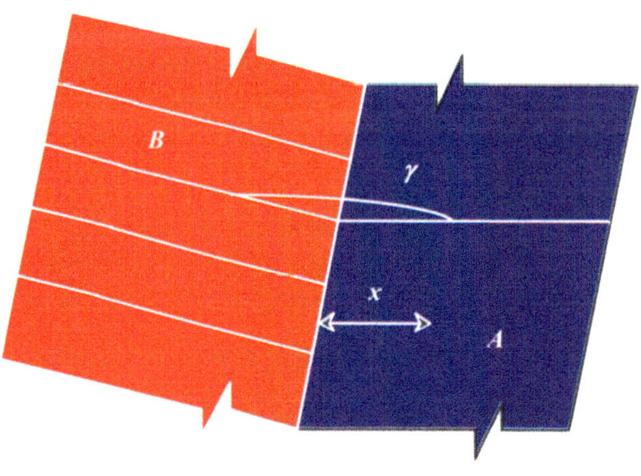

Figure 8.10. *Two semi-infinite planes connected by an opening angle γ (e.g., γ = 45°; adapted from [FunDre92]).*

An exact modeling of ρ_k would have to consider an infinite number of interreflections between the surfaces. For the sake of simplification, the *one-bounce model* for interreflections is observed (see [DreFun90]). With the one-bounce model the color signal $C^A(x,\lambda)$ of a pixel is described at location x on surface A by

$$C^A(x,\lambda) \cong \alpha^A(x)E(\lambda)R^A(\lambda) + \beta^{BA}(x)R^A(\lambda)E(\lambda)R^B(\lambda) \tag{8.1}$$

whereby α^A and β^{BA} designate each of the proportionality factors. For surface B an equation can be set up similar to this.

According to Eq. (8.1), C is comprised of the light reflected from surface A without influence of interreflection, the *no-bounce* color signal, plus the light reflected first from surface B and then from surface A. The latter light is then reflected again from surface B, and so forth. For simplification of the model, the infinite series of reflections is broken off after the first term. The second addend in Eq. (8.1) is called the *one-bounce* contribution to the light signal. In the following, the two components in Eq. (8.1) are indicated as *no-bounce* and *one-bounce*.

The intensity of the direct illumination E depends on the orientation and position of the surface, even when a constant spectrum is assumed. Factor $\alpha^A(x)$ represents this intensity variation and shading in point x.

For the reflection from surface B onto surface A, factor $\beta^{BA}(x)$ represents the relative portion of interreflection on the color signal. It unites all possible influences on the size of the one-bounce component into one number. This number takes into consideration the local surface orientation in point x, the general form of both surfaces, the shade on both surfaces, and the possibility that several points on surface B are hidden from point of view x. In computer graphics the calculation of $\beta^{BA}(x)$ represents a numerically intensive problem. In contrast to this, in image analysis the scaling factors $\alpha(x)$ and $\beta(x)$ can be determined quite efficiently from a given *RGB* image. The sensor model of the color space is observed for this.

According to Eq. (4.1) in Section 4.3, the sensor response ρ_k is represented by

$$\rho_k = \int C(\lambda) S_k(\lambda) d\lambda \tag{8.2}$$

whereby $S_k(\lambda)$ denotes the spectral sensitivity of the component sensor k. By using Eq. (8.2) on $C(\lambda)$ in Eq. (8.1), vectors $\mathbf{\rho}$ arise with vector components ρ_k^A and ρ_k^B on surfaces A and B, which are determined by

$$\rho_k^A(x) \cong \alpha^A(x)\rho_k^{A\_nobounce} + \beta^{BA}(x)\rho_k^{onebounce}$$

and

$$\rho_k^B(x) \cong \alpha^B(x)\rho_k^{B\_nobounce} + \beta^{AB}(x)\rho_k^{onebounce}$$

with

$$\rho_k^{A\_nobounce} = \int E(\lambda)R^A(\lambda)S_k(\lambda)d\lambda$$

$$\rho_k^{B\_nobounce} = \int E(\lambda)R^B(\lambda)S_k(\lambda)d\lambda$$

and

$$\rho_k^{onebounce} = \int E(\lambda)R^A(\lambda)R^B(\lambda)S_k(\lambda)d\lambda\,.$$

All ρ from surface A are determined by a linear combination of $\boldsymbol{\rho}^{A\_nobounce}$ and $\boldsymbol{\rho}^{onebounce}$. Analogously, all ρ from surface B are determined by $\boldsymbol{\rho}^{B\_nobounce}$ and the identical $\boldsymbol{\rho}^{onebounce}$. Apart from the assumption that the color signal of surface A is correctly approximated by Eq. (8.1), all ρ arising from surface A lie on a plane in the ρ space. This holds similarly for all ρ resulting from surface B.

8.2.2 Determination of the One-Bounce Color Portion

Both planes determined by all ρ_k^A and ρ_k^B contain the common color $\boldsymbol{\rho}^{onebounce}$. This is given thus by the intersection of the two planes. A proven method for determining the levels of the planes is by the analysis of the *singular value decomposition* (*SVD*). The *SVD* is used on matrix **R** of all data from surface A. **R** has the dimension $n \times 3$, whereby n is the number of pixels. Additional matrixes are produced by the *SVD*, which divide matrix **R** into

$$\mathbf{R} = \mathbf{UWV}\,.$$

Hereby **U** is an $n \times 3$ matrix, **W** is a 3×3 diagonal matrix with the eigenvalues in the diagonal, and **V** is a 3×3 matrix of the eigenvectors in the *RGB* space. If the third eigenvalue in **W** is much smaller than both the others, then most of the pixels lie on one plane. This is spanned by both main component vectors $\mathbf{v}_1$ and $\mathbf{v}_2$ from the corresponding columns in **V**.

In analogy to a similar procedure in highlight detection (see [TomWan89] and [TonFun88]), the intersection between both color levels produced by surfaces A and B is computed in order to calculate color $\boldsymbol{\rho}^{onebounce}$. If $\mathbf{v}_1^A, \mathbf{v}_2^A$ and

$\mathbf{v}_1^B, \mathbf{v}_2^B$ are the first two main component vectors of both surfaces *A* and *B*, then the intersecting vector is determined by solving the vector equation

$$\mathbf{v}_1^A + a_2\mathbf{v}_2^A - b_1\mathbf{v}_1^B - b_2\mathbf{v}_2^B = \mathbf{O}$$

with the three unknowns a_2, b_1, and b_2. The following applies:

$$\boldsymbol{\rho}^{onebounce} = \mathbf{v}_1^A + a_2\mathbf{v}_2^A .$$

For surface *A*, the use of the one-bounce model for interreflections implies that each ρ from the color specified above and from the no-bounce color signal for surface *A* composes a $\boldsymbol{\rho}^{A\_nobounce}$. Therefore, the question remains how this latter color component for both surfaces *A* and *B* can be determined.

8.2.3 Quarter-Circle Analysis

$\boldsymbol{\rho}^{nobounce}$ is to be determined from a set of measured vectors ρ and the already-calculated $\boldsymbol{\rho}^{onebounce}$. For this, *quarter-circle analysis* can be employed according to [TomWan90]. This corresponds to the procedure for highlight analysis in Section 8.1.6. According to this technique, all measured ρ of surface *A* are projected into the $\mathbf{v}_1^A, \mathbf{v}_2^A$ plane. All ρ are subsequently rotated into a new coordinate system with the axes $\mathbf{a}_1$ and $\mathbf{a}_2$. Thereby $\mathbf{a}_1$ is identical to $\boldsymbol{\rho}^{onebounce}$ and $\mathbf{a}_2$ stands perpendicular to $\mathbf{a}_1$ in the $\mathbf{v}_1^A, \mathbf{v}_2^A$ plane. Described in this new coordinate system and standardized on unit length, each measured vector ρ must lie in the first quadrant of a unit circle (see Fig. 8.11). Therefore, all ρ have positive components a_1 and a_2 along the $\mathbf{a}_1, \mathbf{a}_2$ directions. Limitations for the direction of $\boldsymbol{\rho}^{nobounce}$ must now be determined.

The first condition for $\boldsymbol{\rho}^{nobounce}$ arises from the fact that from all measured ρ, the vector lying closest to the $\mathbf{a}_2$ axis represents the point of the surface on which the interreflection is the smallest. This vector has the largest relation of its components a_2 / a_1. It determines a lower limit for a_2 / a_1 from $\boldsymbol{\rho}^{nobounce}$. An upper limit results from the condition that $\boldsymbol{\rho}^{nobounce}$ can have no negative components. Since $\mathbf{a}_1$ has positive components and stands perpendicular to $\mathbf{a}_2$, it results that $\mathbf{a}_2$ can have negative components. If one moves from $\mathbf{a}_2$ in the direction of $\mathbf{a}_1$, then a point is reached where each component of vector ρ is nonnegative. This point is the upper limit for a_2 / a_1.

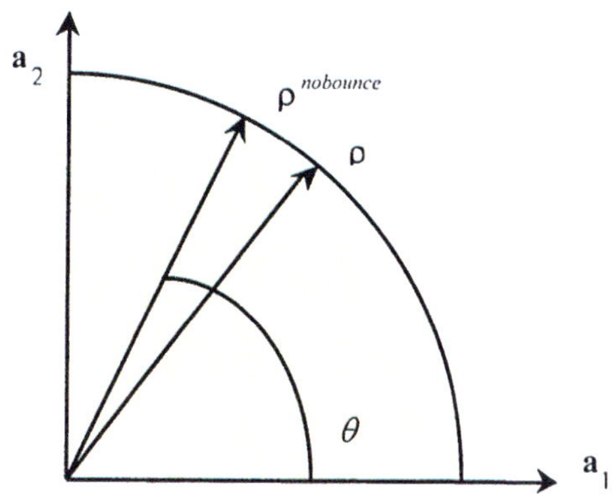

Figure 8.11. *Example of a quarter-circle analysis.*

The bordering vector contains exactly so much of $\mathbf{a}_1$ added to $\mathbf{a}_2$ that the component, which corresponds to the strongest negative component of $\mathbf{a}_2$, becomes exactly zero. The condition for this is

$$a_1\mathbf{a}_1 + a_2\mathbf{a}_2 \geq \mathbf{0},$$

so that the following applies:

$$a_2 / a_1 \leq \left(\max_{k=1,\dots 3} \{-a_{2k} / a_{1k}\} \right)^{-1}.$$

Therefore, the domain of possible $\boldsymbol{\rho}^{nobounce}$ vectors is fixed and vectors ρ from the image can be divided into their components in the direction $\boldsymbol{\rho}^{onebounce}$ and $\boldsymbol{\rho}^{nobounce}$. However, by the previous analysis only one domain was found to have a $\boldsymbol{\rho}^{nobounce}$. Thus, the components are fixed also only within a domain. In their investigations with synthetic images, Funt and Drew [FunDre92] discovered that the error is smallest if $\boldsymbol{\rho}^{nobounce}$ is selected in that place where the relationship a_2 / a_1 is maximum.

One case where the technique of interreflection analysis outlined in this section fails is if one of the two examined surfaces is achromatic. For example, if interreflections between a blue and an ideal white or gray surface arise, then the color without interreflection is equal to the color with interreflection on the blue side, and the colors cannot be separated. In addition, the *SVD* analysis cannot be used if no interreflections are present at all. This happens, for example, if two surfaces lying close together are in one plane or the angle between the surfaces is greater than 180°.

8.2.4 Minimization of Interreflections in Real Color Images

In the previous section a procedure was presented with which a color signal afflicted by interreflection can be split up into its interreflection portion and the color portion not affected by interreflection. Here the one-bounce model for interreflections was used. The procedure is applicable in its original form (see [FunDre92]) only if the image to be analyzed consists exclusively of two planar surfaces suited to each other. However, this constellation will arise only rarely. Therefore, this technique must be expanded for the removal of interreflections in real color images. This extension was presented in [Kos97].

In order to implement the technique described in the previous section it is presupposed that no highlights are present in the image. If the scene to be examined also contains specular objects, then the highlights must be eliminated in a preprocessing step (e.g., with the procedure described in Section 8.1.3). The resulting matte images are then the input images for an interreflection analysis.

Furthermore, the technique for eliminating interreflections described in the previous section supplies false results if no interreflections between the examined surfaces arise at all. Therefore, in a first processing step those areas in which the color signal could be influenced by interreflections in the image must first be detected. To do this, segmentation of the color image into areas with small local color changes is first carried out.

8.2.5 Segmentation with Consideration to Interreflections and Shadows

Segmentation is based on the idea that pixels that represent a monotone object surface differ only minimally in hue in a local neighborhood. This applies also when the color signal is influenced by interreflections or shadows. The observation is made (see [Baj et al. 96] and Section 7.4.2) that:

1. Shadows reduce the brightness of a pixel, but do not change the hue
2. Interreflections increase the brightness of a pixel somewhat and slightly change the hue in a local neighborhood.

In relation to a larger object surface, the hues of the pixels can be substantially changed by arising interreflections. However, this change of hue is less locally observed (i.e., for two pixels situated next to or over one another in the image).

In order to exclude single outliers, the image is first smoothed by means of the vector median regulation (see Section 5.3.2). The image data is then transformed into the *HSI* color space. For the regulation of segments each pixel is compared with its right neighbor in each line. If the difference of both hues is smaller than a given tolerance value, then the pixel to the right is allocated to the same line segment (i.e., a segment within the line). One condition for hue comparison is a certain difference between the values in the red, green, and blue spectral transmissions, as well as a minimum brightness of the pixels to be predetermined. If these two conditions are not met, then these pixels are excluded from further treatment. For these pixels, meaningful values cannot be determined either by presegmentation on the basis of hue or by later *SVD*. A line segment ends if a pixel that can be excluded is found, or the hue difference to the next pixel exceeds the tolerance value.

The minimum brightness is very small to select and should apply only to the areas in the scene that reflect almost no light. Shadow areas in the image should in general meet the minimum brightness. The minimum hue difference applies for each of the two image pixels lying next to each other in the line segment. The single pixels in a line segment can exhibit quite larger hue differences. Subsequently, in each line each pixel from a line segment is compared to the pixel lying underneath it. If the difference of the two hues is smaller than the given tolerance value and the minimum brightness is given, then the pixels are combined into regions.

The calculated regions already can be observed as the result of color image segmentation taking into account interreflections and shadows. Figure 8.12 shows (upper left) an example color image and (upper right) the segmentation result calculated taking into account interreflections and shadows. If the processing goal plans only the partitioning of the color image into homogeneous color gamuts, then this procedure already achieves the goal. However, here the minimization of interreflections in the color image is observed. A complete elimination of the interreflections is usually not possible in real images since, first, the measured color signal is frequently influenced by interreflections on several surfaces, and, second, only approximate values are observed in the one-bounce model used and likewise in the quarter-circle analysis. The processing goal is the calculation of a color image in which the overlay of the color signal by interreflections is as small as possible in order to minimize the influence of interreflection on the results of the processing procedures (e.g., a photometric stereo analysis or a shadow analysis).

8.2.6 Determination of Interreflection Areas

In principle, interreflections can exist between all surfaces or, in this case, all segmented regions. This can occur if they lie close together and in the correct angle to each other and to the source of light. In order to be able to detect interreflections in the image, certain conditions must be met. On the one hand, the distance between the areas concerned in the image must be small (e.g., smaller than 30 pixels), since otherwise the mutual influence of the hues by interreflections is negligibly small. On the other hand, they should be multicolored, since otherwise no hue changes arise and then the interreflection analysis described in the previous section supplies false results.

For the examination of the latter condition a mean hue is determined for each segmented region by arithmetical averaging of all pixels of this region. The difference of the mean hues must possess a certain minimum size. For all pairings of regions the minimal distance to each other and the hue difference is computed. The influence of interreflection on the hue of two surfaces is largest where the surfaces are closest to one another. The influence decreases with increasing distance. Thus, not all pixels need to be included within the regions in the interreflection analysis. It is sufficient to consider only the pixels lying closest together. For this, the pixels with the smallest distance to the other region are determined in each region. Apart from these pixels, the interreflection areas to be investigated can be arranged in the form of a right triangle or a half-ellipsis (see Fig. 8.12).

8.2.7 Analysis of Shadow

All regions between which an interreflection is assumed are examined now for possible shadow domains. At the borders of shadow domains the even process of values of the pixels is disturbed in an area afflicted with interreflections. Thereby in the analysis of the *SVD*, false results would be computed. For the shadow analysis a histogram of the intensity values is computed for all pixels in one area to be examined. If no shadow borders are contained, then the histogram is well balanced. Otherwise, the area contains a light and a dark domain. This becomes apparent in the histogram by two separated maxima. In this case, the difference of the intensities in the maxima is determined and the intensity value is increased by this difference for all pixels that lie in the shadow. Subsequently, all *HSI* values are transformed into the *RGB* space.

8.2.8 Minimization of Interreflections

After increasing the intensity in the shadow areas at the latest, the two areas to be examined for interreflections contain relatively homogeneous *RGB* values. An analysis of the *SVD* can now be accomplished for both matrixes formed in such a way. The determination of the surface colors without the influence of interreflection and the determination of the one-bounce portion of the measured light take place according to the procedure described in Section 8.2.1. In determining the no-bounce color in real color images an additional difficulty can result from the color of a surface being overlaid by several interreflections on multicolor surfaces. In practice, quite good results could be obtained in which that no-bounce color is selected whose hue exhibits the smallest difference to the mean hue of the region determined at the beginning. After the minimization of interreflections in the entire image, the intensity values can be reduced again in the shadow domains, depending on the requirement of a processing procedure.

Figure 8.12 gives an example of the minimization of interreflections in a real color image. At upper right in Fig. 8.12, the segmentation result with consideration to interreflections and shadows is presented. The white points in the regions represent an internal numbering of the regions (i.e., region number three is designated with three points). In the lower left corner of the image (emphasized in the segmentation result by a white rectangle), the measured intensity in the shadow area is too small to be able to accomplish a color analysis. This area is therefore not considered in further processing steps. At bottom left in Fig. 8.12, the partitioning of the image into areas is shown (represented as ellipses), within which an interreflection is assumed.

Figure 8.12. *Original image (upper left), the segmented image (upper right), the areas in the image (represented as ellipses) with possible interreflections (lower left), and the resulting color image after minimization of the interreflections (lower right).*

(a) (b) (c)

Figure 8.13. *Differences between the original image with interreflections (see Fig. 8.12 upper right) and the computed image after minimizing the interreflections (see Fig. 8.12 lower right) in the blue (a), red (b), and green channel (c).*

At lower right the figure shows the resulting image after minimization of interreflections. For better illustration of changes of the surface colors due to interreflections, the differences between the original image with interreflections and the computed image after minimizing interreflections are represented separately in Fig. 8.13 for the three components of the color vectors (in the blue channel and short-wave spectral region, in the red channel and long-wave spectral region, and in the green channel and middle-wave spectral region, respectively).

8.3 COLOR CONSTANCY

The color values acquired by a camera depend among other things on the color of light (or more exactly, the spectral power distribution of the illumination). If the color of light changes, then the measured color values also change. *Color constancy* designates the production of a color classification invariant in relation to lighting changes for the description of surfaces from color images. To achieve this in digital color image processing is a problem that is difficult to solve. This is because the color signal measured with the camera depends on the spectral distribution of the lighting and the light reflected at the surface, as well as object geometry. These characteristics of the scene are usually not known.

In digital color image processing, as previously mentioned in Section 1.2.4, different procedures for the acquisition of color constancy are in widespread use. One procedure consists of estimating the spectral distributions of the reflected light for each visible surface in the scene. If the entering lighting is known, then the spectral reflectance factor of the surface material can be determined. Another technique for color constancy consists of generating a color image of the acquired scene in such a way as it would occur under known lighting conditions. This usually takes place related to basic reference lighting, such as standard illuminant *D*65, which cannot be completely reproduced by any technical source of light (see Section 4.2.2). Figure 8.14 illustrates the latter variant of a color constancy algorithm. A third procedure for the acquisition of color constancy consists of determining characteristics of the colored object surfaces in the color image independently from lighting conditions and canceling the variances due to lighting changes.

From the technical point of view, color constancy plays an important role for a lighting independent color comparison between surfaces or objects. Thus in color object recognition, an approximate color constancy is frequently a condition for the recognition process. In contrast to this, color constancy does not play a role in the static stereo analysis of color images (see Chapter 9). Here it can be always assumed that the two images were generated under identical lighting conditions. However, if a technique of motion stereo analysis is applied to color images or if a colored object in an image sequence is to be tracked, then changes in the color values conditioned by changes of the lighting can lead to errors in the computed results. Certain color constancy is usually also presupposed here.

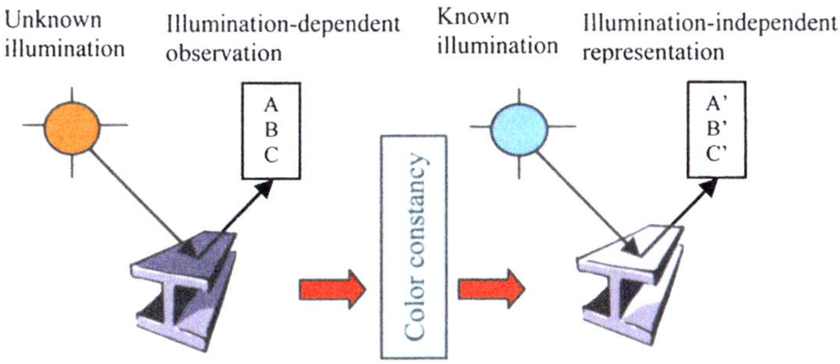

Figure 8.14. *An example of color constancy: The eye (or the color camera) views a scene under unknown lighting. A color constancy algorithm transforms, for example, the scene into a known lighting.*

8.3.1 Mathematical Formulation of the Color Constancy Problem

For a mathematical formulation of the color constancy problem the sensor model from Section 4.3 can be again observed. According to Eq. (4.1) it holds that

$$\rho_k = \int E(\lambda) \cdot R(\lambda) \cdot S_k(\lambda)\, d\lambda\,,$$

whereby $E(\lambda)$ indicates the spectral power distribution of the illumination. $R(\lambda)$ is the spectral reflectance factor of the surface material, $S_k(\lambda)$ is the spectral sensitivity of subsensor k, and ρ_k its sensor response. For each pixel (x, y) there exist altogether p subsensors, $k = 1,\ldots,p$.

Color constancy is not possible without constraints or assumptions about the structure of the scene [Hea et al. 92]. In solving the color constancy problem the assumption is frequently made that the spectral reflectance factor of the surface material can be represented by a linear combination of a finite quantity of basic functions and that the spectral power distribution of the illumination can be represented by a linear combination of another quantity of basic functions. The use of such models is meaningful for the characterization of the kind of scenes for which an algorithm supplies color constancy.

In principle, the color constancy problem can be formulated as an inverse problem in a finite-dimensional linear area and solved by matrix algebra (see [Jai et al. 95]). Let us assume that the spectral reflectance factor of the surface material can be represented by a linear combination of basic functions

$$R(\lambda) = \sum_{i=1}^{n} b_i R_i(\lambda).$$

The number n of basic functions is the number of degrees of freedom in the surface reflection. Based on the data from Cohen [Coh64], this is mostly set at $n = 3$. Cohen discovered in an investigation of 150 Munsell color samples (see Section 3.6.1) that the spectral reflectance factors could be modeled to 99.2% by three suitably selected basic functions. It concludes from the fact that a good approximation can be achieved with three basic functions. Let us assume that the basic functions $R_i(\lambda)$ are known.

Let us assume further that the spectral power distribution of the illumination is represented by a linear combination with m degrees of freedom

$$E(\lambda) = \sum_{j=1}^{m} c_j E_j(\lambda)$$

and the spectral distributions $E_j(\lambda)$ are known. Three or four basic functions are usually sufficient for a good approximation of the spectral power distribution of daylight (see [Jud et al. 64] and [MalWan86]). Four or five basic functions are frequently employed for the description of artificial lighting (see [NovSha90] and [NovSha91]).

The color determination problem can now be formulated in a matrix. The m values of c_j form a column vector **c**, which describes the light $E(\lambda)$. The n values of b_i form a column vector **b**, which describes the spectral reflectance factor $R(\lambda)$. By including the column vector in Eq. 4.1 a matrix model results for each pixel in the image. It holds for

$$\boldsymbol{\rho} = \boldsymbol{\Lambda}_{\mathbf{c}} \mathbf{b} \, ,$$

whereby $\boldsymbol{\rho}$ describes the column vector that is formed by the response of the p subsensors. $\boldsymbol{\Lambda}_{\mathbf{c}}$ is a $p \times n$ matrix and its elements at the kith position are of the form

$$\int E(\lambda) R_i(\lambda) S_k(\lambda) \, d\lambda \, .$$

If the lighting is known, then the lighting matrix $\boldsymbol{\Lambda}_{\mathbf{c}}$ is also known. If the number of subsensors is equal to the number of degrees of freedom of the spectral reflectance factor (i.e., $p = n$), then the solution of the color constancy problem is reduced to inverting the lighting matrix $\boldsymbol{\Lambda}_{\mathbf{c}}$. If p is smaller than n, then the above equation is underdetermined and no clear solution exists. If the lighting is not known, then to solve the problem, the number of subsensors must be greater than the number of degrees of freedom of the spectral reflectance factor.

8.3.2 Techniques for Color Constancy

The classical theory for describing the ability for color constancy in the human visual system is the (monochromatic-based) *retinex theory* (see [LanMcC71], [Hor74], and [Lan86]), whose limits were already discussed in Section 2.4. This approach investigates color constancy behavior from psychophysical experiments. Land studied the psychological aspects of lightness and color perception of human vision and proposed a theory to obtain an analogous performance in machine vision systems. Retinex is not only used as a model of human vision color constancy, but is also used as a platform for digital image enhancement and lightness/color rendition. Land's retinex theory is based on the design of a surround function. Hurlbert [Hur89] proposed a Gaussian surround function by choosing three different sigma values to achieve good dynamic range compression and color rendition. From that point onward, numerous retinex theory implementations were published (see [BarFun98], [CooBag04], [Fun et al. 04], [Kim et al. 03], [Can04], [Rah etal97], [Riz et al. 04]) and efforts were made to optimize the performance of the retinex algorithm by tuning the free parameters [FunCiu04]. The *multiscale retinex* (*MSR*) implementation [Rah et al. 97] intertwined a number of image processing operations and, as a result, the colors are changed in the image in an unpredicted way. Barnard and Funt [BarFun98] presented a way to make MSR operations more clear and to ensure color fidelity.

Many techniques for color constancy differ regarding the basic function and reference sizes used, which are implemented in determining the correspondence between the measured values under unknown lighting and the measured values for a standard illuminant (see Section 4.2.2 for standard illuminant). A brief overview of basic functions and reference sizes is contained in [NovSha90].

The complexity of the color constancy problem can be reduced by using mathematical restrictions. Maloney and Wandell [MalWan86] determine m coefficients for light by measurement on more than m positions in the image. The number of subsensors must hereby always be greater, by at least one, than the number of degrees of freedom of the spectral reflectance factor (i.e., $p \geq n+1$) in accordance with Section 8.3.1. If a commercial color camera that supplies a three-channel *RGB* image is used, then the number of basic functions for the spectral reflectance factor of the surface material is limited to maximally two. This limitation can be overcome as further spectral data is used additionally or instead of the *RGB* values (see [Ho et al. 90] and [TomWan89]).

For the practical use of a color constancy procedure frequently only *RGB* images with three spectral transmissions are available. The underdetermined set of equations can be solved when not merely one single color image, but rather several color images are evaluated for color constancy [TsuOht90]. Here it is assumed that a colored object surface is visible in two color images generated under different lighting conditions. One difficulty of converting this procedure is that an exact allocation must take place between the surface pixels visible in the different images. However, this correspondence problem cannot be solved so easily. Furthermore, Tsukada and Ohta [TsuOht90] assume that the spectral characteristics of the sensor are known, which frequently not the case is.

Further techniques for color constancy consider highlights on inhomogeneous dielectric materials in order to determine the color of the lighting [D'ZmLen86], [HeaBin87], [Lee86], [Lee90], [TomWan89]. According to the dichromatic reflection model (see Section 7.4.1), the highlight range in the image contains the same color as the lighting. If highlights are found in the image, then the color of the lighting is also certain. This technique fails if no highlights are contained in the image or if highlights also arise on noninhomogeneous dielectric materials, such as metals.

The reference sizes used in several techniques for determining correspondence between measured values under unknown illumination and measured values for a standard illuminant are based in general on heuristics of real images. A common heuristic is that the average value of all pixel values in the image results in roughly a gray tone (*gray-world assumption*; see [D'ZmLen86] and [Tak et al. 96]). The averaged intensity values are determined separately (monochromatic-based) for each component of the color vector over the entire image. Subsequently, multiplicative factors are calculated that adjust the relationship between the vector components on a scale of 1:1:1. This procedure is also partly used with automatic film recorders.

A more far-reaching assumption is that the average value of all pixel values in the image results in a certain color. This special color can be selected on the basis of statistic investigations [Ger87], [Ger et al. 87c]. The advantage of this method over the previous technique is that also different brightnesses of incidental lighting become balanced. For both procedures, images can be easily found with which the assumptions lead to a false result. A further difficulty of such procedures results from the general assumption that the transformation between the color areas is nonambiguous. But this is not always the case, as the existence of metamers proves [DreFun92].

The number and the kind of basic functions form a further distinction criterion for a set of color constancy procedures. The spectral reflectance factor can be represented, for example, by a set of Fourier functions [Baj et al. 96], [Wan87], or Legendre polynomes [HeaBin87], [NovSha90]. A set of basic functions is used in exactly the same way, which was determined by a main component analysis of a set of images of natural and artificial objects [Ger87]. Each finite number of basic functions represents only one approximation of the spectral reflectance factor and/or the spectral power distribution of the illumination. All color constancy techniques mentioned use the common assumption that the spectral reflectance factor can be represented by a small set of basic functions.

Takebe, Nakauchi, and Usui [Tak et al. 96] examined the problem of determining the spectral distribution of lighting with an uneven illumination (or partial shadows) of the scene. They solve the color constancy problem by minimizing an energy function that contains a term for distinction between light and shadow areas in the color image. This distinction is based on the assumption by Ruby and Richards [RubRic84] that the signs of differences of at least two components of the color vectors change at material edges, while the signs at shadow edges remain the same. Yet this behavior does not occur in every constellation. In [Ger et al. 92] some examples of color images were given in which shade edges are falsely classified as material changes with this technique.

Some researchers use spectral reference values for converting color values under unknown lighting into color values under known lighting. This procedure is called supervised color constancy. In addition to a white reference surface [Baj et al. 89], the Macbeth ColorChecker (see Section 3.6.2) with its known spectral values is also used as a reference map (*ground truth*) [NovSha90]. First, an acquisition under unknown lighting of the reference map is provided. Subsequently, the color of the lighting is determined from the known spectral values. On the assumption that the lighting does not change, the images to be examined are then generated. The assumption over the nonchanging spectral power distribution of the lighting is, however, usually not allowed for outdoor images.

A different approach for supervised color constancy has been proposed in [Tak et al. 99]. Their method is for restoring surface spectral reflectance using a set of two images: A first image is taken under unknown illuminant, and a second one is taken under known illuminant in addition to the unknown illuminant.

Ohta and Hayashi [OhtHay94] assume for outdoor images in daylight that the spectral power distribution of the illumination should be similar to a CIE illuminant for daylight in a certain color temperature (see Section 4.2.2). They model the daylight lighting of unknown spectral power distribution by three basic functions for daylight as proposed in [Jud et al. 64]. A set of equations is set up for two images of a colored object acquired under different unknown daylight conditions. Here the spectral sensitivities of the subsensors must be known. Ohta and Hayashi ensure this by using a black-and-white camera with three Kodak Wratten filters (see Section 4.2.1). The successive generation of the three components of the color vectors presupposes here a static scene, a static camera, and nonchanging lighting conditions.

All techniques specified so far presuppose that only one lighting source is present. For scenes with varying illumination, the color constancy problem was examined in [Bar et al. 97]. The varying lighting conditions in the scene designate different spectral distributions of the lighting at different positions in the scene (e.g., how this is the case by lighting the scene using a lamp and additional light coming through a window). In this technique, the variation of the lighting is used as an additional restriction for generating an image as it would appear under known lighting conditions.

For the localization of the areas in the image, within which a variation of the lighting conditions is present, a region-growing technique is used that represents an extension of the retinex algorithm. As with the retinex method, here conditions (monochromatic-based) are checked separately for each component. The color constancy procedure is based on the assumption that the transformation between the available image and the image to be provided (under known lighting conditions) can be realized by a diagonal matrix [Fin et al. 93]. For the determination of this diagonal matrix, among other things, the restriction suggested by Forsyth [For90] is used. This states that the values in the *RGB* space, measured with the camera using a firm lighting for a combination of a large set of surfaces, form a convex hull that does not fill out the entire *RGB* space.

A second restriction takes place due to the assumption that only certain spectral distributions (e.g., those of the standard illuminant or a Planck emitter with 2000 K) are plausible for lighting [Fin et al. 94]. A further restriction of the problem takes place due to the assumption that one surface pixel is subject to several illuminations with sufficiently different spectral distributions. The more different the lighting sources in the scene (e.g., 3, 4, or 5), the better the results to be obtained. However, if only one lighting source is present, then (as expected) no good results are achieved with this technique (see [Bar et al. 97]).

Another procedure for producing spectral reflectance factors of surface materials that are invariant to lighting variations was proposed by Mahlmeister, Schmidt, and Sommer [Mah et al. 95]. They first transform the color vectors in the *RGB* space into the uniform CIELAB model (see Section 3.3.1). In the CIELAB model the product, described in Eq. (4.1), of the factors of the illumination's spectral power distribution and the surface material's spectral reflectance factor can be approximated by a sum of the two terms [Mah et al. 95]. On the basis of the assumption [Wan87] that variances caused by illumination are more low-frequency in the image signal than those due to material changes, the CIELAB image can be convolved with a set of scaled and rotated Andersson filters [And92] in such a manner that the filter result contains only the reflection component of the color signal. The lowfrequency portions, which according to the assumption contain the lighting variations, are suppressed in such a way. This technique can be integrated into an object recognition system that will be described later.

A very efficient (video-real-time capable) method for the recognition of colored objects in color images is based on the comparison of color distributions or color histograms of objects in a color space [SwaBal91]. Since this method of object recognition, called *color indexing*, is susceptible to illumination changes, a set of expansion suggestions were submitted that would make the technique more color constant. One procedure would consist of first applying a color constancy algorithm to the image and afterward computing the color histogram. We agree with [FunFin95] in the fact that hereby the efficiency and elegance of the technique are destroyed. It should be a goal rather to amend the procedure in such a manner that certain invariance in relation to lighting changes is ensured.

Funt and Finlayson [FunFin95] suggested using *quotients* of color histograms for object comparisons instead of color histograms. Thereby the results of object recognition can be improved with changing lighting conditions opposite "direct" color indexing. This procedure is susceptible to noise, particularly within badly illuminated scene areas. Another suggestion from Healey and Slater

[HeaSla94] consists of using a small set of *moments* from color histograms for the representation of objects. On the assumption that a linear model can describe the lighting change, they show that some moments of the color distribution are invariant in relation to lighting changes.

A further suggestion in [Fin et al. 96] consists of using six angles of the color distribution instead of a color histogram for the representation of an object. In [Fin et al. 96] some examples of objects were presented with which a better rate of recognition for objects was obtained with this procedure with changing lighting than with the moment formulation of Healey and Slater. An additional suggestion plans to use histograms of color orientations instead of color distributions for representing objects [Mah et al. 96]. Here (in expansion of the technique named above in [Mah et al. 95]), a set of Andersson filters is applied afterward to the single components of the color image in the CIELAB model. Subsequently, orientation and amplitude are determined for each vector component in the CIELAB model. The characteristic vectors obtained in such a way are then registered using several constraints into a color orientation index, for which a histogram (the *color orientation histogram*) is afterward determined. In [Mah et al. 96] the robustness of these color orientation histograms in relation to lighting variations is demonstrated on the basis of some tests and comparisons. Additional investigations included spatial-chromatic [Cin et al. 01], viewpoint-invariant [GevSme99], and ratio-based color indexing [AdjLee01].

In this section, various techniques and procedures for achieving color constancy in digital color image processing were specified and discussed. Each of these techniques is based on assumptions without which the color constancy problem cannot be solved. The choice for or against a procedure will depend also, apart from the estimation of the cost of computation, on the validity of the assumptions that are the basis for the procedure. Further works on color constancy, which will not be covered further here, can be found in [Bar et al. 01], [BraWan86], [Bri92], [DufLum91], [For90], [FunDre88], [Fun et al. 91], [FunHo89], [Geu et al. 01], [Hwa et al. 93], [LanMcC71], [Lan86], [Len et al. 99], [MarRiz00], [NagGri98], [Orw et al. 01], and [SatIke93]. An extensive investigation of techniques for solving the color constancy problem and some results with real images are summarized in [Fin96]. More recent work of his group is presented in [Fin et al. 01], [FinHor01], and [FinXu02]. An overview of color constancy algorithms is on hand in [Aga et al. 06].

Here it should be noted that some authors apply their color constancy algorithm exclusively to images similar to Mondrian scenes and publish only these results. The use of Mondrian images is understandable and welcome from the tradition of psychological experiments (see Section 2.4). Other scenes should be also included in the investigations in order to enhance the applicability of the technique more clearly for other questions of digital color image processing. Since color constancy is of importance for many areas of digital image processing, additional research work will concern itself with this subject in the future.

8.4 REFERENCES

[AdjLee01] D.A. Adjeroh, M.C. Lee. On ratio-based color indexing. *IEEE Transactions on Image Processing* **10** (2001), pp. 36 –48.

[Aga et al. 06] V. Agarwal, B.R. Abidi, A. Koschan, M.A. Abidi. An Overview of Color Constancy Algorithms. *J. of Pattern Recognition Research* **1** (2006), pp. 42-54.

[And92] M.T. Andersson. Controllable multidimensional filters and models in low level computer vision. Ph.D. Thesis, Linköping University, Sweden, 1992.

[Baj et al. 89] R. Bajcsy, S.W. Lee, A. Leonardis. Image segmentation with detection of highlights and interreflections using color. Technical Report GRASP LAB 182 MS-CIS-89-39, Dept. of Computer and Information Science, University of Pennsylvania, 1989.

[Baj et al. 96] R. Bajcsy, S.W. Lee, A. Leonardis. Detection of diffuse and specular interface reflections and inter-reflections by color image segmentation. *Int. J. of Computer Vision* **17** (1996), pp. 241-272.

[BarFun98] K. Barnard and B. Funt. Investigation into multiscale retinex, *Proc. Color Imaging in Multimedia*, Derby, UK, 1998, pp. 9-17.

[Bar et al. 97] K. Barnard, G. Finlayson, B. Funt. Color constancy for scenes with varying illumination. *Computer Vision and Image Understanding* **65** (1997), pp. 311-321.

[Bar et al. 01] K. Barnard, F. Ciurea, B. Funt. Sensor sharpening for computational color constancy. *J. Optical Society of America. A*, Optics, Image Science, and Vision **18 (**2001), pp. 2728-2743.

[BraWan86] D.H. Brainard, B.A. Wandell. Analysis of the Retinex theory of color vision. *J. Optical Society America* **3** (1986), pp. 1651-1661.

[Bri92] M.H. Brill. Image segmentation by object color: A unifying framework and connection to color constancy. In: G. Healey, S.A. Shafer, L.B. Wolff (eds.). *Physics-Based Vision: Principles and Practice Color*. Jones and Bartlett, Boston, 1992, pp. 109-115.

[Can04] J. J. McCann. Capturing a black cat in shade: Past and present of retinex color appearance models. *J. of Electronic Imaging* **13** (2004), pp. 36-47.

[Cin et al. 01] L. Cinque, G. Ciocca, S. Levialdi, A. Pellicanò, R. Schettini. Color-based image retrieval using spatial-chromatic histograms. *Image and Vision Computing* **19** (2001), pp. 979-986.

[Coh64] J. Cohen. Dependency of the spectral reflectance curves of the Munsell color chips. *Psychonomic Science* **1** (1964), pp. 369.

[CooBag04] T.J. Cooper, F.A. Baqai. Analysis and extensions of the Frankle-McCann retinex algorithm. *J. of Electronic Imaging* **13** (2004), pp. 85-92.

[DreFun90] M.S. Drew, B.V. Funt. Calculating surface reflectance using a single-bounce model of mutual reflection. *Proc. 3rd Int. Conference on Computer Vision*, Osaka, Japan, 1990, pp. 394-399.

[DreFun92] M.S. Drew, B.V. Funt. Natural metamers. *Computer Vision, Graphics, and Image Processing: Image Understanding* **56** (1992), pp. 139-151.

[DufLum91] P.A. Dufort, C.J. Lumsden. Color categorization and color constancy in a neural network model of V4. *Biological Cybernetics* **65** (1991), pp. 293-303.

[D'ZmLen86] M. D'Zmura, P. Lennie. Mechanisms of color constancy. *J. Optical Society of America A* **3** (1986), pp. 1662-1672.

[Fin et al. 93] G.D. Finlayson, M.S. Drew, B.V. Funt. Diagonal transform suffice for color constancy. *Proc. 4th Int. Conference on Computer Vision*, Berlin, Germany, 1993, pp.164-171.

[Fin et al. 94] G.D. Finlayson, M.S. Drew, B.V. Funt. Spectral sharpening: sensor transformations for improved color constancy. *J. Optical Society of America A* **11** (1994), pp. 1553-1563.

[Fin et al. 96] G.D. Finlayson, S.S. Chatterjee, B.V. Funt. Color angular indexing. *Proc. 4th European Conference on Computer Vision*, Cambridge, England, 1996, Vol. II, pp. 16-27.

[Fin96] G.D. Finlayson. Color in perspective. *IEEE Transactions on Pattern Analysis and Machine Intelligence* **18** (1996), pp. 1034-1038.

[Fin et al. 01] G.D. Finlayson, S.D. Hordley, P.M. Hubel. Color by correlation: A simple, unifying framework for color constancy. *IEEE Transactions on Pattern Analysis and Machine Intelligence* **23** (2001), pp. 1209–1221.

[FinHor01] G.D. Finlayson, S.D. Hordley. Color constancy at a pixel. *J. Optical Society of America A*, Optics, Image Science, and Vision **18** (2001), pp. 253-264.

[FinXu02] G.D. Finlayson, R. Xu. Non-iterative comprehensive normalization. *Proc. 1st European Conference on Color in Graphics, Imaging, and Vision*, Poitiers, France, April 2002, pp. 159-163.

[Fol et al. 95] J.D. Foley, A. van Dam, S.K. Feiner, J.F. Hughes. *Computer Graphics: Principles and Practice*. 2nd ed., Addison-Wesley, 1995.

[For90] D.A. Forsyth. A novel algorithm for color constancy. *Int. J. of Computer Vision* **5** (1990), pp. 5-35.

[Fun et al. 91] B.V. Funt, M.S. Drew, J. Ho. Color constancy from mutual reflection. *Int. J. of Computer Vision* **6** (1991), pp. 5-24.

[FunCiu04] B. V. Funt and F. Ciurea. Parameters for retinex. *Proc. 9th Congress of the International Color Association*, Rochester, New York, 2004.

[FunDre88] B.V. Funt, M.S. Drew. Color constancy computation in near-Mondrian scenes using a finite dimensional linear model. *Proc. Int. Conference on Computer Vision and Pattern Recognition*, Ann Arbor, Michigan, 1988, pp. 544-549.

[FunDre92] B.V. Funt, M.S. Drew. Color space analysis of mutual illumination. In: G. Healey, S.A. Shafer, L.B. Wolff (eds.). *Physics-Based Vision: Principles and Practice Color*. Jones and Bartlett, Boston, Massachusetts, 1992, pp. 385-410.

[Fun et al. 04] B. V. Funt, F. Ciurea, and J. McCann. Retinex in MATLAB. *J. Electronic Imaging* **13** (2004), pp. 48-57.

[FunFin95] B.V. Funt, G.D. Finlayson. Color constant color indexing. *IEEE Transactions on Pattern Analysis and Machine Intelligence* **17** (1995), pp. 522-529.

[FunHo89] B.V. Funt, J. Ho. Color from black and white. *Int. J. of Computer Vision* **3** (1989), pp. 109-117.

[Ger87] R. Gershon. The use of color in computational vision. Ph.D. Thesis, Technical Report RBCV-TR-87-15, Dept. of Computer Science, University of Toronto, Ontario, Canada, 1987.

[Ger et al. 87a] R. Gershon, A. D. Jepson, J. K. Tsotsos. The use of color in highlight identification. *Proc. 10th Int. Joint Conference on Artificial Intelligence*, Milan, Italy, 1987, pp. 752-754.

[Ger et al. 87b] R. Gershon, A. D. Jepson, J. K. Tsotsos. Highlight identification using chromatic information. *Proc. 1st Int. Conference on Computer Vision*, London, England, 1987, pp. 161-170.

[Ger et al. 87c] R. Gershon, A.D. Jepson, J.K. Tsotsos. From [R,G,B] to surface reflectance: Computing color constant descriptors in images. *Proc. 10th Int. Joint Conference on Artificial Intelligence*, Milan, Italy, 1987, pp. 755-758.

[Ger et al. 92] R. Gershon, A.D. Jepson, J.K. Tsotsos. Ambient illumination and the determination of material changes. In: G.E. Healey, S.A. Shafer, L.B. Wolff (eds.). *Physics-Based Vision: Principles and Practice Color*, Jones and Bartlett, Boston, Massachusetts, 1992, pp. 101-108.

[Geu et al. 01] J.-M. Geusebroek, R. van den Boomgaard, A.W.M. Smeulders, H. Geerts. Color invariance. *IEEE Transactions on Pattern Analysis and Machine Intelligence* **23** (2001), pp. 1338 –1350.

[GevSme99] T. Gevers, A.W.M. Smeulders. Content-based image retrieval by viewpoint-invariant color indexing. *Image and Vision Computing* **17** (1999), pp. 475-488.

[Hea et al. 92] G. Healey, S.A. Shafer, L.B. Wolff (eds.). *Physics-Based Vision: Principles and Practice Color*. Jones and Bartlett, Boston, Massachusetts, 1992.

[HeaBin87] G. Healey, T.O. Binford. The role and use of color in a general vision system. *Proc. Image Understanding Workshop*, Vol. II, Los Angeles, California, 1987, pp. 599-613.

[HeaSla94] G. Healey, D. Slater. Global color constancy: Recognition of objects by use of illumination-invariant properties of color distributions. *J. Optical Society of America A* **11** (1994), pp. 3003-3010.

[Ho et al. 90] J. Ho, B.V. Funt, M.S. Drew. Separating a color signal into illumination and surface reflectance components: Theory and applications. *IEEE Transactions on Pattern Analysis and Machine Intelligence* **12** (1990), pp. 966-977.

[Hor74] B.K.P. Horn. Determining lightness from an image. *Computer Graphics and Image Processing* **3** (1974), pp. 277-299.

[Hur89] A. C. Hurlbert. The Computation of Color. Ph.D Dissertation, Massachusetts Institute of Technology, September, 1989.

[Hwa et al. 93] P.-W. Hwang, Y.-S. Chen, F.-H. Cheng, W.-H. Hsu. Color recovery from biased illumination: Color constancy. *Proc. Int. Conference on Computer Vision and Pattern Recognition*, New York, 1993, pp. 631-632.

[Jai et al. 95] R. Jain, R. Kasturi, B.G. Schunck. *Machine Vision*. McGraw-Hill, Singapore, 1995.

[Jan91] Y. Jang. Identification of interreflection in color images using a physics-based reflection model. *Proc. Int. Conference on Computer Vision and Pattern Recognition*, Maui, Hawaii, 1991, pp. 632-637.

[Jud et al. 64] D.B. Judd, D.L. MacAdam, G. Wyszecki. Spectral distribution of typical daylight as a function of correlated color temperature. *J. Optical Society of America* **54** (1964), pp. 1031.

[Kim et al. 03] R. Kimmel, M. Elad, D. Shaked, R. Keshet, and I. Sobel. A variational framework for retinex. *Int. J. of Computer Vision* **52** (2003), pp. 7-23.

[Kle et al. 96] R. Klette, A. Koschan, K. Schlüns. *Computer Vision: Räumliche Information aus digitalen Bildern*. Vieweg, Braunschweig/Wiesbaden, Germany, 1996.

[Kle et al. 98] R. Klette, K. Schlüns, A. Koschan. *Computer Vision: Three-Dimensional Data from Images*. Springer, Singapore, 1998.

[Kli93] G.J. Klinker. A physical approach to color image understanding. A.K. Peters, Wellesley, Massachusetts, 1993.

[Kli et al. 88] G.J. Klinker, S.A. Shafer, T. Kanade. Image segmentation and reflection analysis through color. *Proc. Image Understanding Workshop*, Cambridge, Massachusetts, 1988, Vol. II, pp. 838-853.

[Kli et al. 90] G.J. Klinker, S.A. Shafer, T. Kanade. A physical approach to color image understanding. *Int. J. of Computer Vision* **4** (1990), pp. 7-38.

[Kos97] A. Koschan. Segmentation of color images for the minimization of interreflections. *Proc. 4th Int. Workshop on Systems, Signals and Image Processing*, M. Domanski, R. Stasinski (eds.), Poznan, Poland, 1997, pp. 191-194.

[Lan86] E.H. Land. Recent advances in retinex theory. *Vision Research*, **26** (1986), pp. 7-21.

[LanMcC71] E.H. Land, J.J. McCann. Lightness and retinex theory. *J. Optical Society of America* **61** (1971), pp. 1-11.

[Lee86] H.-C. Lee. Method for computing the scene-illuminant chromaticity from specular highlights. *J. Optical Society of America A* **3** (1986), pp. 1694-1699.

[LeeBaj92] S.W. Lee, R. Bajcsy. Detection of specularity using color and multiple views. *Proc. 2nd European Conference on Computer Vision*, Santa Margherita Ligure, Italy, 1992, pp. 99-114.

[Lee90] H.-C. Lee. Illuminant color from shading. *Proc. SPIE* **1250**, Perceiving, Measuring and Using Color (1990), pp. 236-244.

[Len et al. 99] R. Lenz, P. Meer, M. Hauta-Kasari. Spectral-based illumination estimation and color correction. *Color Research & Application* **24** (1999), pp. 98-111.

[LinLee96] S. Lin, S.W. Lee. Detection of specularity using stereo in color and polarization space. *Proc. 13th Int. Conference on Pattern Recognition*, Vienna, Austria, 1996, Vol. I, pp. 263-267.

[LinLee97] S. Lin, S.W. Lee. Detection of specularity using stereo in color and polarization space. *Computer Vision and Image Understanding* **65** (1997), pp. 336-347.

[Mah et al. 95] U. Mahlmeister, B. Schmidt, G. Sommer. Preattentive colour features by steerable filters. *Proc. 17th DAGM-Symposium Mustererkennung*, G. Sagerer, S. Posch, F. Kummert (eds.), Bielefeld, Germany, 1995, pp. 464-472.

[Mah et al. 96] U. Mahlmeister, H. Pahl, G. Sommer. Color-orientation indexing. *Proc. 18th DAGM-Symposium Mustererkennung*, B. Jähne et al. (eds.), Heidelberg, Germany, 1996, pp. 3-10.

[MalWan86] L.T. Maloney, B.A. Wandell. Color constancy: A method for recovering surface spectral reflectance. *J. Optical Society of America A* **3** (1986), pp. 1651-1661.

[MarRiz00] D. Marini, A. Rizzi. A computational approach to color adaptation effects. *Image and Vision Computing* **18** (2000), pp. 1005-1014.

[NagGri98] K. Nagao, W.E.L. Grimson. Using photometric invariants for 3D object recognition. *Computer Vision and Image Understanding* **71** (1998), pp. 74-93.

[Nay et al. 91a] S.K. Nayar, K. Ikeuchi, T. Kanade. Surface reflection: Physical and geometrical perspectives. *IEEE Transactions on Pattern Analysis and Machine Intelligence* **13** (1991), pp. 611-634.

[Nay et al. 91b] S.K. Nayar, K. Ikeuchi, T. Kanade. Shape from interreflection. *Int. J. of Computer Vision* **6** (1991), pp. 173-195.

[Nay et al. 93] S.K. Nayar, X.-S. Fang, T. Boult. Removal of specularities using color and polarization. *Proc. Int. Conference on Computer Vision and Pattern Recognition*, New York, 1993, pp. 583-590.

[Nay et al. 97] S.K. Nayar, X.-S. Fang, T. Boult. Separation of reflection components using color and polarization. *Int. J. Computer Vision* **21** (1997), pp. 163-186.

[NayGon92] S.K. Nayar, Y. Gong. Colored interreflections and shape recovery. *Proc. Image Understanding Workshop*, San Diego, California, 1992, pp. 333-343.

[Nov92] C.L. Novak. Estimating scene properties by analyzing color histograms with physics-based models. Ph.D. Thesis, Technical Report CMU-CS-92-222, School of Computer Science, Carnegie Mellon University, Pittsburgh, Pennsylvania, 1992.

[NovSha90] C.L. Novak, S.A. Shafer. Supervised color constancy using a color chart. Technical Report CMU-CS-90-140, School of Computer Science, Carnegie Mellon University, Pittsburgh, Pennsylvania, June 1990.

[NovSha91] C.L. Novak, S.A. Shafer. Supervised color constancy for machine vision. *Proc. SPIE* **1453**, Human Vision, Visual Processing, and Digital Display II, Santa Clara, California, 1991, pp. 353-368.

[NovSha92] C.L. Novak, S.A. Shafer. Anatomy of a color histogram. *Proc. Int. Conference on Computer Vision and Pattern Recognition*, Champaign, Illinois, 1992, pp. 599-605.

[OhtHay94] Y. Ohta, Y. Hayashi. Recovery of illuminant and surface colors from images based on the CIE daylight. *Proc. European Conference on Computer Vision*, J.-O. Eklundh (ed.), Stockholm, Sweden, 1994, pp. 235-246.

[Orw et al. 01] J. Orwell, P.M. Remagnino, G.A. Jones. Optimal color quantization for real-time object recognition. *Real-Time Imaging* **7** (2001), pp. 401-414.

[Rah etal97] Z. Rahman, D. Jobson, G.A.Woodell. A multiscale retinex for bridging the gap between color images and the human observation of scenes. *IEEE Transactions on Image Processing* **6** (1997), pp. 965-976.

[Riz et al. 04] A. Rizzi, C. Gatta, D. Marini. From retinex to automatic color equalization: Issues in developing a new algorithm for unsupervised color equalization. *J. Electronic Imaging* **13** (2004), pp. 75-84, 2004.

[RubRic84] J.M. Rubin, W.A. Richards. Color vision and image intensities: Representing material changes. AI Memo 764, MIT Artificial Intelligence Laboratory, Cambridge, Massachusetts, 1984.

[SatIke93] Y. Sato, K. Ikeuchi. Temporal colorspace analysis of reflection. *Proc. Int. Conference on Computer Vision and Pattern Recognition*, New York, 1993, pp. 570-576.

[SchKos00] K. Schlüns, A. Koschan. Global and local highlight analysis in color images. *Proc. 1st Int. Conference on Color in Graphics and Image Processing*, Saint-Etienne, France, 2000, pp. 300-304

[SchTes95a] K. Schlüns, M. Teschner. Fast separation of reflection components and its application in 3d shape recovery. *Proc. 3rd Color Imaging Conference*, Scottsdale, Arizona, 1995, pp. 48-51.

[SchTes95b] K. Schlüns, M. Teschner. Analysis of 2d color spaces for highlight elimination in 3d shape reconstruction. *Proc. Asian Conference on Computer Vision*, Vol. II, Singapore, 1995, pp. 801-805.

[Sha et al. 90] S.A. Shafer, T. Kanade, G.J. Klinker, C.L. Novak. Physics-based models for early vision by machine. *Proc. SPIE* **1250**, Perceiving, Measuring and Using Color, 1990, pp. 222-235.

[Sha85] S.A. Shafer. Using color to separate reflection components. *Color Research Applications* **10** (1985), pp. 210-218.

[Sha92] S.A. Shafer. Color interreflection. In: G. Healey, S.A. Shafer, L.B. Wolff (eds.). *Physics-Based Vision: Principles and Practice Color.* Jones and Bartlett, Boston, Massachusetts, 1992, pp. 349-350.

[SwaBal91] M.J. Swain, D.H. Ballard. Color indexing. *Int. J. of Computer Vision* **7** (1991), pp. 11-32.

[Tak et al. 96] K. Takebe, S. Nakauchi, S. Usui. A computational model for color constancy by separating reflectance and illumination edges within a scene. *Neural Networks* **9** (1996), pp. 1405-1415.

[Tak et al. 99] K. Takebe, N. Ito, S. Nakauchi, S. Usui. A digital system with color constancy from a couple of images. *Proc. Int. Conference on Systems, Man, and Cybernetics*, Tokyo, Japan, 1999, pp. 936-942.

[TomWan89] S. Tominaga, B.A. Wandell. Standard surface-reflectance model and illuminant estimation. *J. Optical Society of America A* **6** (1989), pp. 576-584.

[TomWan90] S. Tominaga, B.A. Wandell. Component estimation of surface spectral reflectance. *J. Optical Society of America A* **7** (1990), pp. 312-317.

[TsuOht90] M. Tsukada, Y. Ohta. An approach to color constancy using multiple images. *Proc. 3rd Int. Conference on Computer Vision*, Osaka, Japan, 1990, pp. 385-389.

[Wan87] B.A. Wandell. The synthesis and analysis of color images. *IEEE Transactions on Pattern Analysis and Machine Intelligence* **9** (1987), pp. 2-13.

[WolBol91] L.B. Wolff, T.E. Boult. Constraining object features using a polarization reflectance model. *IEEE Transactions on Pattern Analysis and Machine Intelligence* **13** (1991), pp. 635-657.

9 STATIC STEREO ANALYSIS IN COLOR IMAGES

Objects can be viewed from two different locations in order to determine their positions in a three-dimensional area on the basis of geometrical relations. The process of transforming an image pair into a three-dimensional representational form of a visible object surface area is named after the spatial vision of humans: *stereo vision*. Stereo analysis techniques are likewise important for digital image processing since they can be applied to a number of different tasks. This is especially true when techniques that work with a laser beam cannot be employed. Another advantage of stereo analysis techniques is that, except for a second camera, no additional special devices are needed.

In static stereo analysis, it is essential that no object or camera movements occur within the time interval of the acquisition of both stereo images. If it is guaranteed that both images are taken at exactly the same time, then this requirement is also fulfilled if the cameras are installed on a mobile robot, or if the objects in the scene move. The difference between static and dynamic stereo analysis (see Chapter 10) is that motion analysis is not part of static stereo vision. The goal of static stereo analysis is to determine depth information on the basis of geometrical relations.

Although very promising results have already been obtained with stereo analysis techniques, a substantially more robust solution is often necessary for its practical application. One possibility for developing such a procedure consists of the search for new mathematical techniques. Another consists of efficient and complete utilization of all available information. The second possibility covers, for example, the evaluation of color information contained in the scene for stereo analysis.

This procedure is quite plausible since red pixels in the left image do not correspond to blue pixels in the right image even when their intensity values are identical or similar. Experiments in color stereo analysis have shown that the quality of the produced results always improves when analyzing color information instead of gray-level information (see [BroYan89], [JorBov88], [JorBov91], [Kos93], and [Kos94]). Color images are taken with commercial color CCD cameras (see Section 4.1).

In principle, a stereo analysis technique consists of the following six processing steps:

1. *Image acquisition.* The result of this process is influenced above all by the resolution of the camera sensor, the scanning frequency used, and the technical characteristics of the lighting source(s).
2. *Camera calibration.* This concerns determining the inner geometry of the camera and the relations between the camera coordinates and world coordinates.
3. *Feature extraction.* Depending on the selected formula, significant image characteristics, such as edges and their orientation, are determined in the image.
4. *Correspondence analysis (stereo matching).* This is the process of automatic determination of corresponding elements in both images.
5. *Depth map estimation.* The corresponding depth value in the scene is calculated apart from two corresponding elements in the right and left image.
6. *Interpolation* or *approximation* of the visible surface(s).

The processing steps of image acquisition and colorimetric camera calibration were described in Chapter 4. The geometrical camera calibration is identical for the use of black-and-white and color cameras. Two geometrical calibration techniques for two different camera models are discussed in detail in [Kle et al. 98]. An overview of edge detection techniques in color images was presented in Chapter 6.

This chapter focuses on the process of correspondence analysis in stereo images by the evaluation of color information. The fundamentals necessary for representing this processing step are likewise identical with those for gray-level techniques. For a detailed representation please refer to [Kle et al. 98]. There several assumptions and *constraints* that are frequently used in stereo analysis procedures (in gray-level and in color images) are introduced. In this chapter, fundamentals are presented only to the extent they are necessary for understanding the correspondence analysis in color stereo images. These basics are also contained in [Kle et al. 98]. The basic geometry of the stereo image acquisition system is described in the following.

9.1 GEOMETRY OF A STEREO IMAGE ACQUISITION SYSTEM

Two cameras with identical effective focal length f are ordered such that the distance between their optical centers $\mathbf{O}_L$ and $\mathbf{O}_R$ is equal to b (see Fig. 9.1). The line between the two optical centers $\mathbf{O}_L$ and $\mathbf{O}_R$ is called the *base line* and the distance b *base distance*. Let the angle between the two optical axes be 2θ. The coordinate systems $X_L Y_L Z_L$ and $X_R Y_R Z_R$ for the left and right camera are defined as in Fig. 9.1. The coordinate system XYZ is defined such that the Z-axis exactly halves the angle between the Z_L-axis and the Z_R-axis. Coordinate system XYZ can be transformed into coordinate system $X_L Y_L Z_L$ through

a rotation by the angle θ clockwise (mathematically negative direction) about the Y-axis

$$\begin{pmatrix} \cos(\theta) & 0 & \sin(\theta) \\ 0 & 1 & 0 \\ -\sin(\theta) & 0 & \cos(\theta) \end{pmatrix}$$

and a translation by $b/2$ to the left

$$\begin{pmatrix} X - b/2 \\ Y \\ Z \end{pmatrix}.$$

Altogether the following thus applies:

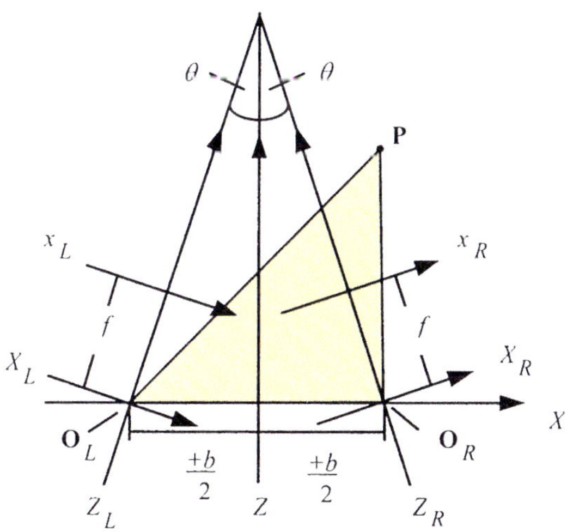

Figure 9.1. *Geometry of a static stereo system (after [Kle et al. 98]).*

$$\begin{pmatrix} X_L \\ Y_L \\ Z_L \end{pmatrix} = \begin{pmatrix} \cos(\theta) & 0 & \sin(\theta) \\ 0 & 1 & 0 \\ -\sin(\theta) & 0 & \cos(\theta) \end{pmatrix} \cdot \begin{pmatrix} X - b/2 \\ Y \\ Z \end{pmatrix}. \tag{9.1}$$

Analogously the coordinate system XYZ can be transformed into the coordinate system $X_R Y_R Z_R$. The following applies:

$$\begin{pmatrix} X_R \\ Y_R \\ Z_R \end{pmatrix} = \begin{pmatrix} \cos(\theta) & 0 & -\sin(\theta) \\ 0 & 1 & 0 \\ \sin(\theta) & 0 & \cos(\theta) \end{pmatrix} \cdot \begin{pmatrix} X + b/2 \\ Y \\ Z \end{pmatrix}. \tag{9.2}$$

A point **P** = (X, Y, Z) in 3D space is assumed to be represented in the image planes of both cameras at points (x_L, y_L) and (x_R, y_R). Assuming central projection, the relation between the points is defined by

$$x_L = \frac{f \cdot X_L}{Z_L}, \quad y_L = \frac{f \cdot Y_L}{Z_L} \quad \text{and}$$
$$x_R = \frac{f \cdot X_R}{Z_R}, \quad y_R = \frac{f \cdot Y_R}{Z_R}. \tag{9.3}$$

The Euclidian distance

$$\sqrt{(x_L - x_R)^2 + (y_L - y_R)^2}$$

between the corresponding pixels $\mathbf{p}_L = (x_L, y_L)$ in the left image and $\mathbf{p}_R = (x_R, y_R)$ in the right image is indicated as *disparity*. If two corresponding pixels (x_L, y_L) and (x_R, y_R) are determined in the left image and the right image, then the three-dimensional position of point (X, Y, Z) can be calculated from Eqs. (9.1), (9.2), and (9.3).

The search for corresponding pixels in the focal planes can be simplified substantially if knowledge of the underlying *epipolar geometry* of image acquisition is used. The terms *epipole, epipolar plane*, and *epipolar line* are introduced for the description of geometrical relations. The *epipolar surface* (or epipolar plane) is stretched by two straight lines. One line is determined by the ray of view that goes through the center of the lens and a point in the focal plane of one of the cameras. The other line, the base line, is, as already mentioned, represented by the connecting line between the centers of the lens of both cameras (see Fig. 9.2).

All object points in the scene that lie on this epipolar plane are represented in the second focal plane at points along a straight line, the *epipolar line*. It is the intersecting line between the epipolar plane and the focal plane of the second camera. The base line intersects in general both focal planes. These intersecting points are called *epipoles* and are the representations of the center of the lens of the other camera in the focal plane. The epipole of a focal plane lies likewise on the epipolar plane. Point **P** in the scene is represented in the left focal plane at point $\mathbf{p}_L$ and in the right focal plane at point $\mathbf{p}_R$ (see Fig. 9.2).

A special case for camera geometry is when angle θ, between the optical axes of the camera, is equal to zero (i.e., if the cameras are shifted parallel to each other and the focal planes are coplanar). In this case, $Z_L = Z_R = Z$ and the calculations of the depth value can be further simplified. For two corresponding points (x_L, y_L) and (x_R, y_R) in the left and right image respectively, the depth value Z is determined by

$$Z = \frac{f \cdot b}{x_L - x_R}.$$

Since the focal length f of both cameras and base distance b between the centers of their lenses are established values in the above-named equations, the depth is inversely proportional to the disparity of the x-coordinates $(x_L - x_R)$ in both images. Since this special ordering of the cameras leads to a substantial simplification of the calculations, it is used in a large number of stereo systems and is defined as *standard stereo geometry*. Under consideration of the geometrical relations represented above, stereo analysis can thus be reduced to the detection of corresponding pixels in the left and the right image.

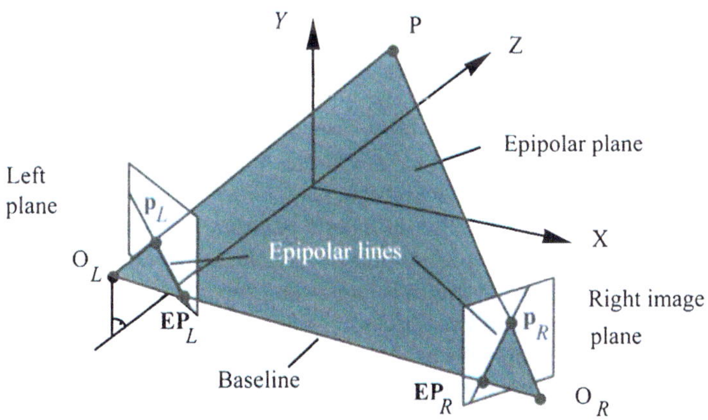

Figure 9.2. *Epipoles, epipolar plane, and epipolar lines.* $\mathbf{O}_L$ *and* $\mathbf{O}_R$ *are the optical centers of the binocular camera systems.* $\mathbf{p}_L$ *and* $\mathbf{p}_R$ *are the projections of pixel P into the respective focal planes.* $\mathbf{EP}_L$ *and* $\mathbf{EP}_R$ *denote the respective epipoles.*

9.2 AREA-BASED CORRESPONDENCE ANALYSIS

In correspondence analysis in stereo images, the assumption is often made that corresponding pixels in both images have similar intensity or color values (*photometric compatibility constraint*). This assumption is, however, not sufficient to establish a nonambiguous correspondence between pixels, since usually a large number of identical intensity or color values arise in both images. Therefore, several neighboring pixels in a window (e.g., of the size 5×5 or 8×8 pixels) are combined into one block. The relation between the pixels results from a similarity between the intensity or color values the pixels have within the blocks. This technique is referred to in the literature as *area-based stereo*.

This technique can be applied to all pixels of an image or only to selected pixels along extracted features, such as edges in an image. Thus, the techniques can be used with or without feature extraction. The fundamental difference to the feature-based technique, described later in Section 9.3, is that stereo matching is affected due to the correlation between the function values of the image functions within the selected working window. Therefore, this procedure is called *area-based correspondence analysis* in the following.

In the practical use of this technique, it may be noted that high-frequency components in the stereo images are especially susceptible to correlation errors. A preprocessing of the stereo images is therefore recommended for improving the results. Apart from that, the quality of the results depends on the similarity used, as well as the selection of a suitable window size.

In Section 9.2.2, an area-based, vector-valued technique is presented that produces a *dense disparity map* (i.e., a disparity value is determined for all pixels in the stereo image).

9.2.1 Dense Disparity Maps by Block Matching

A common technique for coding image sequences is based on the method of *motion analysis*. In this technique, a current image is determined from the temporally previous image by using motion vectors for the pixels. A motion vector describes the change of the position of a pixel between two temporally successive images (see Fig. 9.3a). Coding of the image sequence is achieved when only that motion vector is stored whose length is larger than zero. Thus, only those pixels of an image are viewed in which two successive images differ.

If the technique of motion analysis is used for correspondence analysis in stereo images, then the temporal change between the images corresponds to the difference in the views of both cameras. The motion vector is not determined, but rather the disparity between the two images. In using standard stereo geometry the disparities (see Section 9.1) must be determined only horizontally (i.e., in one row) (see Fig. 9.3b).

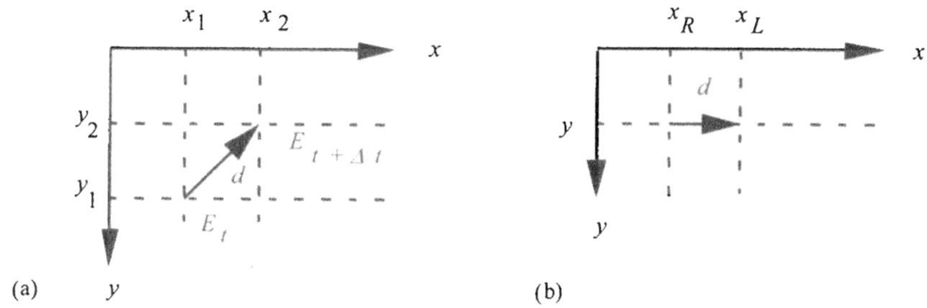

Figure 9.3. *(a) Motion vector of a pixel between the points of time t and t +Δt; (b) Representation of a disparity vector (after [Kle et al98]).*

While in motion analysis the search domain is generally small and the search direction is unknown, in determining disparity the search direction is known and the search domain is generally large. For this reason, software and hardware developed for motion analysis cannot as a rule be implemented for determining disparity in stereo images.

As with the technique described in Section 9.2.1, this technique is based on a similarity comparison of the distribution of gray and color values between two equal-sized blocks ($n \times m$ matrixes) in the left and right images (*area-based stereo*). The similarity comparisons are implemented here not only along the edges, but rather for all pixels of the stereo image. It is assumed that all pixels within a block have the same disparity value. Thus, only *one* disparity value must be determined for each block. This technique is called *block matching*.

The technique is first described for gray-level images and later expanded for application in color images. The technique consists of several steps. In the first processing step an image (e.g., the right image) is divided into a constant number of equal sized blocks. The search for a corresponding block in the left image is carried out only for the established blocks of the right image. As measurement for the similarity of the two blocks, the *MSE* (*mean square error*) between the intensity values of the pixel can be used within the corresponding blocks.

The intensity functions of the left and right images are again indicated as E_L and E_R. The similarity (or more exact: dissimilarity) is defined for an offset δ, which indicates the difference $(x_L - x_R)$ between the column positions in the right and left images, and a block size of $n \times m$ pixels by

$$MSE(x, y, \delta) = \frac{1}{n \cdot m} \sum_{i=0}^{n-1} \sum_{j=0}^{m-1} \left| E_R(x+i, y+j) - E_L(x+i+\delta, y+j) \right|^2 . \qquad (9.4)$$

Here (x, y) indicates in each case the upper left corner of a block in the right image (in accordance with Fig. 9.3). The disparity D between the blocks is defined by the distance between the positions (column difference) of the blocks that exhibit minimum deviation. In addition, the search domain is limited horizontally in the left image by a disparity limit d_{max} (i.e., a maximally permissible disparity value is established). Within the search domain the $n \times m$ -sized block is shifted pixel by pixel. The shift value δ, for which the *MSE* function assumes its minimum, determines the block disparity value D. However, the disparity value is by this definition only then clearly determined if the *MSE* function in the search domain has a clear minimum. In cases in which no clear minimum exists, an additional decision criterion is used.

Under the assumption that the disparity values of neighboring blocks differ only slightly, all disparity values for which the *MSE* function assumes a minimum are compared with the values of the neighboring blocks. The disparity with the smallest difference from the disparity of the neighboring block is selected.

Using the similarity described above, a correct correspondence analysis is to some degree only limitedly possible in image functions that contain regular textures or areas with small gray-level change (homogeneous regions). The selection of a suitable block size is of greater importance for the quality of the results. Experimental examinations have produced in this connection the best results using a square block size of $n = m = 8$ pixels.

The result produced using the block-matching technique is a disparity matrix, in which all blocks of fixed size have an identical value. By using a pixel selection technique, this result can be further refined so that a disparity value can be given for *each* pixel. This technique is comprised of three processing steps:

1. Use of the median operator on the disparity values of the blocks
2. Pixel selection
3. Use of the median operator on the disparity values that were determined for each individual pixel

The median operator determines the value in a set of values that would take the middle position in a sorted sequence of values. As a first step, the median operator is used on the block disparities within a 3×3 -block environment in order to eliminate individual outliers in disparities. The disparity for each pixel (x', y') of a block is subsequently determined using the disparity values of this and neighboring blocks.

For determining the disparity of an individual pixel at position (x', y'), the differences $DIFF(k)$ between the intensity value of the right image at (x', y') and the intensity values of the left image at positions $(x' + D(k), y')$ are formed for all disparities $D(k)$ $(1 \leq k \leq 9)$ from the 3×3 -block environment (see Fig. 9.4):

$$DIFF(k) = \left| E_R(x', y') - E_L(x' + D(k), y') \right| \quad \text{with } k = 1,...,9. \qquad (9.5)$$

The disparity value $DISP(x', y')$ is defined by that value $D(k)$ for which the difference $DIFF(k)$ assumed its minimum.

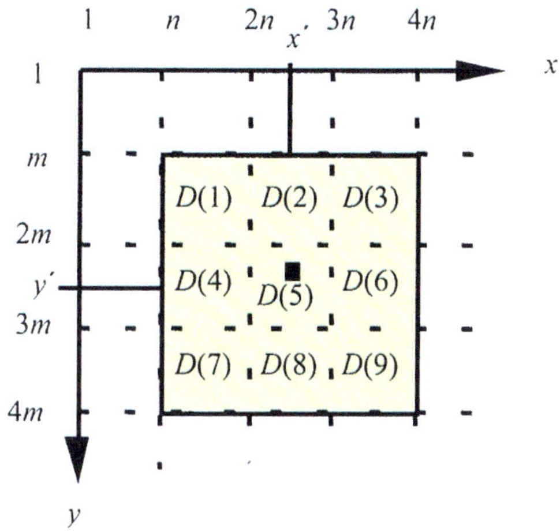

Figure 9.4. *Pixel selection of the disparity value at position (x', y') using block disparities D(k) in the 8-neighborhood.*

By using pixel selection on each pixel, a disparity matrix in original image size results, in which *each* matrix element contains a disparity value (*dense disparity map*). In conclusion, the median operator is used on the disparity values calculated by pixel selection.

When comparing and evaluating 10 techniques for correspondence analysis in gray-level stereo images, the block-matching technique turned out to be the best variant in comparison with the other examined techniques. The exactness of the calculated results was in this comparison always just as good as or better than with more time-consuming techniques [Kos92]. For this reason the block-matching technique, proven in the correspondence analysis in gray-level stereo images, is selected for an expansion to color stereo analysis. This expansion is described in the following.

9.2.2 Chromatic Block Matching for Color Stereo Analysis

In each technique of digital color image processing the selection of a suitable color space and coordinate system for the representation of color vectors is an important criterion. In the following, color stereo analysis in the *RGB* color space, *HSI* color space, and $I_1I_2I_3$ color space (see Chapter 3) is observed. However, in principle, color stereo analysis can occur in any color space. Here it should be taken into consideration that the analysis result could be influenced by the selection of the color space.

For an expansion of the block-matching technique to color stereo analysis the similarity criterion defined for gray-level images must be adjusted for color images. For this, the differences between intensity values are to be replaced by color differences in the respective color space. In Section 3.5 color distance measurements for the *RGB* and *HSI* color spaces were described. These color distance measurements can be implemented for the calculation of color differences. Four distance measurements for color stereo analysis are examined as an example in the *RGB* color space and $I_1I_2I_3$ color space. However, any other difference measurement can be used. The four selected distance measurements describe the difference between two color vectors $\mathbf{F}_1 = (u_1, u_2, u_3)^T$ and $\mathbf{F}_2 = (v_1, v_2, v_3)^T$ by

$$\Delta_1(\mathbf{F}_1, \mathbf{F}_2) = \sqrt{(u_1 - v_1)^2 + (u_2 - v_2)^2 + (u_3 - v_3)^2}\ ,$$

$$\Delta_2(\mathbf{F}_1, \mathbf{F}_2) = |u_1 - v_1| + |u_2 - v_2| + |u_3 - v_3|,$$

$$\Delta_3(\mathbf{F}_1, \mathbf{F}_2) = |u_1 - v_1|^2 + |u_2 - v_2|^2 + |u_3 - v_3|^2,\ \text{and}$$

$$\Delta_4(\mathbf{F}_1, \mathbf{F}_2) = \max\{|u_1 - v_1|, |u_2 - v_2|, |u_3 - v_3|\}.$$

These color distance measurements are used for the standardized color values (defined in accordance with Eq. (3.1)) as well as similarly for the nonstandardized color values.

If, for example, the *RGB* color space with the color distance measurement Δ_3 is selected, then the similarity of two blocks can be assessed by expanding the *MSE* function presented in Eq. (9.4). $\mathbf{C}_L$ and $\mathbf{C}_R$ denote the left and the right color image, respectively. The similarity MSE_{color} can be established for an offset δ, which gives the difference $(x_L - x_R)$ between the column positions in the right and left images, and a block size of $n \times m$ pixels by

$$MSE_{color}(x, y, \delta) = \frac{1}{n \cdot m} \sum_{i=0}^{n-1} \sum_{j=0}^{m-1} \Delta_3(\mathbf{C}_R(x+i, y+j), \mathbf{C}_L(x+i+\delta, y+j)) \tag{9.6}$$

$$= \frac{1}{n \cdot m} \sum_{i=0}^{n-1} \sum_{j=0}^{m-1} |r_R(x+i, y+j) - r_L(x+i+\delta, y+j)|^2$$
$$+ |g_R(x+i, y+j) - g_L(x+i+\delta, y+j)|^2\ .$$
$$+ |b_R(x+i, y+j) - b_L(x+i+\delta, y+j)|^2$$

Here (x, y) indicates again the upper left corner of a block in the right image (in accordance with Fig. 9.3). The other techniques for determining block disparities are carried out similarly to those of the gray-level technique.

For the evaluation of color information in pixel selection, the color distance measurement must likewise be taken into consideration. The disparity of a single pixel at position (x', y') is determined in which the differences $DIFF_{color}(k)$ between the color value of the right image at position (x', y') and the color values of the left image at positions $(x'+D(k), y')$ are formed for all disparities $D(k)$ $(1 \leq k \leq 9)$ from the 3×3 block neighborhood (see Fig. 9.4). During use of the distance measurement Δ_3, Eq. (9.5) changes to

$$DIFF_{color}(k) = \Delta_3\left(\mathbf{C}_R(x', y') - \mathbf{C}_L(x'+D(k), y')\right) \text{ with } k = 1,...,9. \quad (9.7)$$

The disparity value $DISP(x', y')$ is defined similarly to the gray-level variant by that value $D(k)$ for which the difference $DIFF_{color}(k)$ assumes its minimum. The dense disparity map is then completely produced if pixel selection is used on each pixel in the entire image.

If, for example, the *HSI* color space is selected instead of the *RGB* color space, then the color distance Δ_3 is to be replaced in Eqs. (9.6) and (9.7) by the color distance measurement Δ_{HSI}, which was presented in Section 3.5.2. The sequence of events for this technique can occur similarly to those of the *RGB* color space. The principal algorithm for the chromatic block-matching technique with pixel selection is presented in Fig. 9.5.

The degree of improvement of the matching results in evaluating color information in the images depends on the number of different colors in the image. If a stereo image contains only a few different colors or mostly achromatic colors, then only a small improvement is to be expected by the evaluation of color information. In contrast to this, the more varying the colors in the image, the better the improvement of the matching results.

For an evaluation of this technique, six real test images were generated. The color stereo images represent natural opaque objects of differing complexity and color. All possible configurations of the algorithm were implemented on these stereo images (altogether 875 tests; see [Kos93]). Using the left stereo images and the calculated dense disparity matrices, the right stereo images were reconstructed. The *displaced frame difference* between the reconstructed and the original right stereo images was subsequently calculated. Furthermore, the number M_{DIFF} of false disparity values and the mean error of the color values (mean value of the Euclidian distances between the original and the reconstructed color vectors) were calculated.

Since only the presence and not the absence of matching errors can be assessed from the evaluation of displaced frame difference, the actual disparity values were determined, in addition, manually as *ground truth* for a color stereo pair "Shawl" (see Fig. 9.6). Most false matches occurred in the image background, which was to be expected since the image background represents a homogeneous color distribution in the test images. In contrast, the best results occurred near the object edges in the image. The matching results were examined with regard to the use of different block sizes, color spaces, and color distances.

Block sizes: The five block sizes 4×4, 6×6, 8×8, 10×10, and 12×12 pixels were examined for the three color spaces and corresponding color distance measurement. The selection of the "best" block size depends on the image function and not the color space or color distance measurement. Some results of the examination are represented in Tables 9.1, 9.2, and 9.3. The details, with regard to the percentage of correct matches, refer to the area visible in both stereo images.

```
begin
  subdivide right image into blocks of size n×m;
  for (each block in right image) do begin                    {block matching}
    initialize min and D;
    set (x, y) according to left upper corner of the relevant block;
    for δ := 1 to d_max do begin
      calculate MSE_color(x,y,δ);
      if MSE_color(x,y,δ) < min then
        min := MSE_color(x,y,δ);   D := δ
      else
        if MSE_color(x,y,δ) = min then
          D := disparity with smallest difference to
          disparity value of neighboring block
        end {if}
      end {if};
      save block disparity value
    end {for}
  end {for};
  filter block disparity value with median operator;
  for (each pixel in right image) do begin                     {pixel selection}
    for k := 1 to 9 do begin calculate DIFF_color(k);
      determine value D for the DIFF_color(k) is minimum;
      DISPARITY (pixel position) := D
    end {for}
  end {for};
  filter matrix DISPARITY with median operator
end
```

Figure 9.5. *Pseudo code for the chromatic block-matching technique with pixel selection.*

Block Size	4×4	6×6	8×8	10×10	12×12
Mean Error	10.2	10.9	10.4	11.1	13.0

Table 9.1. *Mean error of color values in six reconstructed images using different block sizes.*

Block Size	4×4	6×6	8×8	10×10	12×12
Mean Error	6.1	4.9	3.5	3.7	5.8

Table 9.2. *Mean error of the disparity value in the image "Shawl" using different block sizes.*

Block Size	4×4	6×6	8×8	10×10	12×12
%	73.8	79.4	83.2	85.2	83.5

Table 9.3. *Percentage of correct matches in the image "Shawl" using different block sizes.*

Figure 9.6. *Gray-level representation of the color stereo pair "Shawl" (cp. also Fig. 5.4).*

In summary, it can be said that good results were always attained using blocks of size 8×8 pixels, while mean errors of the disparity values rose using blocks of a smaller size (4×4) or larger size (12×12) in this examination.

Color spaces: The algorithm was tested using the *RGB*, $I_1I_2I_3$, and *HSI* color spaces, as well as a gray-level representation of color images. The chromaticities, standardized by intensities, and the nonstandardized color values were examined in the *RGB* color space. The exactness of the matching results was always higher in the assessment of color information than in the evaluation of gray-level information. The mean errors of the color values and disparity values are reduced by about 20 - 25% by assessing color information. Using the (by intensity) standardized chromaticities, the accuracy of the matching results was lower than in gray-level images. The exactness of the matches calculated with the *HSI* color space laid approximately 0.3% below the accuracy attained with the *RGB* and $I_1I_2I_3$ color spaces. It is not yet clear whether this is due to the use of the *HSI* color space or the distance measurement Δ_{HSI}. In the results represented in Tables 9.4, 9.5, and 9.6, the accuracy in the $I_1I_2I_3$ color space lies somewhat above that in the *RGB* color space. In an examination of additional test images, however, the accuracy in the *RGB* color space lay somewhat above that in the $I_1I_2I_3$ color space. Thus, the accuracy of the mean can be seen as almost equal for both color spaces.

Color Distance Measurement: The color differences were determined in the *RGB* and $I_1I_2I_3$ color spaces by the distance measurements Δ_1, Δ_2, Δ_3, and Δ_4, as well as in the *HSI* color space by means of Δ_{HSI}. In addition, the distance measurements were examined for (by intensity) standardized and nonstandardized values in the *RGB* color space. In general, the mean errors of false disparities using standardized values were roughly twice as great as those using nonstandardized values. Apart from that, the selection of color distance measurement for nonstandardized values had no deciding influence in this investigation on the matching results in all color spaces. Tables 9.7, 9.8, and 9.9 show some results. For a reduction of calculation time, the use of the distance measurement Δ_3 is recommended in the *RGB* color space. Further investigations are necessary in order to support this recommendation.

Color Space	*RGB*	$I_1I_2I_3$	*HSI*	*rgb*	Gray Value
Mean Error	8.7	8.7	9.8	14.6	11.2

Table 9.4. *Mean error of color values in six reconstructed images using different color spaces.*

Color Space	*RGB*	$I_1I_2I_3$	*HSI*	*rgb*	Gray Value
Mean Error	4.3	4.2	.6	5.3	5.6

Table 9.5. *Mean error of disparity values in the image "Shawl" using different color spaces.*

Color Space	*RGB*	$I_1I_2I_3$	*HSI*	*rgb*	Gray Value
%	82.5	82.4	80.9	79.8	78.7

Table 9.6. *Percentage of correct matches in the image "Shawl" using different color spaces.*

Measurement	Δ_1	Δ_2	Δ_3	Δ_4	Δ_{HSI}
Mean Error	11.1	11.2	11.2	11.2	9.8

Table 9.7. *Mean error of color values in six reconstructed images using different color distance measurements.*

Measurement	Δ_1	Δ_2	Δ_3	Δ_4	Δ_{HSI}
Mean Error	4.9	4.9	4.8	4.8	4.6

Table 9.8. *Mean error of disparity values in the image "Shawl" using different color distance measurements.*

Measurement	Δ_1	Δ_2	Δ_3	Δ_4	Δ_{HSI}
%	80.9	80.9	81.3	80.9	80.9

Table 9.9. *Percentage of correct matches in the image "Shawl" using different color distance measurements.*

9.2.3 Hierarchical Block Matching in a Color Image Pyramid

A faster and more accurate correspondence analysis in color stereo images is attained by a hierarchical production of the chromatic block-matching technique with an image pyramid [Kos et al. 96]. The idea of using pyramid models in image analysis was proposed for edge detection by Tanimoto and Pavlidis [TanPav75]. The important characteristic of pyramid models is that they can be implemented efficiently [Kro96]. In the calculation of a *quad pyramid*, each level is determined by a reduction of the resolution by the factor four from the nearest lower level. The color values of the pixel are determined by calculating the mean values in each color channel.

It is noted that a color distortion appears by calculating the mean values in the color channels (see Section 5.3.2). However, this is not important for processing since only estimated values for the disparities are determined in the upper levels of the pyramid. The final disparity values for the original color images are calculated in the lowest level (here level number four). The example in Fig. 9.7 shows an image data pyramid of the stereo image "Andreas", presented later in Section 9.11, in five resolutions of 32×32 pixels to 512×512 pixels.

The chromatic block-matching technique presented in Section 9.2.2 is used first on the smallest image pair in order to attain the first estimates of the disparities in the images. These results are used as starting values $D(0)$ for determining the disparities in level (1).

By using a modified block-matching algorithm on the images in level $(s + 1)$, the disparities $D(s + 1)$ in level $(s + 1)$ can be calculated in accordance with the disparities $D(s)$ that were determined in the previous level. The search area for the disparity determination of each block in level $(s+1)$ is established by the disparity of the corresponding block in level (s) and a tolerance factor D_T. The tolerance factor determines the width and size of the reduced search interval $[D_{MIN}, D_{MAX}]$ and regulates the smoothness of the disparity map. The following applies:

$$D_\Delta(s) = 2^{(s-1)} \cdot D_T,$$

$$D_{MIN}(s) = \begin{cases} D(0) - D_\Delta(s) & \text{for } s = 1 \\ D_{MIN}(s-1) - D_\Delta(s-1) & \text{for } s > 1 \end{cases},$$

$$D_{MAX}(s) = \begin{cases} D(0) - D_\Delta(s) & \text{for } s = 1 \\ D_{MAX}(s-1) - D_\Delta(s-1) & \text{for } s > 1 \end{cases}.$$

If a lesser value for the tolerance factor D_T is determined, then the difference between the final disparities and the average disparities found in level (0) is very small. This corresponds to a small variation of disparities over the entire image. A larger tolerance factor produces a larger search area. This reduces the influence of the disparities on the current results determined in the previous level. Figure 9.8 illustrates the establishment of the search area with the tolerance factor $D_T = 3.0$.

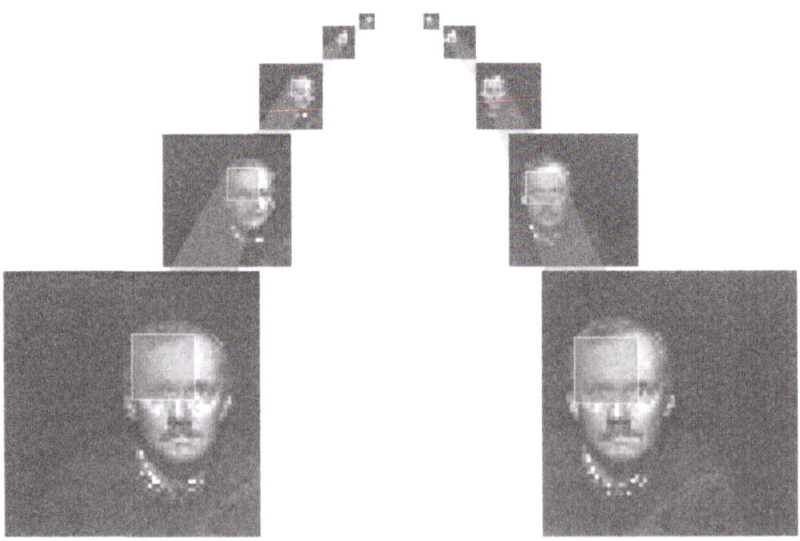

Figure 9.7. *Example of an image data pyramid.*

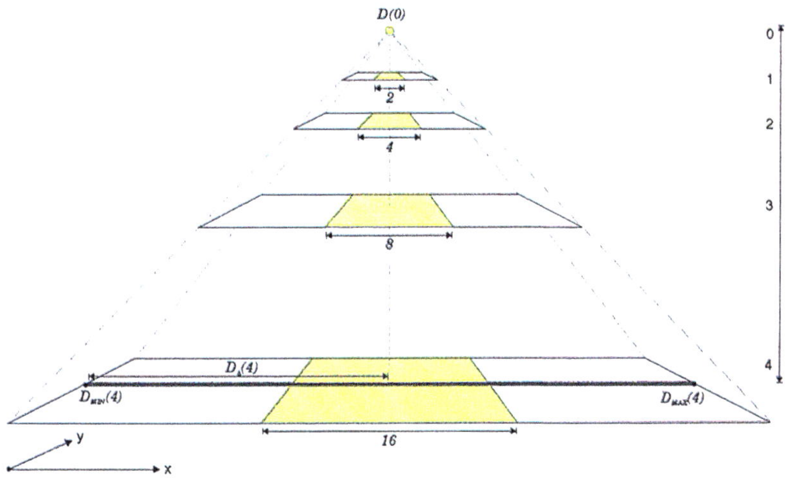

Figure 9.8. *Definition of a search area with the tolerance factor $D_T = 3.0$.*

Apart from modifying the search area in each level of the pyramid, the block-matching algorithm is used as described in Section 9.2.2. A comparison of the results of hierarchical and nonhierarchical variants of the block-matching technique has revealed that the hierarchical variant is more robust and more efficient than the nonhierarchical variant, and in general the average error (in pixels) is reduced by an additional 5% (see [Kos et al. 96] and [KosRod97]).

Nevertheless, the color variant of the block-matching technique requires more computation time and more memory as opposed to the gray-level technique. While the need for memory plays a subordinated role in falling memory costs, the calculation time can be decisive for using the technique. Here the decision for the faster versus the more exact technique must be made depending on the application. In vision system-supported navigation and three-dimensional object recognition, high demands are placed on mathematical accuracy as well as computation speed. For these application areas a parallelization of color stereo analysis algorithms is useful and necessary.

Due to the separate calculation of the disparity values for each image line, the block-matching technique can be well parallelized. A parallel algorithm for the block-matching technique and an examination of the change of the computing time of the algorithm using several processors (PUs) is given in [KosRod95].

As an example, the calculation times are given in Table 9.10 (without post-processing) for sequential and parallel implementations of the nonhierarchical and hierarchical block-matching techniques with 12 R8000 processors (75 MHz) on an SGI Power Challenge from Silicon Graphics. The techniques were used on color images of differing sizes, since the calculation times needed for the correspondence analysis do not linearly depend on the image size.

Image size in pixels	Nonhierarchical block matching		Hierarchical block matching	
	1 PU	10 PUs	1 PU	10 PUs
256 ×256	0.96 sec.	0.13 sec.	0.25 sec.	0.03 sec.
768 × 566	10.14 sec.	1.07 sec.	3.51 sec.	0.64 sec.

Table 9.10. *Calculation times for sequential and parallel implementations of the nonhierarchicl and hierarchical block-matching technique (without postprocessing).*

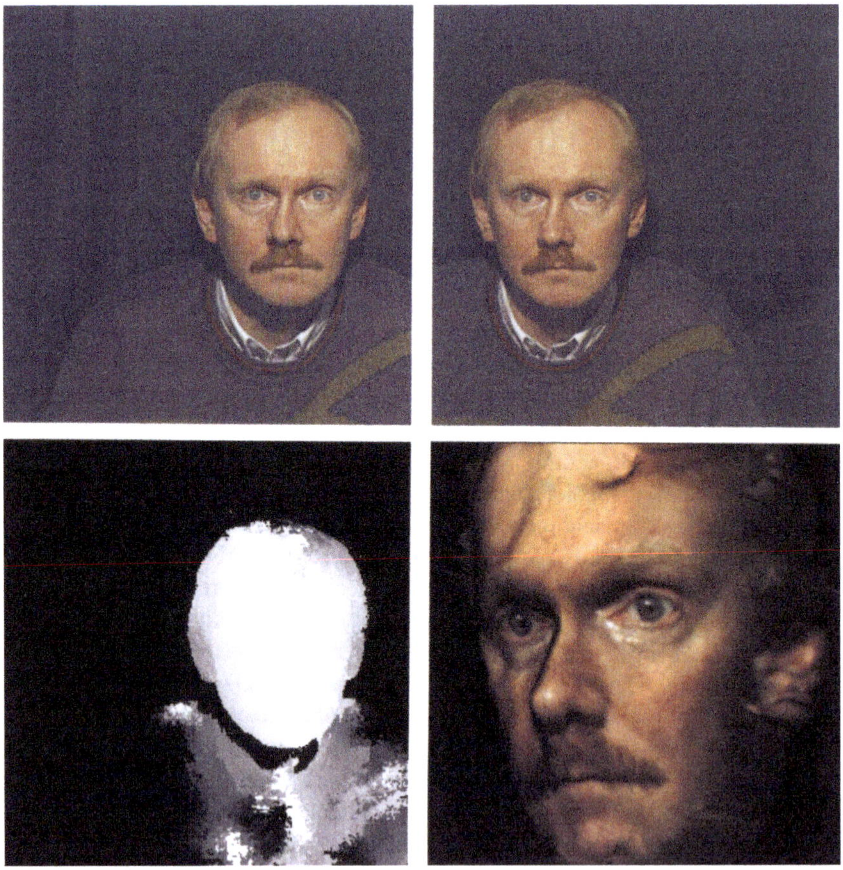

Figure 9.9. *Above: The color stereo pair "Andreas." Lower left: gray-level coded representation of the dense disparity map for the stereo image "Andreas" calculated using the color variant of the block-matching technique with pixel selection. Lower right: Face reconstructed using this dense disparity map overlaid with texture (reprinted from [Kle et al. 96] with permission from Vieweg).*

For a color image of the size 256×256 pixels, processing in video real time (24 images per second) is nearly achieved with the hierarchical block-matching technique using 10 processors. In Fig. 9.9 the dense disparity map for a portrait, determined using the color variant of the block-matching technique, is presented as an example. The representation shows, in addition, the color stereo image and the reconstruction of the scene overlaid with texture. By including color information in the matching process, the (if also imperfect) reconstruction can be substantially improved.

Most of the errors in determining disparity arise in homogeneous image areas. There, no clear matching between the pixels can occur from the image values. These ambiguities can be partially resolved if a colored pattern is projected into the scene. By overlay with a colored pattern, the object surfaces obtain a texture identifiable (for the most part) in both images. This variant is described in the following section.

In conclusion, it should also be mentioned that other techniques for calculating dense disparity maps from color stereo images were implemented. These are, however, in general very time consuming. In one case, for example, the technique of "simulated annealing" is used for optimization [JorBov92]. In another case, for example, a statistical criterion and an iterative technique with a convergence speed not predictable in advance were used [Oku et al. 92]. In contrast to this, the hierarchical variant of the chromatic block-matching technique presented in this section illustrates a very efficient technique for calculating dense disparity maps from color stereo images.

9.2.4 Stereo Analysis with Color Pattern Projection

Most stereo analysis techniques cannot calculate correct dense disparity maps within homogeneous image areas. From investigations it is known [Vri et al. 92] that the variability of objects can be improved by a suitable regulation of the illuminating color. Thus, the search for an "optimal" illuminating color for the recognition of objects in a structured environment [MurNay94] is a promising, yet very difficult, task. Fortunately, no optimal illuminating color need be found to improve the matching results in stereo images.

Kanade and his colleagues [Kan et al. 95] project a sinus-shaped varying gray-level pattern into the scene to be treated. By this, they do achieve an improvement of the matching results, but because of the limited dynamic areas of the cameras, they still get many false matches. These mismatches arise predominantly on dark object surfaces where the projected gray-level pattern shows only a small relative lightness contrast (i.e., where a dark pattern is projected onto a dark surface). Yet if a color-coded pattern is projected into the scene, then the results can be substantially improved. The technique of *stereo analysis with color pattern projection* consists of projecting a color pattern onto the objects to be examined and subsequent use of a technique of static stereo analysis.

By overlaying the projected color-coded pattern with the unknown object colors, mixed colors can result in the camera image, in which an identification of the original color of the coded pattern is no longer possible. The mixed colors produced by overlaying are identically visible in both camera images. By a correspondence analysis in both images, the color strips (most of them) can be assigned from the visible colors in the images without previous knowledge of the object colors or the projected color pattern being needed. In Fig. 9.10, the principal structure of a stereo analysis system with color pattern projection is represented.

The selection of the projected color pattern is of great importance for the clearest possible matching of the pixels in both images. In principle, the color pattern can be generated either such that between neighboring strips the greatest possible difference exists (*discontinuous illumination*), or that between the strips continuous transitions arise. In Fig. 9.11 the course of the values of a color vector component is represented schematically within a subpattern with discontinuous or continuous illumination. In discontinuous illumination and uniform distance change in the scene, the color strips can be especially well differentiated in the camera image. However, at depth discontinuities in the scene (i.e., at object edges) discontinuous illumination can lead to a small color difference between neighboring strips in the image. In contrast, in continuous illumination the color differences between neighboring strips in the image are generally small; however, at object edges they are just as large (see [KnoSas95]).

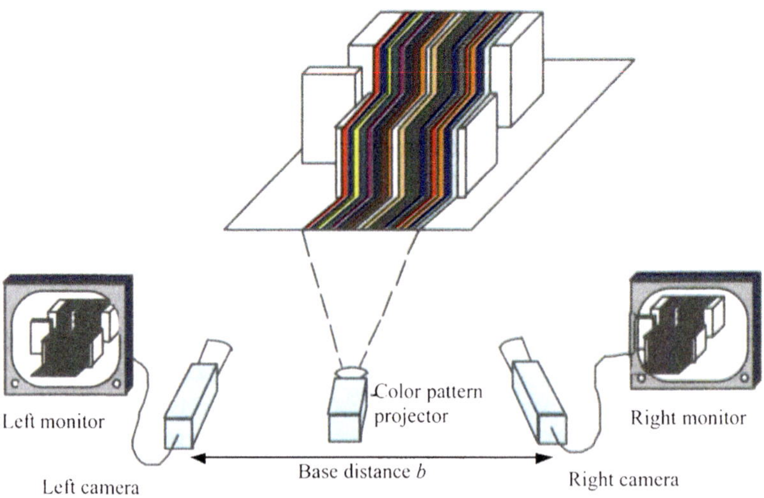

Figure 9.10. *Principal structure of a stereo analysis with color pattern projection.*

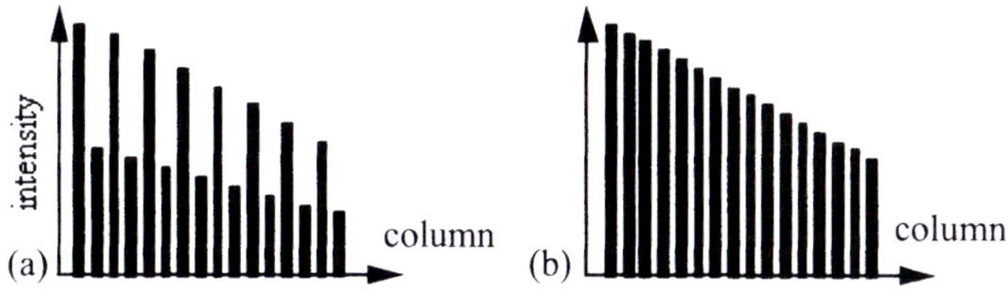

Figure 9.11. *Intensity distributions for contrasting illumination (a) and continuous illumination (b) (reprinted from [Kle et al. 96] with permission from Vieweg).*

The greater the color distance between neighboring color strips in the images, the easier it is to determine the corresponding pixels in both images. The small color distance with continuous illumination is thus a principal disadvantage. In order to be able to achieve optimal results with "discontinuously" produced color strips, the projected strips must correspond exactly to one pixel width in the camera image, which is practically impossible. Therefore, either several strips on one pixel or one strip on several pixels is reproduced, which in both cases leads to a reduction of the color difference between neighboring strips in the image.

Furthermore, a slide projector is employed as a rule for the projection of the color strip pattern into the scene, which can be focused only for a determined distance area. Through this, color strips lying outside this area overlap and are unclearly represented, which likewise leads to a reduction of the color difference between neighboring strips in the image. Naturally the problem of overlapping of light level arises with continuous illumination. However, this effect can be reduced to a great extent for continuous illumination by an image smoothing, while an image smoothing with discontinuous illumination can lead to a significant change of the color code.

The problem of overlapping can be solved in principle with discontinuous illumination by using a vector median operator (see Section 5.3.2) in the three-dimensional color space. This topic will not be discussed here. The generation of a color strip pattern with continuous illumination is recommended due to the simpler handling.

Based on these preexaminations a color pattern in the *RGB* space can be generated whereby no limits of creativity are set for color selection. For example, only one of the respective three components of the color vector can be set not equal to zero in a strip, which can lead to a very small relative lightness contrast between neighboring strips. Two components of the color vectors can be varied and the third component is set at zero.

Another possibility consists of setting the three components of the color vector in such a relation that the sum of the values established in the individual vector components results in a predetermined size. The last two color codes enable a relatively simple color correction of images with the differing spectral sensitivity of both cameras.

Using qualitative high-quality three-chip-CCD color cameras a subsequent color correction is as a rule not necessary. Thus, all three components of the color vector can be varied at the same time in order to further increase the ability of neighboring strips to be differentiated. For example, the values in the color components can be modeled by sawtooth or sine functions. Figure 9.12 shows the diagram of a subpattern within which the values of the three color components are determined by sine functions, each shifted by $2/3\pi$. The entire color code is formed by periodical arrangement of slightly different subpatterns.

The color spectrum image S_{RGB} with the three vector components S_R, S_G, and S_B in the *RGB* color space can be produced with the equations

$$S_R = \sin\left(\frac{1}{n} \cdot \pi\right) \cdot \left(\frac{G_{\max}}{2} - 1\right) + \frac{G_{\max}}{2},$$

$$S_G = \sin\left(\left(\frac{2}{3} + \frac{i}{n}\right) \cdot \pi\right) \cdot \left(\frac{G_{\max}}{2} - 1\right) + \frac{G_{\max}}{2}, \quad \text{for } i \in \{0, \cdots, n\},$$

$$S_B = \sin\left(\left(\frac{4}{3} + \frac{i}{n}\right) \cdot \pi\right) \cdot \left(\frac{G_{\max}}{2} - 1\right) + \frac{G_{\max}}{2}.$$

With this, i indicates the column position of the color vector in the color spectrum image and $G_{\max} + 1$ denotes the maximum permissible value in each component of the color vector.

The color strip pattern is projected into the scene and a stereo image is produced from two different positions. For the improvement of the color quality in both images a white balance for both cameras should be carried out before beginning the acquisition (see Section 4.5.3). Using standard stereo geometry (see Section 9.1), the correspondence analysis is simplified considerably since corresponding pixels lie on both lines in both images. In order to attain the best possible depth values and to estimate the color code as well as possible, a stereo analysis technique that well analyzes the available color information and calculates dense disparity maps should be used.

One technique that fulfills the requirements listed above is, for example, the block-matching technique for color stereo analysis described in Section 9.2.3. The expansion to the procedures described there consists merely in the fact that (e.g., by a slide projector) a color strip pattern is projected onto the object to be reconstructed. The technique described in Section 9.2.3 can be subsequently used (without modification) on the stereo images taken from two different positions.

In the following, the improvement to be achieved by projecting a colored pattern onto the object surface during surface reconstruction is illustrated using two examples.

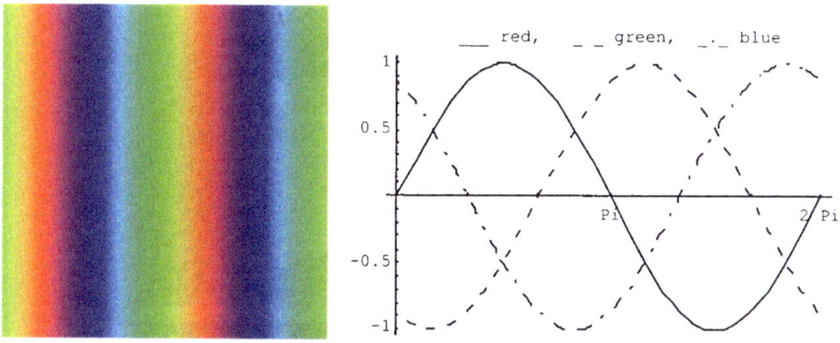

Figure 9.12. *Sketch of the intensities of the three color components in a subpattern (S_{RGB}) (reprinted from [Kle et al. 96] with permission from Vieweg).*

The difference is most clearly recognizable when the surfaces of an object in the scene, such as the surfaces of the synthetically generated cube in Section 9.15, are monotone and homogeneous. In this case, the dense disparity map determined by the block-matching technique contains many false disparity values (see Fig. 9.13, lower left) due to the many ambiguities in the image values. In contrast, these ambiguities can be generally solved by overlay with a projected color pattern and determining more accurate disparity values (see Fig. 9.13, lower right).

At the top of Fig. 9.15, a gray-level representation without overlapping color code of a stereo image pair "Tom" is presented, and in the center the same scene with overlapping of a color code is presented. The dense disparity maps calculated by the block-matching technique are visualized for both stereo image pairs in Fig. 9.15e and f. Errors in the disparity map calculated for the stereo image without overlapping color code are visible, for example, in the head and feet areas of the figure. For better distinction of both disparity maps, two sections from each are shown enlarged. The errors are shown dark in the lefthand example in Fig. 9.14. The right example in Fig. 9.14 shows a shaded representation of the reconstructed scene (from the disparity map generated with projected color code).

In conclusion, it should be indicated that first, in stereo analysis with color pattern projection, no knowledge of the projected color code is needed, and second, with the exception of very dark or black surfaces, no further restrictions are placed on the object colors. Stereo analysis techniques have difficulties in general in determining correct dense disparity maps in homogeneous image areas. This problem can be remedied considerably by projecting the color strip pattern into the scene, since homogeneous object surfaces can be distinguished as textured color areas in the camera images. Another advantage of stereo analysis with color pattern projection is that only one single image pair needs to be taken. Therefore, short processing times are possible and the technique can also be used with moving or nonrigid objects.

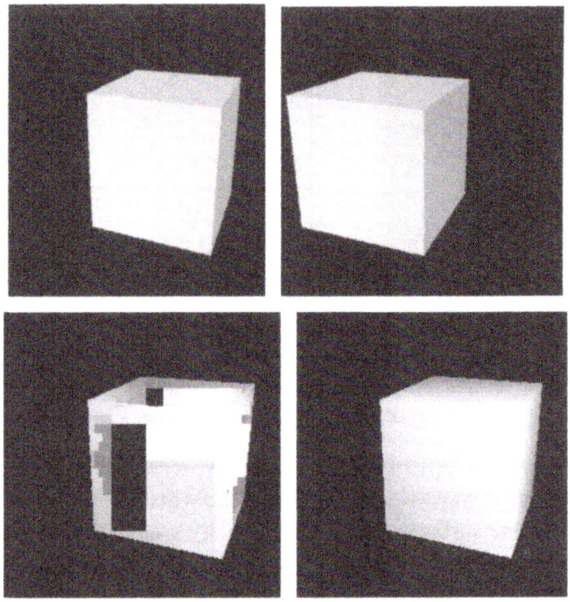

Figure 9.13. *A synthetic stereo image "Cube" (above) and the disparity maps that were calculated by the block-matching technique without projected color pattern (lower left) and with projected color pattern (lower right).*

Figure 9.14. *Left: Difference (scaled) between the two disparity maps in Fig. 9.14. Right: Shaded representation of the scene "Tom" reconstructed with projected color code (reprinted from [Kle et al. 96] with permission from Vieweg).*

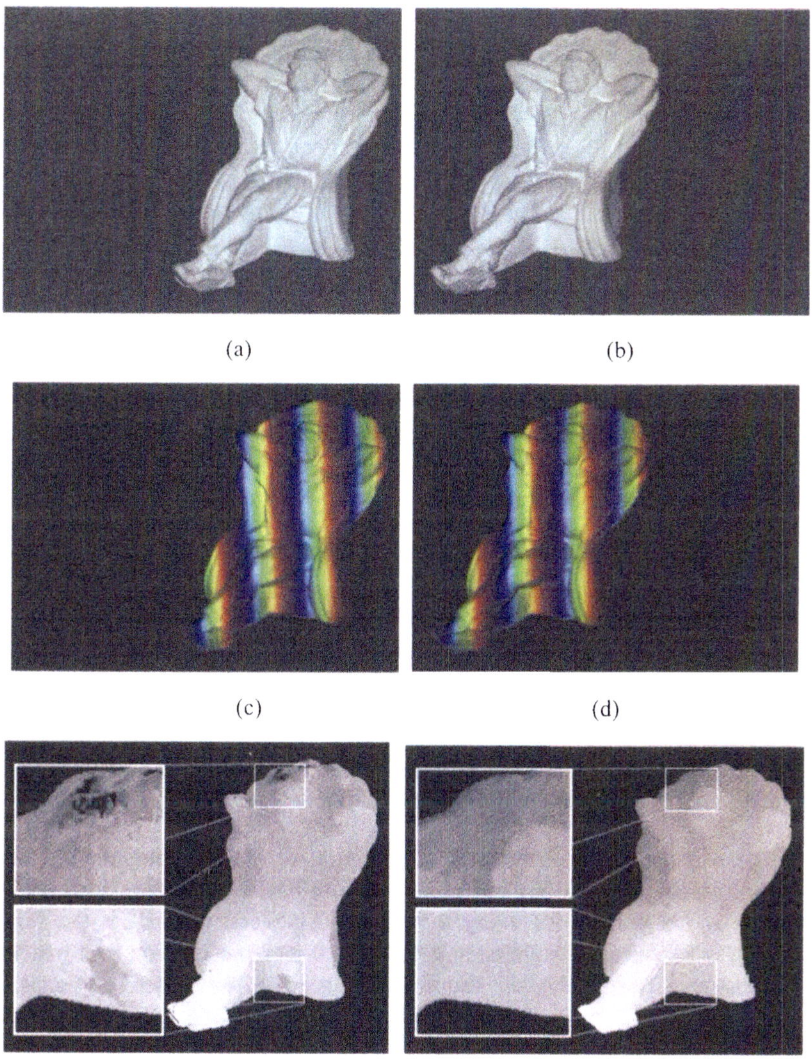

(a) (b) (c) (d) (e) (f)

Figure 9.15. *(a) Left and b) right color stereo image "Tom." (c) and (d) Stereo image with projection of a color code. (e) disparity map calculated without projected color code; (f) disparity map calculated with projected color code (each with enlarged sections, reprinted from [Kle et al. 96] with permission from Vieweg).*

9.3 FEATURE-BASED CORRESPONDENCE ANALYSIS

In the previous sections techniques for determining dense disparity maps from color stereo images were introduced. These dense disparity maps are necessary, for example, for the most accurate shape analysis and surface reconstruction. However, for many other applications (e.g., in robotics) the knowledge of a few distance values is sufficient for solving the problem. In this case a dense disparity map does not absolutely have to be determined. It suffices, as in the case of the technique presented in Section 9.2.1, to determine only some selected disparity values.

If correspondence analysis in stereo images is not carried out between the intensity or color values of the image functions but rather between selected image features, then this technique is indicated as *feature-based correspondence analysis* (*feature-based stereo*). Especially distinctive sections of the image, such as edges or pixels along edges, are referred to as *features*. The matching occurs, as a rule, due to selected characteristics of these features, such as on the basis of orientation or the length of the edges. Feature-based techniques have the following advantages over area-based techniques:

1. Ambiguities in correspondence analysis using feature-based techniques are considerably less than with area-based techniques, since the number of possible candidates is considerably less for a correspondence.
2. Stereo matchings are less sensitive in relation to photometric variations occurring during image generation, since features represent significant details of the scene.
3. Often, the determination of disparities can occur substantially more accurately since the position of the features can be calculated with subpixel accuracy (i.e., more accurately than the camera resolution).

Regarding the first characteristic, reduction of required processing time is often the reason given for implementing a feature-based technique. However, this statement is not correct for every technique. In several techniques the feature extraction is very time consuming and even much more costly than later matching of the features. By reducing the features to be matched in the stereo images, the correspondence analysis is indeed considerably simplified and thus, as a rule, the time needed for matching is also reduced. From this it cannot be concluded for every case that the entire stereo analysis achieved using a feature-based correspondence analysis is faster than using an area-based correspondence analysis.

The second characteristic of a feature-based technique is of great importance for correspondence analysis. Photometric variations during image generation directly influence the result of an area-based technique. In contrast, they have only a small influence on the physical origin of features and thus also on the results of a feature-based technique.

The third advantage describes the possibility of a considerable increase in accuracy of disparity determination. This can substantially more precisely calculate surface object points in a three-dimensional space. If, in addition to that, the geometry of three-dimensional object surfaces should be described as accurately as possible, then a calculation with subpixel accuracy is absolutely necessary. This accuracy can be achieved only with certain area-based techniques.

Next is shown how the results of edge-based stereo analysis can be improved by evaluating color information.

9.3.1 Edge-Based Correspondence Analysis

Few techniques exist for an edge-based correspondence analysis in color stereo images. This can lie first in the fact that an edge generally contains no color information, or second, that edge-based techniques are frequently selected for reasons of reducing computation time. As a rule, the analysis of color images needs, more computation time than a corresponding analysis of gray-level images. Furthermore, due to the correlation between the color channels, at least 90% of the edges are identical in gray-level and color images (see Chapter 6). Exceptions arise when neighboring objects in the scene demonstrate the same brightness but differing hues. In these cases no edges can be detected in the gray-level image. By using a "good" color edge finder, the number of edges found can increase up to 10%. Since only edges that were detected can be matched, a certain improvement of the results is to be expected.

In addition to this quantitative argument, it should now be shown how the ambiguities during correspondence analysis in stereo images are reduced by using color information in the edge-matching process. Thus, the false matches of edges that do not correspond to each other can be reduced. Up to now, only monochromatic-based techniques have been published for this. In order to clarify the still widely held way of thinking about monochromatic-based techniques, the following citation is presented: "Perhaps the best way to extract chromatic edges is to detect zero crossings from LoG-filtered chromatic images" [JorBov91, pp. 101]. The "chromatic" images show here the individual components of the color signal. Here a rethinking of vector-valued color signals is necessary. Techniques in edge-based correspondence analysis that, for example, include the eigenvalues or eigenvectors of the derivatives of vector-valued color stereo images in the matching process do not exist to our knowledge. So far only monochromatic-based techniques have been published, and these shall be explained in the following. An example of a stereo analysis technique proposed by Jordan and Bovik [JorBov88] is outlined for this method.

For simplification of stereo analysis, a camera arrangement is assumed in the image acquisition according to the standard stereo geometry. Due to the position of the epipolar lines, the correspondence search can again be limited to a line-by-line search. The zero crossings in both LoG-filtered gray-level stereo images (with

$\sigma = 1.41$, see Section 6.2) form the features for edge-based stereo matching. In addition, a maximum-appearing disparity value (disparity limit) of 20% of the image width is assumed.

Starting from a zero crossing in the left image, the following three criteria in the fundamental matching algorithm (for gray-level stereo images) are used for the selection of a matching candidate in the right image:

1. The zero crossing in the right image lies within the search area
2. It has the same contrast conditions
3. It has approximately the same orientation (±30°), as the zero crossing in the left image.

After a set of possible matching candidates are determined for all zero crossings in the left image with the above-named algorithm, all those matching pairs are chosen in which a noncontradicting definite match is possible. These definite combinations are matched and all other pairings are rejected. Here it should be noted that a definite match with the above-named criteria is not also compulsorily a correct match. In order to reduce the number of false positive matches, color information should be made full use of. This can be included, for example, in the matching algorithm as follows:

1. The zero crossings are determined individually for each component of the color vectors, and the gray-level algorithm is carried out separately for each component.
2. Color information is used for characterizing the zero crossings in the gray-level image.

The first steps correspond principally to those from Section 9.2.1. It is very numerically intensive since the matching must be performed several times. The second step is more efficient. Here color information shall be characterized on one zero crossing.

For a description of this kind, in which color information varies at the position of a zero crossing in the LoG-filtered gray-level image, Jordan and Bovik [JorBov88] suggest using three standardized displaced frame difference images: red-minus-green (D_{rg}), green-minus-blue (D_{gb}), and blue-minus-red (D_{br}). Each single (monochrome) displaced frame difference image originates by calculating the differences of values in the selected vector components of the color image. If, for example, the standardized red-minus-green displaced frame difference image is observed with

$$D_{rg}(x,y) = r(x,y) - g(x,y) = \frac{R(x,y) - G(x,y)}{R(x,y) + G(x,y) + B(x,y)},$$

then it is clear that the gradient of this displaced frame difference image describes at each zero crossing the way and direction in which this difference varies most strongly. Thus, the sign and orientation are used as attributes of the gradient. The sign of the direction variation along the horizontal line defines the sign of the gradient of the smoothed displaced frame difference image. The following applies:

$$\frac{\delta}{\delta x}(GAUSS * D_{rg}) = \left(\frac{\delta}{\delta x} GAUSS\right) * D_{rg},$$

whereby $GAUSS(x, y)$ indicates a two-dimensional Gaussian function and $*$ the convolution operation. For a given pixel, a positive sign of the gradient signals a relative increase of values in the "red" component as opposed to the values in the "green" component (from left to right), while a negative sign signals a reduction relating to this. The orientation of the gradient is given for the smoothed difference image by

$$\tan^{-1}\left(\frac{\frac{\delta}{\delta y}(GAUSS * D_{rg})}{\frac{\delta}{\delta x}(GAUSS * D_{rg})}\right).$$

The definitions can be determined similarly for the green-minus-blue and blue-minus-red difference images.

Based on the above-named ideas, the matching algorithm can be expanded for an additional criterion. For a zero crossing in the left image, in the matching algorithm (for color stereo images) the following four criteria are used for the selection of a matching candidate in the right image:

1. The zero crossing in the right image lies within the search area
2. It has the same contrast sign
3. It has approximately the same orientation as the zero crossing in the left image (±30°)
4. The signs of the gradients are equal for each standardized difference image in both candidates.

Otherwise, this is similar to the steps used in the matching algorithm for gray-level images (i.e., the definite matches are selected and declared valid). In three selected test images additional use of the fourth criterion resulted in a moderate increase in the number of correct matches of roughly 5% as opposed to the gray-level algorithm according to [JorBov88]. This relatively small change results, above all, from the simple matching pattern, with which only a certain number of edge pixels in stereo images can be matched.

Jordan and Bovik attained considerably better results in a further investigation [JorBov91]. Instead of viewing the difference images, they proposed a monochromatic-based expansion from the disparity gradient limit for color images [Pol et al. 85]. The conditions to be met, which Jordan and Bovik call *chromatic gradient constraints*, are determined for two gray-level images and two color channels belonging to them by

$$
\begin{aligned}
&\left| E_{L_x}\left(x - \frac{\delta - b}{2}, y \right) - E_{R_x}\left(x + \frac{\delta + b}{2}, y \right) \right| \\
&\leq \frac{1}{2} \cdot \left| E_{L_x}\left(x - \frac{\delta - b}{2}, y \right) + E_{R_x}\left(x + \frac{\delta + b}{2}, y \right) \right| + 2\sqrt{2}\sigma_x
\end{aligned}
\tag{9.8}
$$

and

$$
\begin{aligned}
&\left| E_{L_y}\left(x - \frac{\delta - b}{2}, y \right) - E_{R_y}\left(x + \frac{\delta + b}{2}, y \right) \right| \\
&\leq \frac{1}{2} \cdot \left| E_{L_x}\left(x - \frac{\delta - b}{2}, y \right) + E_{R_x}\left(x + \frac{\delta + b}{2}, y \right) \right| + 2\sqrt{2}\sigma_y
\end{aligned}
\tag{9.9}
$$

Here E_{L_x} and E_{R_y} indicate the partial derivatives with respect to x of the left and right stereo images, respectively; E_{L_y} and E_{R_y} the partial derivatives with respect to y of the left and right stereo images respectively; δ a value that assumes the column difference (the disparity); b the base line; and σ_x and σ_y the standard derivatives for an additive noise in the gray-level images derived from x and y. The above-given criterion 4 is replaced in [JorBov91] by the criterion $4'$:

$4'$. The matching candidates meet inequalities (9.8) and (9.9) in each component of the color signal.

Using the additional criterion $4'$ instead of the gray-level algorithm, some percentages of changes in the results are presented in Table 9.11 from [JorBov91]. The large increase of on average about 240% more correct matches, using criterion $4'$ as opposed to the gray-level algorithm, can be largely attributed to the fact that only very small edge pixels can be definitely and correctly matched with the gray-level algorithm.

Measured Quantity	Percentage Change
Total number of matching candidates	- 28.5
Number of definite matches	+191.3
Percentage number of correct matches	+246.3

Table 9.11. *Percentage changes of results using additional criterion $4'$ instead of the gray-level algorithm gathered from four color stereo images according to [JorBov91].*

Another monochromatic-based procedure for including color information in an edge-based correspondence analysis in stereo images was proposed by Jarvis [Jar95]. He used a structure with four coplanar arranged cameras whose optical axes are aligned parallel to each other. For feature extraction, the Marr-Hildreth operator [MarHil80] is used separately on the individual color channels, and the results are combined. An explanation of how this combination is implemented is not presented in [Jar95].

For the matching of the edges, the color values are viewed each time only on one side of the edge (*onesidedness*), since on the basis of differing observer positions of the cameras, areas along the object edges must not be uniformly visible in all images. However, it is very probable that a possible covering exists only on one side of the edge. The values in the individual color components are inspected for similarity separately on the left and right of the possibly corresponding edges along the scanlines in the images. Instead of the individual inspection in the vector components, a color distance measurement could also have been employed here for the color vectors.

Grimson and his colleagues [Gri et al. 94] use edge-based stereo analysis by means of an active stereo system in model-based recognition and localization of a colored object in the scene. In the model description of the object to be searched, geometry information as well as mean values and standard derivatives in the individual components in the *HSV* color space are noted for the object surfaces. For the localization process, intensity edges in the images are first determined. Areas in the color stereo images that demonstrate similar color values are subsequently searched. Only edges that lie in one of the image areas presegmented in this manner are examined in edge-based stereo analysis. Color information is made full use of here for a rough preselection of the edges to be matched for a known object.

In conclusion, it should be noted that a feature-based correspondence analysis in color stereo images must not be inevitably implemented for edges. Also conceivable are, for example, matches between regions that represent surfaces with similar characteristics. Nguyen and Cohen [NguCoh92] propose for this a monochromatic-based technique in which they apply a gray-level algorithm separately to the components of the color stereo images and subsequently combine the results. A vector-valued approach could consist of determining cluster centers or vector medians as features for color vectors belonging to one region in the stereo images and then matching these on the basis of similarity. A principal problem in region-based matching is that for each matched feature (i.e., for each matched region) only one single disparity value can be calculated. Thus, only approximate depth values can be determined for the pixels of a region. A subsequent interpolation of the disparity or depth values is difficult and hardly possible since, as a rule, no knowledge of the position of the object pixels, represented by a region in the image, is present in the scene.

9.3.2 General Ideas

In the previous section it was shown that evaluating color information could reduce ambiguities in edge-based correspondence analysis in stereo images. Furthermore, by using a good color edge finder (e.g., the Cumani operator, see Section 6.1.2), the number of edges found in the color image is increased by up to 10%. Since only edges that were detected can be matched, this can mean a definite improvement of results. For a subsequent surface reconstruction, the errors of a single reconstructed object edge can have substantial consequences.

One challenge and at the same time new possibility of color image processing is an at least partial classification of image value edges from their physical causes (see Section 6.3). This is especially important for an edge-based correspondence analysis in stereo images [Kos97]. Image value edges can be subdivided into orientation edges, reflection edges, illumination edges, highlight edges, and occlusion edges. In static stereo analysis only orientation edges, reflection edges, and illumination edges may be taken into consideration. Highlight edges and occlusion edges may not be matched since they depend on observer position. The example in Fig. 9.16 clarifies that occlusion edges and highlight edges do not represent the same physical location in the scene. In addition, no illumination edges may be matched with dynamic stereo analysis (see Chapter 10).

So far a (partially) automatic classification of image value edges in color images is possible only under certain preconditions and only at great computational cost (see Section 6.3 and Chapter 8). If, for example, it is known that the objects in the scene consist of nonhomogeneous dielectric materials, then highlight areas in stereo images can be detected with the technique for spectral differencing (see Section 8.1.5) or can be eliminated with a technique for highlight

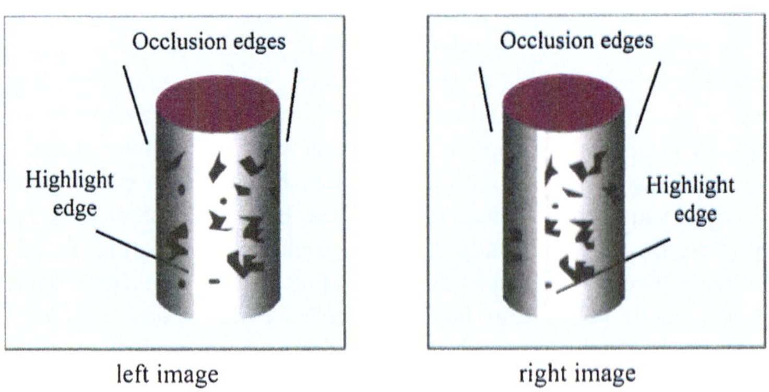

Figure 9.16. *Example of observer location dependent cover and highlight edges in stereo images (adopted from [Kle et al. 96]).*

analysis, for example, the Klinker-Shafer-Kanade technique (see Section 8.1.1) or the Schlüns-Teschner technique (see Section 8.1.4). Further advances in the classification of edges can be directly used for improving an edge-based correspondence analysis in color stereo images.

9.4 REFERENCES

[BroYan89] D.C. Brockelbank, Y.H. Yang. An experimental investigation in the use of color in computational stereopsis. *IEEE Transactions on Systems, Man, and Cybernetics* **19** (1989), pp. 1365-1383.

[Gri et al. 94] W.E.L. Grimson, A.L. Ratan, P.A. O'Donnell, G. Klanderman. An active visual attention system to "play where's Waldo". *Proc. Image Understanding Workshop*, Monterey, California, 1994.

[Jar95] R. Jarvis. Synchronised search multiple camera edge based stereopsis ranging with colour/intensity verification. *Proc. 2nd Asian Conference on Computer Vision*, Singapore, 1995, Vol. II, pp. 316-320.

[JorBov88] J.R. Jordan III, A.C. Bovik. Computational stereo vision using color. *IEEE Control Systems Magazine*, June 1988, pp. 31-36.

[JorBov91] J.R. Jordan III, A.C. Bovik. Using chromatic information in edge-based stereo correspondence. *Computer Vision, Graphics, and Image Processing: Image Understanding* **54** (1991), pp. 98-118.

[JorBov92] J.R. Jordan III, A.C. Bovik. Using chromatic information in dense stereo correspondence. *Pattern Recognition* **25** (1992), pp. 367-383.

[Kan et al. 95] S.B. Kang, J.A. Webb, C.L. Zitnick, T. Kanade. A multibaseline stereo system with active illumination and real-time image acquisition. *Proc. 5th Int. Conference on Computer Vision and Pattern Recognition*, Cambridge, Massachusetts, 1995, pp. 88-93.

[Kle et al. 96] R. Klette, A. Koschan, K. Schlüns. *Computer Vision: Räumliche Information aus digitalen Bildern*. Vieweg, Braunschweig/Wiesbaden, Germany, 1996.

[Kle et al. 98] R. Klette, K. Schlüns, A. Koschan. *Computer Vision: Three-Dimensional Data from Images*. Springer, Singapore, 1998.

[KnoSas95] A. Knoll, R. Sasse. An active stereometric triangulation technique using a continuous colour pattern. In: W. Straßer, F. Wahl (eds.), *Graphics and Robotics*. Springer, Berlin, 1995, pp. 191-206.

[Kos92] A. Koschan. Methodic evaluation of stereo algorithms. *Proc. 5th Workshop 1992 on Theoretical Foundations of Computer Vision*, R. Klette, W.G. Kropatsch (eds.), Buckow, Germany, 1992, pp. 155-166.

[Kos93] A. Koschan. Chromatic block matching for dense stereo correspondence. *Proc. 7th Int. Conference on Image Analysis and Processing*, S. Impedovo (ed.), Capitolo, Monopoly, Italy, 1993, pp. 641-648.

[Kos94] A. Koschan. How to utilize color information in dense stereo matching and in edge-based stereo matching. *Proc. Int. Conference on Automation, Robotics, and Computer Vision*, Singapore, 1994, Vol. 1, pp. 419-423.

[Kos96] A. Koschan. Using perceptual attributes to obtain dense depth maps. *Proc. IEEE Southwest Symposium on Image Analysis and Interpretation*, San Antonio, Texas, 1996, pp. 155-159.

[Kos97] A. Koschan. Improving robot vision by color information. *Proc. 7th Int. Conference on Artificial Intelligence and Information-Control Systems of Robots*, I. Plander (ed.), Smolenice Castle, Slovakia, 1997, pp. 247-258.

[KosRod95] A. Koschan, V. Rodehorst. Towards real-time stereo employing parallel algorithms for edge-based and dense stereo matching. *Proc. IEEE Workshop on Computer Architectures for Machine Perception*, Como, Italy, 1995, pp. 234-241.

[KosRod97] A. Koschan, V. Rodehorst. Dense depths maps by active color illumination and image pyramids. In: F. Solina, W.G. Kropatsch, R. Klette, R. Bajcsy (eds.), *Advances in Computer Vision*, Springer, Wien, 1997, pp. 137-148.

[Kos et al. 96] A. Koschan, V. Rodehorst, K. Spiller. Color stereo vision using hierarchical block matching and active color illumination. *Proc. 13th Int. Conference on Pattern Recognition*, Vienna, Austria, 1996, Vol. I, pp. 835-839.

[Kro96] W.G. Kropatsch. Properties of pyramidal representations. *Computing Supplements* **11** (1996), pp. 99-111.

[MarHil80] D. Marr, E. Hildreth. Theory of edge detection. *Proc. Royal Society of London* **B207** (1980), pp. 187-217.

[MurNay94] H. Murase, S.K. Nayar. Illumination planning for object recognition in structured environments. *Proc. Int. Conference on Computer Vision and Pattern Recognition*, Seattle, 1994, pp. 31-38.

[NguCoh92] H.H. Nguyen, P. Cohen. Correspondence from color shading. *Proc. 11th Int. Conference on Pattern Recognition,* The Hague, Netherlands, August/September 1992, Vol. I, Conf. A. Computer Vision and Applications. pp. 113-144.

[Oku et al. 92] M. Okutomi, O. Yoshizaki, G. Tomita. Color stereo matching and its application to 3-d measurement of optic nerve head. *Proc. 11th Int. Conference on Pattern Recognition*, The Hague, Netherlands, August/September 1992, Vol. I, pp. 509-513.

[Pol et al. 85] S.B. Pollard, J.E.W. Mayhew, J.P. Frisby. PMF: A stereo correspondence algorithm using a disparity gradient limit. *Perception* **14** (1985), pp. 449-470.

[TanPav75] S. Tanimoto, T. Pavlidis. A hierarchical data structure for picture processing. *Computer Graphics and Image Processing* **4** (1975), pp. 104-119.

[Vri et al. 92] M. Vriesenga, G. Healey, K. Peleg, J. Sklansky. Controlling illumination color to enhance object discriminability. *Proc. Int. Conference on Computer Vision and Pattern Recognition*, Champaign, Illinois, 1992, pp. 710-712.

10 DYNAMIC AND PHOTOMETRIC STEREO ANALYSES IN COLOR IMAGES

In the previous chapter, stereo analysis based on geometrical considerations for an assumed static scene was introduced. For stereo analyses, motions of objects in the scene or photometric changes due to varying lighting constellations can be observed. This is the subject of this chapter.

Evaluating projected motion vectors can support the calculation of gradient vectors and depth values. Under ideal circumstances, a movement in the three-dimensional space (*3D motion*) corresponds to a projected 2D motion in an acquired image sequence. These projected motions may be represented in the image plane as a *field of local displacement vectors*. Only in very limited cases it is possible to calculate almost error-free fields of local displacement vectors from image sequences. The main problems of a technical realization of a dynamic stereo analysis are the correct measurement of projected motions and the evaluation of partially distorted fields of local displacement vectors. Fields of local displacement vectors can be approximately calculated by the optical flow (defined in Section 10.1).

A detailed representation of the dynamic and photometric stereo analysis for gray-level images is given in [Kle et al. 98]. The basic principles for the calculation of the optical flow in gray-level image sequences are viewed only to the extent of how they are necessary for understanding the expansion to include color image sequences. The representation of the basics was adapted from [Kle et al. 98]. A technique for calculating the optical flow in color image sequences is described in the following. Furthermore, it is shown that in contrast to a gray-level technique, a photometric stereo analysis for evaluating color information is also possible for moving objects or non-Lambertian surfaces.

10.1 OPTICAL FLOW

Local displacement fields can be calculated approximately by the *optical flow*. This represents, graphically speaking, the course of changes of image values (image irradiances) measured with the camera in an image sequence from image E_i to image E_{i+1}. It is assumed that relative or absolute object movements cause these changes in the image values. Under this assumption, the optical flow is an approximation of the local translation field.

In Section 10.1.1, general solutions for calculating the optical flow in gray-level image sequences are discussed. In Section 10.1.2, the Horn-Schunck constraint, derived from them, is extended for calculating the optical flow in color image sequences.

10.1.1 Solution Strategy

Optical flow can be analyzed for any image sequences including those that contain nonrigid objects, such as persons in street scenes or clouds in meteorological images. The optical flow can be calculated for such applications (and also for *active* cameras that should follow the course of motion) to aid motion estimation. The term *active camera* relates here to cameras that can pan, tilt, and zoom as opposed to fixed cameras. Active vision systems seek to dynamically and intelligently gather selective scene information An introduction to the rapidly developing area of *active vision* or *animate vision* is presented in [Alo93].

Optical flow cannot be identified generally with a field of local displacements [Kle et al. 98]. In the case of a sphere without a texture that is rotating in front of a camera, no optical flow can be seen; however, a movement of surface pixels occurs (i.e., the field of local translations is defined accordingly). As opposed to this, the optical flow can, in fact, be observed in the case of a sphere with a uniform textured surface in a fixed position under changing lighting conditions. However, no movement of the surface pixels takes place.

Furthermore, the *aperture problem* must be taken into account. This is characterized by the fact that often no indication or only inadequate information about the motion vectors can be gained from a locally restricted image region in an image sequence. In Fig. 10.1, the aperture problem in motion analysis is illustrated. The optical flow is also affected by camera instability, by changing illumination, or by different surface appearance for different viewing directions (see Fig. 10.2). Thus, it is clear in advance that optical flow allows only approximating local displacement fields.

Let the image irradiance of image E_i in pixel $\mathbf{p} = (x, y)$ be described by $E(x, y, t_i)$, for $i = 0,1,2,\ldots$, where $t_i = i \cdot \delta t_{const}$ applies and δt_{const} is the assumed constant time difference between two acquisitions of the image sequence.

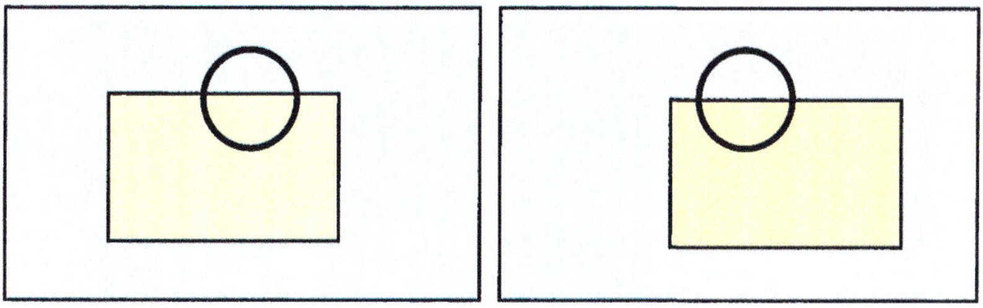

Figure 10.1. *Illustration of the aperture problem in motion analysis: In sole (local) observation of the circle area in both images of an image sequence, the movement of the object (movement of the rectangle toward the upper-right corner) cannot be unambiguously detected (i.e., the motion component toward the right cannot be detected).*

Let

$$\mathbf{u}_i(x,y) = \left(u_i(x,y), v_i(x,y)\right)^T$$

be the optical flow of image E_i to image E_{i+1} that characterizes the changes of image irradiances from image E_i to image E_{i+1}. Here we also assume that the equation

$$E\left(x + u_i(x,y), y + v_i(x,y), t_{i+1}\right) = E(x,y,t_i)$$

or

$$E\left(x + u_i(x,y), y + v_i(x,y), t_i + \delta t_{const}\right) = E(x,y,t_i)$$

has to be satisfied. Such supposed *image value fidelity* of the optical flow can only slightly support the design of an algorithm for calculating vectors $(u,v)^T$ since the individual gray levels $0, 1, \ldots, G_{\max}$ are always affected by noise. For calculating optical flow, additional assumptions or *constraints* are formulated in order to support a solution.

10.1.2 Horn-Schunck Constraint for Color Image Sequences

For a gray-level image function $E(x,y,t_i)$, it can be assumed that it can be locally represented for a small step $(\delta x, \delta y, \delta t)$ by a Taylor expansion

$$\begin{aligned} &E(x+\delta x, y+\delta y, t_i+\delta t) \\ &\quad = E(x,y,t_i) + \delta x \cdot \frac{\delta E}{\delta x}(x,y,t_i) + \delta y \cdot \frac{\delta E}{\delta y}(x,y,t_i) + \delta t \cdot \frac{\delta E}{\delta t}(x,y,t_i) + e \end{aligned}$$

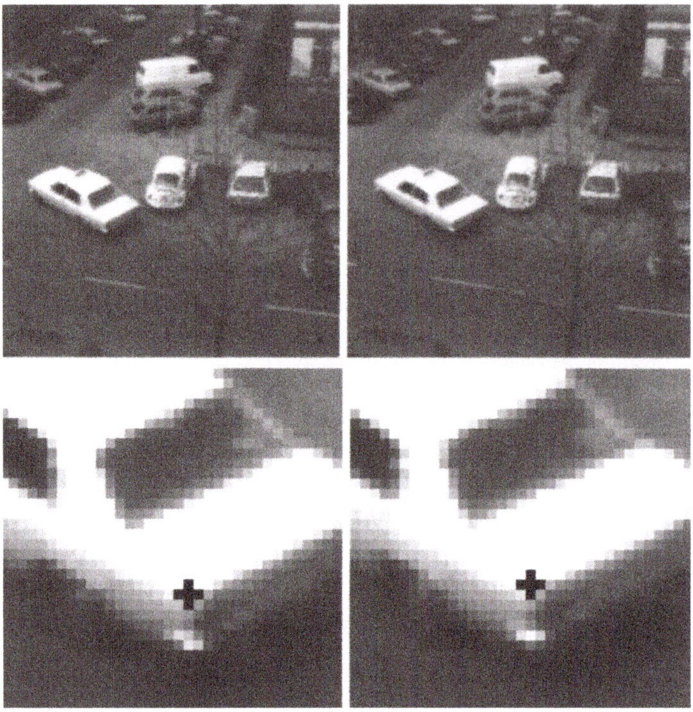

Figure 10.2. *The portrayal of a particular object pixel should be "pursued" in the course of the image sequence. Above, two original images of the image sequence "Hamburg Taxi" are presented, and below, an enlarged section, each with the same marked surface pixel. A comparison of these windows shows minor variations in the local gray value distributions around the marked surface point that can be true due to various causes during image acquisition (reprinted from [Kle et al. 96] with permission from Vieweg).*

(see [Kle et al. 98]). The Taylor expansion of $E(x,y,t_i)$ is viewed especially for a particular step $(u_i(x,y), v_i(x,y), \delta t_{const})$. From an assumed image value fidelity of the optical flow follows

$$0 = u_i(x,y)\cdot\frac{\delta E}{\delta x}(x,y,t_i) + v_i(x,y)\cdot\frac{\delta E}{\delta y}(x,y,t_i) + \delta t_{const}\cdot\frac{\delta E}{\delta t}(x,y,t_i) + e$$

and in simplified form (for $e = 0$ and $\delta t_{const} = 1$)

$$0 = u_i(x,y)\cdot\frac{\delta E}{\delta x}(x,y,t_i) + v_i(x,y)\cdot\frac{\delta E}{\delta y}(x,y,t_i) + \frac{\delta E}{\delta t}(x,y,t_i) + e\,.$$

This equation (see [HorSch81]) is denoted as the *Horn-Schunck constraint*.

The Horn-Schunck constraint follows from the assumed image value fidelity of the optical flow and from the assumption that these vectors describe only small steps, for which the linearity assumption for $E(x,y,t_i)$ is justified (i.e., for which $e=0$ can be assumed). Thus, all pixels are treated equally, unaffected by the distance of the projected object pixels in the image plane. This assumed situation corresponds to the projection model of parallel projection. The normalization $\delta t_{const} = 1$ can be accepted without loss of generality.

The time scale factor $\delta t_{const} = 1$ is used in the following, due to the formal simplification. It is, therefore,

$$t_i = i \quad \text{for} \quad i = 0,1,2,\ldots \quad \text{or} \quad t = 0,1,2,\ldots,$$

which are the discrete points in time to observe, at which the individual images of the image sequence are acquired.

For a definite point in time of the image sequence, the Horn-Schunck constraint assumes the form

$$u \cdot E_x + v \cdot E_y = -E_t \quad \text{or} \quad \left(E_x, E_y\right) \cdot (u,v)^T = -E_t, \tag{10.1}$$

where u and v depend on (x,y), and the derivatives E_x, E_y, and E_t of the image function can be dependent on (x,y,t). The derivatives of the image function can be determined or approximated with a technique introduced in Chapter 6. Each of the values of u and v are limited for a special value pair (x,y) by this linear dependency (see Fig. 10.3), if $E_x \neq 0$ or $E_y \neq 0$ is valid. On these lines in the uv space (*velocity space*) only one pixel can be selected as the solution (u,v). The uncertainty of a solution (u,v) on the lines established by $u \cdot E_x + v \cdot E_y = -E_t$ corresponds to the aperture problem illustrated in the previous section.

Now a transfer of the Horn-Schunck constraint is observed for the calculation of the optical flow in color image sequences. With this, the Horn-Schunck constraint (10.1) can be used in a sequence of color images

$$\mathbf{C}_i(x,y,t) = \left(R_i(x,y,t),\ G_i(x,y,t),\ B_i(x,y,t)\right),\ i = 0,1,2,\ldots$$

on two of the three vector components of a color image [MarFli90] or on each of the three vector components of a color image [Li91]. In the first procedures, a system of equations results from two equations with two unknowns, and in the latter procedures that are presented here, an overdetermined system of equations from three equations with two unknowns. The *Horn-Schunck constraint for color images* can thus be described by

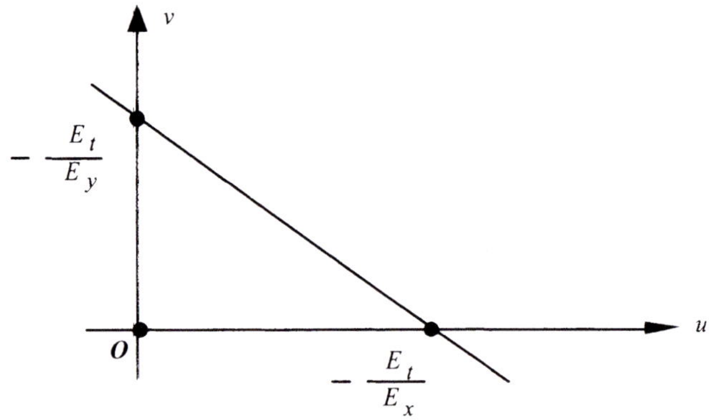

Figure 10.3. *The Horn-Schunck constraint limits the possible value range of the optical flow to a line in the* uv *space. The intersections of this line with the u-axis and the v-axis are equal to* $(-E_t / E_x, 0)$ *or* $(0, -E_t / E_y)$*, respectively.*

$$R_x \cdot u + R_y \cdot v = -R_t,$$

$$G_x \cdot u + G_y \cdot v = -G_t, \text{ and}$$

$$B_x \cdot u + B_y \cdot v = -B_t$$

[Li91]. The indixes x and y indicate again the respective partial derivatives of the functions, for example,

$$R_x = \frac{\partial R}{\partial x} \text{ and } R_y = \frac{\partial R}{\partial y}.$$

In vector notation, the equations above can be described by $\mathbf{J} \cdot \mathbf{u} = -\mathbf{C}_t$ with

$$\mathbf{J} = \begin{pmatrix} R_x & R_y \\ G_x & G_y \\ B_x & B_y \end{pmatrix}, \mathbf{u} = \begin{pmatrix} u \\ v \end{pmatrix}, \text{ and } \mathbf{C}_t = \begin{pmatrix} R_t \\ G_t \\ B_t \end{pmatrix}.$$

A solution to this equation results from the conversions

$$\mathbf{J} \cdot \mathbf{u} = -\mathbf{C}_t$$

$$(\mathbf{J}^T \cdot \mathbf{J}) \cdot \mathbf{u} = -\mathbf{J}^T \cdot \mathbf{C}_t$$

$$(\mathbf{J}^T \cdot \mathbf{J})^{-1}(\mathbf{J}^T \cdot \mathbf{J}) \cdot \mathbf{u} = -(\mathbf{J}^T \cdot \mathbf{J})^{-1}\mathbf{J}^T \cdot \mathbf{C}_t$$

$$\mathbf{u} = -(\mathbf{J}^T \cdot \mathbf{J})^{-1} \cdot \mathbf{J}^T \cdot \mathbf{C}_t \ . \tag{10.2}$$

A complete solution to the optical flow is determined by Eq. (10.2), the unique solution of which is represented by the intersection of three lines in the uv-space. Figure 10.4 illustrates the geometric context.

The aperture problem in determining the optical flow remains likewise unsolved with this method of evaluating color image sequences. The definite solvability of Eq. (10.2) can be tested with the help of the determinants of the matrix $(\mathbf{J}^T \cdot \mathbf{J})$. If $\det(\mathbf{J}^T \cdot \mathbf{J}) = 0$, then no definite solution exists, and the aperture problem remains unsolved. The matrix $(\mathbf{J}^T \cdot \mathbf{J})$ is given by

$$\mathbf{J}^T \cdot \mathbf{J} = \begin{pmatrix} R_x^2 + G_x^2 + B_x^2 & R_x R_y + G_x G_y + B_x B_y \\ R_x R_y + G_x G_y + B_x B_y & R_y^2 + G_y^2 + B_y^2 \end{pmatrix}. \tag{10.3}$$

From this follows

$$\left\| \mathbf{J}^T \cdot \mathbf{J} \right\| = \left(R_x^2 + G_x^2 + B_x^2\right)\left(R_y^2 + G_y^2 + B_y^2\right) - \left(R_x R_y + G_x G_y + B_x B_y\right). \tag{10.4}$$

Conversion results in the following equation:

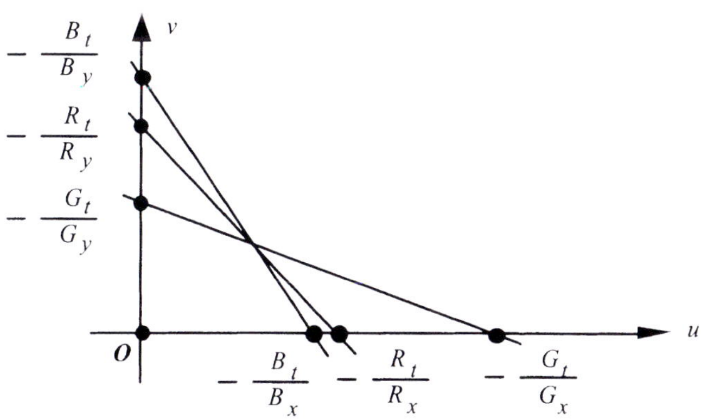

Figure 10.4. *Three noncollinear lines, fixed by the Horn-Schunck constraint, for the three vector components of the color signal form a definite solution to the optical flow in the uv space.*

$$\left\| \mathbf{J}^T \cdot \mathbf{J} \right\| = \left(R_x G_y - R_y G_x\right)^2 + \left(R_x B_y - R_y B_x\right)^2 + \left(G_x B_y - G_y B_x\right)^2. \quad (10.5)$$

The sum of the quadratic equation is then equal to zero if each of these summands is eliminated, that is, if

$$R_x G_y = R_y G_x \text{ and } R_x B_y = R_y B_x \text{ and } G_x B_y = G_y B_x$$

applies. This is then the case if, for example,

$$R_x = G_x = B_x = 0 \text{ or}$$

$$R_y = G_y = B_y = 0 \text{ or}$$

$$R_x = R_y = G_x = G_y = 0 \text{ or}$$

$$R_x = R_y = B_x = B_y = 0 \text{ or}$$

$$G_x = G_y = B_x = B_y = 0$$

or with $R_y \neq 0, G_y \neq 0$, and $B_y \neq 0$

$$\frac{R_x}{R_y} = \frac{G_x}{G_y} = \frac{B_x}{B_y}$$

applies. Thus, the aperture problem remains unsolved if the gradients in all three color components have the same direction. This is, however, more likely the norm than the exception, due to the correlation between the components of the color signal. The aperture problem in color image sequences can be geometrically illustrated with this technique in the *uv* space by three collinear lines fixed by the Horn-Schunck constraint. Finally, we would like to mention that aperture ambiguity may be reduced when combining range and color information [LukFis05]. Moreover, photometric invariants, which are independent of shadow, shading, and specular reflectance changes, may be used to improve the robustness of optical flow estimation [WeiGev04] (compare Section 6.3).

10.2 PHOTOMETRIC STEREO ANALYSIS

Apart from determining the distance and motion values for objects in the scene, the objects' surface orientation is also frequently important for stereo analysis. The three-dimensional surface orientations (or more accurately, the surface standards) can be identified with the help of image values measured with a camera under differing lighting constellations (or more accurately, image irradiance values). This technique is called *photometric stereo analysis* (see [Woo80]) since it is based only on the evaluation of measured intensities. A detailed representation of the photometric stereo analysis can be found in [Kle et al. 98].

In photometric stereo analysis, the evaluation of vector-valued color images offers two substantial advantages over the evaluation of gray-level images:

1. By using color information, photometric stereo analysis can also be implemented for nonstatic scenes with moving objects.
2. By the treatment of physical phenomena, as for example, highlights in color images, the photometric stereo analysis (under certain conditions) can also be implemented for non-Lambertian surfaces in the scene.

Both expansions of the classical gray-level technique of photometric stereo analysis are discussed in the following.

10.2.1 Photometric Stereo Analysis for Nonstatic Scenes

For photometric stereo analysis, two or three images of the scene are viewed. Here the positions of the objects in the scene and the camera position for all images are generally accepted as static. The scene is lit in each shot from another direction. The principal structure of a photometric stereo analysis system with three lighting positions is presented in Fig. 10.5. The directions $\mathbf{s}_1$, $\mathbf{s}_2$, and $\mathbf{s}_3$, toward the lighting source, and the direction $\mathbf{v}$, toward the camera, are assumed as being known. In image E_1, the lighting source is in lighting position 1, in image E_2 in lighting position 2, and in image E_3 in lighting position 3.

Thus, a triplet of intensities can be assigned to all surface pixels without having to execute correspondence analysis in the images, as in static stereo analysis (see Chapter 9). It is assumed that the objects in the scene have Lambertian surfaces. By using the Lambertian reflection law, the three equations

$$\begin{aligned} E_1 &= E_{01}\rho \cdot \mathbf{n}^{0T}\mathbf{s}_1^0 \,, \\ E_2 &= E_{02}\rho \cdot \mathbf{n}^{0T}\mathbf{s}_2^0 \,, \text{ and} \\ E_3 &= E_{03}\rho \cdot \mathbf{n}^{0T}\mathbf{s}_3^0 \end{aligned} \tag{10.6}$$

can be formulated. Here, for $i = 0,1,2,\ldots$, E_i denotes the intensity values measured with the ith camera (or more exact, the image irradiance value), E_{0i} the irradiances of the lighting sources, ρ the albedo in the observed surface pixel, $\mathbf{n}^0$ the surface normal, and $\mathbf{s}_i^0$ the normalized illumination angle. The irradiances and the illumination angles of the lighting sources are assumed constant in the scene.

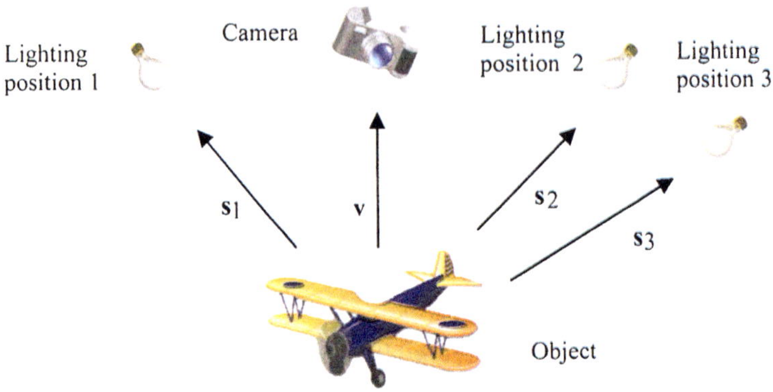

Figure 10.5. *Principal structure of a photometric stereo analysis system with three lighting positions.*

A matrix can also represent the above equation. For this, the image values E_1, E_2, and E_3 are combined to one vector

$$\mathbf{E} = (E_1,\ E_2,\ E_3)^T$$

and the irradiances E_{01}, E_{02}, and E_{03} to a diagonal matrix

$$\mathbf{D} = diag(E_{01},\ E_{02},\ E_{03}) = \begin{pmatrix} E_{01} & 0 & 0 \\ 0 & E_{02} & 0 \\ 0 & 0 & E_{03} \end{pmatrix}. \tag{10.7}$$

Furthermore, vectors $\mathbf{s}_1^0$, $\mathbf{s}_2^0$, and $\mathbf{s}_3^0$ of the illumination directions can be represented as a matrix

$$\mathbf{S} = \begin{pmatrix} s_{1x}^0 & s_{1y}^0 & s_{1z}^0 \\ s_{2x}^0 & s_{2y}^0 & s_{2z}^0 \\ s_{3x}^0 & s_{3y}^0 & s_{3z}^0 \end{pmatrix}.$$

With the three quantities **E**, **D**, and **S**, the equation for the image values can be transferred into the form

$$\mathbf{E} = \rho \cdot \mathbf{D} \cdot \mathbf{S} \cdot \mathbf{n}^0 . \tag{10.8}$$

Here a linear transformation exists between **E** and $\mathbf{n}^0$. If the matrixes **D** and **S** are invertible, then the unit vector of the surface normal scaled with the albedo ρ can be determined with

$$\rho \cdot \mathbf{n}^0 = \mathbf{S}^{-1} \cdot \mathbf{D}^{-1} \cdot \mathbf{E} . \tag{10.9}$$

If the three vectors $\mathbf{s}_1$, $\mathbf{s}_2$, and $\mathbf{s}_3$ are not complanar, then an inverse matrix $\mathbf{S}^{-1}$ exists for matrix **S**. Matrix **D** is invertible if all irradiances are equal to zero. On the basis of the length of the vector

$$\mathbf{S}^{-1} \cdot \mathbf{D}^{-1} \cdot \mathbf{E} \tag{10.10}$$

albedo ρ can then be reconstructed.

This standard technique for photometric stereo analysis can be expanded for nonstatic scenes in which three color light sources are used, whose spectral power distributions are nearly disjunctive. If these three light sources illuminate the scene simultaneously, then only one color image is sufficient for photometric stereo analysis (see [Sch92] and [Dre97]). The three color channels of the image contain the three necessary intensity images. A photometric stereo analysis can in principle also be implemented with this constellation, for example, for objects that move at any speed on a conveyer belt in a factory.

Color light sources can be implemented by introducing color filters with known spectral transmissions (see Section 4.2.1) into the light beam of a white light source. The main problem in the technical transformation of this lighting is that the spectral transmissions of the color filter should be suited to the spectral sensitivities of the component sensors of the color camera. The spectral sensitivities of the camera sensors arc, however, as a rule not known and must be determined. Furthermore, it should be considered that overlapping spectral transmissions of color filters, in view of overlapping spectral sensitivities of component sensors arising in commercial CCD cameras, can cause errors in the results [Sch92].

10.2.2 Photometric Stereo Analysis for Non-Lambertian Surfaces

According to the fundamentals of the Lambertian reflection law, photometric stereo analysis can be successfully implemented for gray-level images only on Lambertian surfaces. Highlights on the surfaces directly cause a distortion of the results in determining the surface normals. This limited applicability of photometric stereo analysis can be overcome (at least partially) by the analysis of three color images (see [ChrSha93], [ChrSha94], [DreKon94], [Dre97], [Sch93], [SchTes95ab], and [SchWit93]).

If the objects affected by highlights consist of inhomogeneous dielectric materials, then the dichromatic reflection model (see Section 7.4.1) can be employed for the description of surface reflection. In this connection, surface reflection can be modeled by a diffuse (Lambertian) and a specular component. By analyzing the dichromatic plane (see Section 7.4.1), both of these reflection components can be generally separated and highlight elimination can be executed (see Section 8.1).

In photometric stereo analysis the multi-image technique, introduced in Section 8.1.6, and the Schlüns-Teschner technique, from Section 8.1.4, were successfully employed for matte image generation (see [SchTes95ab]). The matte images determined in this way are subsequently transformed into intensity images, and in a second processing step, a gray-level variant of photometric stereo analysis is implemented. In this technique no additional lighting sources that would lead to a reduction of the (from all illuminated light sources) reconstructed image area are necessary. Furthermore, no additional assumptions on the roughness of the materials are necessary since reflection analysis can be implemented, unaffected by the geometric reflection component m_S (compare Section 7.4.1).

While the transformation of the variants of the photometric stereo analysis, introduced in the previous section, is marked for nonstatic scenes with technical border problems, no knowledge of spectral sensitivities of the camera sensors is needed for the variants introduced in this section. In summary, the following holds: If the scene to be analyzed contains only objects with Lambertian surfaces, then the technique for photometric stereo analysis introduced in this section is practically identical to the classic gray-level variant. However, if the scene also contains objects with non-Lambertian surfaces consisting of inhomogeneous dielectric materials, then highlight elimination can be implemented, taking into account the dichromatic reflection model. The surface normals can be determined from the matte images extracted in this way.

10.3 REFERENCES

[Alo93] Y. Aloimonos. *Active Perception*. Lawrence Erlbaum, Hillsdale, New Jersey, 1993.

[ChrSha93] P.H. Christensen, L.G. Shapiro. Determining the shape of multi-colored dichromatic surfaces using color photometric stereo. *Proc. Int. Conference on Computer Vision and Pattern Recognition*, New York, 1993, pp. 767-768.

[ChrSha94] P.H. Christensen, L.G. Shapiro. Three-dimensional shape from color photometric stereo. *Int. J. of Computer Vision* **13** (1994), pp. 213-227.

[DreKon94] M.S. Drew, L.L. Kontsevich. Closed-form attitude determination under spectrally varying illumination. *Proc. Int. Conference on Computer Vision and Pattern Recognition*, Seattle, Washington, 1994, pp. 985-990.

[Dre97] M.S. Drew. Photometric stereo without multiple images. *Proc. SPIE* **3016**, Human Vision and Electronic Imaging, 1997, pp. 369-380.

[HorSch81] B.K.P. Horn, B.G. Schunck. Determining optical flow. *Artificial Intelligence* **17** (1981), pp. 185-203.

[Kle et al. 96] R. Klette, A. Koschan, K. Schlüns. *Computer Vision: Räumliche Information aus digitalen Bildern*. Vieweg, Braunschweig/Wiesbaden, Germany, 1996.

[Kle et al. 98] R. Klette, K. Schlüns, A. Koschan. *Computer Vision: Three-Dimensional Data from Images*. Springer, Singapore, 1998.

[Li91] H. Li. Optical flow from a color image sequence. Technical Report LiTH-ISY-I-1270, Linköping University, Dept. of Electrical Engineering, Linköping, Sweden, 1991.

[LukFis05] T.C. Lukins, B. Fisher. Colour constrained 4D flow. *Proc. British Machine Vision Conference*, Oxford, UK, 2005.

[MarFli90] V. Markandey, B.E. Flinchbaugh. Multispectral constraints for optical flow computation. *Proc. 3rd Int. Conference on Computer Vision*, Osaka, Japan, 1990, pp. 38-41.

[Sch92] K. Schlüns. Colourimetric stereo. *Proc. 5th Workshop on Theoretical Foundations of Computer Vision*, R. Klette, W.G. Kropatsch (eds.), Akademie Verlag, Berlin, Germany, 1992, pp. 181-190.

[Sch93] K. Schlüns. Photometric stereo for non-Lambertian surfaces using color information. *Proc. 5th Int. Conference on Computer Analysis of Images and Patterns*, D. Chetverikov, W.G. Kropatsch (eds.), Budapest, Hungary, 1993, pp. 444-451.

[SchWit93] K. Schlüns, O. Wittig. Photometric stereo for non-Lambertian surfaces using color information. *Proc. 7th Int. Conf. on Image Analysis and Processing*, S. Impedovo (ed.), Capitolo, Monopoly, Italy, 1993, pp. 505-512.

[SchTes95a] K. Schlüns, M. Teschner. Fast separation of reflection components and its application in 3d shape recovery. *Proc. 3rd Color Imaging Conference*, Scottsdale, Arizona, 1995, pp. 48-51.

[SchTes95b] K. Schlüns, M. Teschner. Analysis of 2d color spaces for highlight elimination in 3d shape reconstruction. *Proc. Asian Conference, on Computer Vision*, Vol. II, Singapore, 1995, pp. 801-805.

[WeiGev04] J. van de Weijer, T. Gevers. Robust optical flow from photometric invariants. *Proc. IEEE Int. Conference on Image Processing*, Singapore, 2004, pp. 251–255.

[Woo80] R.J. Woodham. Photometric method for determining surface orientations from multiple images. *Optical Engineering* **19** (1980), pp. 139-144.

11 COLOR-BASED TRACKING WITH PTZ CAMERAS

The problem of tracking people and recognizing their actions in video sequences is of increasing importance to many applications. Examples include video surveillance, human/computer interaction, and motion capture for animation, to name a few (see [Dom et al. 06], [Har et al. 00], [Kak et al. 07], [Li et al. 02], [McK et al. 99], [PläFua01], [Rob et al. 06], [WuHua02], [WuYu06], [XioDeb04]). This chapter presents case studies of color use in an automated video tracking and location system. The system could be utilized in any situation where detection of adverse flow motion with subsequent video tracking would be beneficial. Examples of these situations include federal buildings, courthouses, large office buildings, military bases, and national laboratories. Guidance of a robot arm employing dynamic imaging and motion trajectory analysis of workers in hazardous environments are also potential applications of the system's tracking aspect.

With the number of people traveling by plane today, security in airport terminals is of great concern. Whenever a suspicious individual is identified or a threat is suspected, the entire section of the airport where the threatening activity is taking place must be cleared for investigation. Knowing or recording the activity of a "bolter" (a person entering a secured area without clearance, usually through an exit lane) at all times would limit the investigation to a smaller area of the airport and/or facilitate an effective and risk-favorable apprehension of the violator.

The system examined in this chapter represents a paradigm shift from the current analog, disconnected, and human-intensive security and surveillance systems to a digital, networked, and fully automated system. This system is a camera-based security system consisting of a network of cooperating cameras controlled by computer vision software. Automatic target acquisition is performed via cooperating fixed and pan-til-zoom (PTZ) cameras, while tracking is achieved solely via PTZ cameras. Several algorithms were proposed in the past to extend camera views to track objects in large areas. Lee et al. [Lee et al. 00] proposed a method to align the ground plane across multiple views to build common coordinates for multiple cameras. Dellaert and Collins [DelColl99] proposed a fast image registration algorithm between the image from a pan/tilt camera and background images from a database. Omnidirectional cameras were also used to extend the field of view to 360 degree [Nic et al. 00]. But the reality remains that most tracking algorithms cater only to the case of fixed cameras and are generally based on adaptive background generation and subtraction [Har et al. 00], [Hor et al. 00], [Lee et al. 00].

Another issue facing automatic tracking in public areas is occlusion and features' robustness. Tracking algorithms based on gray-level images [PläFua01], shape information [BlaIsa98], and color [McK et al. 99] have been proposed before. But despite the various levels of accuracy in modeling objects to be tracked, the assumption must be made in some applications that the object is nonrigid or deformable. In order to represent a nonrigid object (people), active shape models (ASMs) are very efficient compact models in which the shape variety of an object class is taught in a training phase. In this chapter, a hierarchical robust approach to an enhanced ASM is proposed to realize an efficient color video tracking system.

11.1 THE BACKGROUND PROBLEM

Generally, an object-tracking algorithm is composed of three main functions:

1. Background modeling
2. Moving object detection
3. Object tracking

Moving objects can be identified by the difference between the background and the current images. The simplest means is by using an image without any motion, but this method cannot deal with the problem of changing illumination. In the method used here, a pixel that does not change for N consecutive frames is used to update the background image. After a background image is generated, moving pixels can be extracted by comparing the background to the current images. A threshold value, which is determined experimentally, is used to extract moving pixels, and morphological operations, such as opening and closing, are used to remove noise when detecting moving pixels.

For tracking purposes, features such as color distribution, height of objects, and motion information are extracted. Color distribution includes the mean and variance of the U and V components [Lev97] for each segment. Height is an important feature because the height of a human body generally does not change much, while the width may vary depending on the position and viewing direction [Son et al. 99]. The final feature, or motion information, includes the direction of change and the acceleration in previous frames for each object area.

In the tracking process, correlation is computed for each feature and the best-matched region is determined as the position of the current target object. For seamless tracking by multiple cameras in a spacious public area, such as an airport concourse, the proposed system should switch the current camera to another that covers a better viewing area. An important criterion in determining a better viewing direction is the distance between the object being tracked and each camera. Generally speaking, a closer camera provides a better image of the object, if it includes the whole object with minimal occlusion.

Two features, the height ratio and motion information, can be used to determine the best viewing camera. For different cameras, the captured images with the same object may have different object sizes because the distances between the cameras and the object are different. The height ratio can be computed by considering actual positions and zooming ratios for each camera. For this research, a simulated environment based on Knoxville's McGhee-Tyson airport was used for determining the height ratio.

Tracking results using two stationary cameras and one PTZ camera are shown in Fig. 11.1. As shown in Fig. 11.1a, the tracking algorithm was applied to each image from stationary cameras, and the detected object is highlighted with a rectangular box. According to the position of the detected person, the PTZ camera changes parameters (PTZ factors) to get the best view, with the result as shown in the third image. Figures 11.1b and c show results of the camera handover task. When the person disappears in the first viewing area and appears in the second viewing area, the proposed system could successfully hand over the object between cameras.

In the 3D scene model shown in Fig. 11.2, the area highlighted by the circle in Fig 11.2a represents the position of an object, and Figs. 11.2b and c show different views covering the person from cameras 2 and 3. Using this model, the person's height and location and the corresponding PTZ parameters, such as panning angle, tilt angle, and zoom ratio, can be verified. Based on this result, we could successfully test the camera handover strategy and perform the virtual-reality experiment.

Figure 11.3 illustrates the transformed background shown in the first row. Mosaic building is accomplished in the first two columns and then updating is performed. In the second row are the images from the PTZ camera, and the third row shows the detected moving regions as white pixels on a black, stationary background. Since we did not generate a new background for the current position, error in the motion-detection process shows up as motion in objects that are obviously stationary. This can be resolved by generating a background for the new position in advance and combining that background with the background transformed from the previous frame, as in column 5 of Fig. 11.3.

(a) Images at the 18th frame

(b) Images at the 75th frame

(c) Images at the 77th frame

Figure 11.1. *Tracking results using two stationary cameras and one PTZ camera. The first and second columns respectively show images captured by the first and second stationary cameras. The third column shows images captured by the PTZ camera.*

11.2 METHODS FOR TRACKING

Tracking and recognizing nonrigid objects in video image sequences are complex tasks. Using color information as a feature to describe a moving object or person can support these tasks. The use of four-dimensional templates for tracking objects in color image sequences was suggested in [Bro et al. 94]. However, if the observation is accomplished over a long period of time and with many single objects, then both the memory requirements for the templates in the database and the time requirements for the search of a template in the database increase. In contrast to this, active shape models (ASMs) represent a compact model for which the form variety and the color distribution of an object class are taught in a training phase [Coo et al. 95].

(a) (b) (c)

Figure 11.2. *Three-dimensional scene model of Knoxville's McGhee Tyson Airport. (a) Top view of the east wing of the concourse. (b) Closeup view from camera 2. (c) Closeup view from camera 3.*

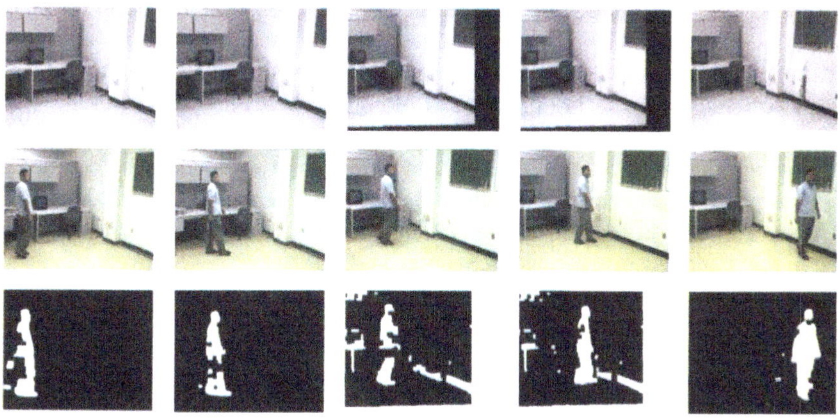

Figure 11.3. *Background generation for PTZ camera using mosaicking; background images (top), PTZ view (middle), and motion detected (bottom).*

Several systems use skin color information for tracking faces and hands (see, e.g., [ComRam00], [Li et al. 00], and [MarVil02]). The basic idea is to limit the search complexity to one single color cluster representing skin color, and to identify pixels based on their membership in this cluster. Several problems affect these approaches. First, skin colors cannot be uniquely defined and, in addition, a person cannot be identified when seen from behind. Here tracking clothes instead of skin is more appropriate [Roh et al. 00].

Second, color distributions are sensitive to shadows, occlusions, and changing illuminations. Addressing the problem occurring with shadows and occlusions, Lu and Tan assume that the only moving objects in the scene are persons [LuTan01]. This assumption does not hold for many applications. Most of the approaches mentioned above cannot be easily extended to multicolored objects other than persons.

A very efficient technique for the recognition of colored objects is color indexing [SwaBal91]. Based on comparisons between color distributions, an object in the image is assigned to an object stored in a database. This technique usually needs several views of the object to be recognized, which is not always ensured when tracking people in a road scene, for example. Furthermore, color indexing partly fails with partial occlusions of the object. Active shape models do not need several views of an object, since by using energy functions they can be adapted to the silhouette of an object represented in the image. However, the outlier problem, which can occur particularly with partial object occlusion, represents a difficulty for these models.

11.2.1 Active Shape Models

For tracking a human target in video, detecting the shape and position of the target is the fundamental task. Since the shape of a human object is subject to deformation and random motion in the two-dimensional image space, ASM is one of the best-suited approaches in the sense of both accuracy and efficiency.

ASM falls into the category of deformable shape models with a priori information about the object. ASM-based object tracking models the contour of the silhouette of an object, and the set of model parameters is used to align different contours in each image frame. An extension of traditional ASMs to color active shape models is presented in Section 11.4.

11.2.2 Automatic Target Acquisition and Handover from Fixed to PTZ Camera

When a breach occurrence is detected, the fixed camera in charge of monitoring the direction of motion triggers an alarm and provides the position of the target in the world coordinate system. The PTZ camera then uses that position information to determine its pan-and-tilt angles and lock on the target for subsequent tracking. The pan-and-tilt angles for the PTZ camera are respectively given as a function of the coordinates (x_t, y_t, h_t) of the target

$$\theta = \sin^{-1} \frac{x_t}{\sqrt{x_t^2 + y_t^2}},$$
$$\delta = \cos^{-1} \frac{\sqrt{x_t^2 + y_t^2}}{\sqrt{x_t^2 + y_t^2 + (h_c - h_t)^2}}. \tag{11.1}$$

Handover is considered complete only when the PTZ camera is able to extract the moving target from its background and lock on it. This step is achieved using the same principle of direction of motion; only this time the motion being searched for is top-down motion instead of left to right. A GUI view of a typical

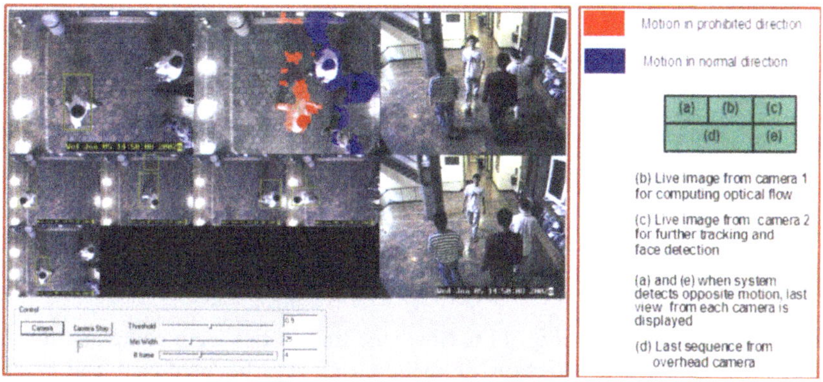

Figure 11.4. *GUI view for the two-camera system.*

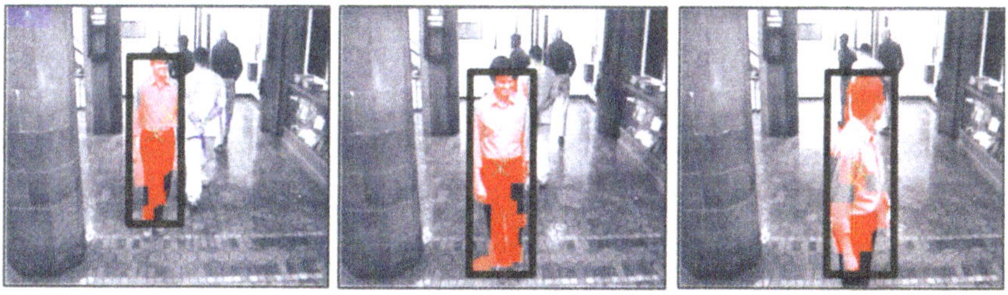

Figure 11.5. *Sequence of frames from the PTZ camera showing achieved handover (successive target views).*

image captured from the two-camera system is shown in Fig. 11.4, whereas Fig. 11.5 shows successive target views from the PTZ camera.

11.2.3 Color and Predicted Direction and Speed of Motion

Image distortions caused by PTZ cameras make the tracking task difficult. Features that are robust to these distortions are needed for the tracking task. Color information of the target can be such a feature. When color constancy is preserved, the color distribution of interesting regions can be used to track objects. Color indexing [SwaBal91 is one of the techniques used to find similar color targets in consecutive frames. The video from the overhead camera is first analyzed to detect and extract breaches. Each extracted region is used to build a color histogram model. Once the histogram models are acquired, the nearest and most similar color regions are searched through histogram intersection. The results are trajectories of the objects that caused the alarm. Experimental results using the histogram

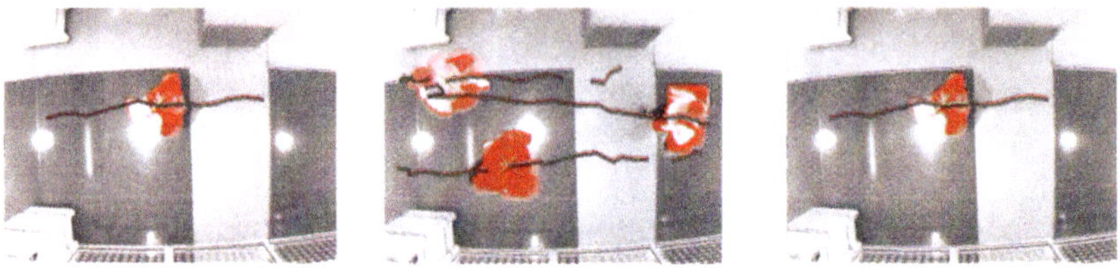

Figure 11.6. *Tracking results using color indexing.*

intersection are shown in Fig. 11.6. Since the trajectories were computed for each frame, the speed and direction of motion can also be predicted and used to compute the internal parameters of the PTZ camera, such as pan and tilt angles. The PTZ camera is then automatically controlled to view the predicted location and to extract the top-down motion caused by the breach. A verification process will then follow to check whether the extracted regions are effectively caused by the breach.

11.3 TECHNICAL ASPECTS OF TRACKING

This section provides more detailed information on various technical considerations associated with color-based tracking.

11.3.1 Feature Extraction for Zooming and Tracking

Three features are selected for automatic zooming and face tracking. The first feature is the mean location (x_c, y_c) of hue values, which are located between $f(x_i)_{Low-th}$, and $f(x_i)_{Hi-th}$, within the detected region-of-interest (ROI)

$$x_c = \frac{\sum_x H(x,y)}{E_H}, y_c = \frac{\sum_y H(x,y)}{E_H}, \tag{11.2}$$

where $H(x,y)$ represents the pixel location of an effective hue value and E_H the number of selected pixels having effective hue values. The second feature is the area of the detected ROI, and the third is the effective pixel ratio, R_{ROI}, within the detected ROI. The mean location x_c and y_c indicates the direction of the moving object and the second feature A_{ROI} determines the optimum zooming ratio; the third feature, R_{ROI}, is used for fault detection in zooming and tracking. The second and the third features can be formulated as

$$A_{ROI} = \text{Width}_{ROI} \times \text{Height}_{ROI} \text{ and } R_{ROI} = \frac{E_H}{A_{ROI}}. \quad (11.3)$$

Automatic zooming is performed using the A_{ROI} feature. There are two experimentally selected limiting values for automatic zooming, Tele and Wide. If A_{ROI} is greater than Wide, the zoom lens turns wide for zooming down, and vice versa. Figure 11.7 presents experimental results of the proposed face tracking algorithms using only the pan/tilt function. The result of 4-channel automatic zooming with face tracking is shown in Fig. 11.8.

In Figs. 11.7, 11.8, and 11.9, segmented face regions are shown in black, and the histogram of the face region is overlaid on each image.

The effective pixel ratio indicates the error probability of zooming and tracking. If this value is smaller than a prespecified value, a new candidate area for the moving object must be detected having the latest $f(x_i)_{Low-th}$ and $f(x_i)_{Hi-th}$ values. This dynamic change of the ROI is necessary for correct tracking. This process is shown in Fig. 11.9.

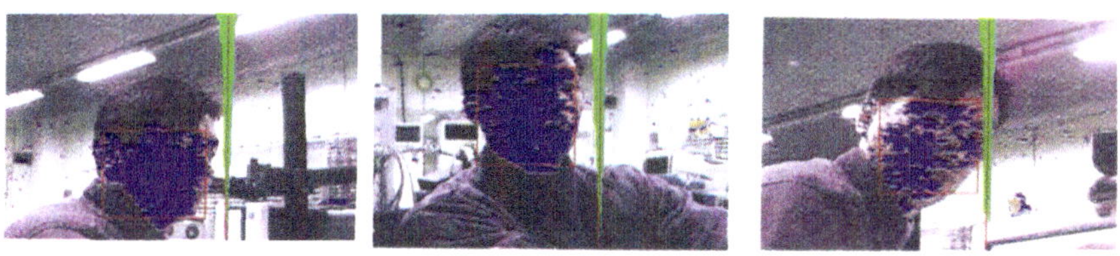

Figure 11.7. *Single face tracking.*

Figure 11.8. *Automatic zooming with face tracking.*

Figure 11.9. *Dynamic change of ROI.*

In order to avoid the undesired extension of the tracked region to neighboring faces, an ellipse fitting is performed every 30 frames, utilizing the generalized Hough transform on the edge image of the rectangular region that is searched based on color distribution. The ellipse-fitting procedure followed by region search can make the detection more robust. Figure 11.10 shows the ellipse-fitting result on the edge image.

Figure 11.10. *Occlusion between two faces (top); Sobel edge within the ROI (bottom-left); ellipse fitting (bottom-right).*

Since accurate identification of a human face is more important than just tracking a moving object, an efficient method to detect the face region and a resulting high-resolution acquisition are needed. The proposed intelligent surveillance system with built-in automatic zooming and tracking algorithms can efficiently detect high-resolution face images and stably track the face. One major contribution of this work is the development of real-time, robust algorithms for automatic zooming and tracking and an intelligent surveillance system architecture using multiple PTZ cameras with seamless interface.

Although face recognition systems work well with "in-lab" databases and ideal conditions, they have exhibited many problems in real applications. (So far, no face-recognition systems, tested in airports, have spotted a single person who is wanted by authorities.) Variations exist in unconstrained environments including pose, resolution, illumination, and age differences. They make face recognition a very difficult problem. Detection of faces from a distance and in crowds is also a challenging task.

In order to increase the performance of face detection and recognition, a combination of robust face detection and recognition is necessary. An incorporated face recognition system using other imaging modalities such as thermal imagery and 3D face modeling, which provide more features and that are invariant to changes in poses, should be developed to be successfully used for surveillance [Kim et al. 03].

11.3.2 Color Extraction from a Moving Target

Entering restricted areas, for any reason, usually requires permission and security procedures, such as walking through a metal detector. One serious security violation is walking in a restricted direction, such as going through an exit lane in an airport to gain access to the passenger concourse and bypassing the checkpoint. To detect an intruder who walks the wrong way through a crowded exit lane, an overhead camera would be used to avoid occlusions. The video from the overhead camera is used to detect intrusions by computing motion direction for every pixel using an optical-flow-based approach. When the system detects intruders, the system sounds an alarm to alert a nearby human operator and sends this information to another system connected to a PTZ camera for target handover and tracking.

The overhead camera has only a fixed view, so the system will lose the breach event after the person who caused the breach walks outside the area of current view. The PTZ cameras will be used to track the target using a color-based tracking algorithm. When a breach occurrence is detected, the fixed camera in charge of monitoring the direction of motion triggers an alarm and provides the position of the target in the world coordinate system. The PTZ camera then uses that position information to determine its pan-and-tilt angles and lock on the target for subsequent tracking. The geometry of the system is shown in Fig. 11.11, and the angles θ and δ can be computed by

$$\theta = \cos^{-1}\frac{x_t}{\sqrt{x_t^2 + y_t^2}}, \quad \delta = \frac{\sqrt{x_t^2 + y_t^2}}{\sqrt{x_t^2 + y_t^2 + (h_c - h_t)^2}}, \tag{11.4}$$

where x_t and y_t are given by

$$x_t = x_i - x_d = (W - i)a - x_d,$$

$$y_t = y_i + y_d = (H - j)b + y_d,$$

where W is the width, H the height, (i, j) the location of the target in the image coordinate system, a and b are the lengths per pixel for horizontal and vertical directions respectively, and the rest of the parameters are shown in Fig. 11.11. These values can be computed by measuring two spots in the scene and finding the corresponding points in the image.

The target detection system sends information on a target that violates the direction of flow to the client. The client then extracts the top-down motion to lock on the target approaching the PTZ camera. Two assumptions are made: (1) The

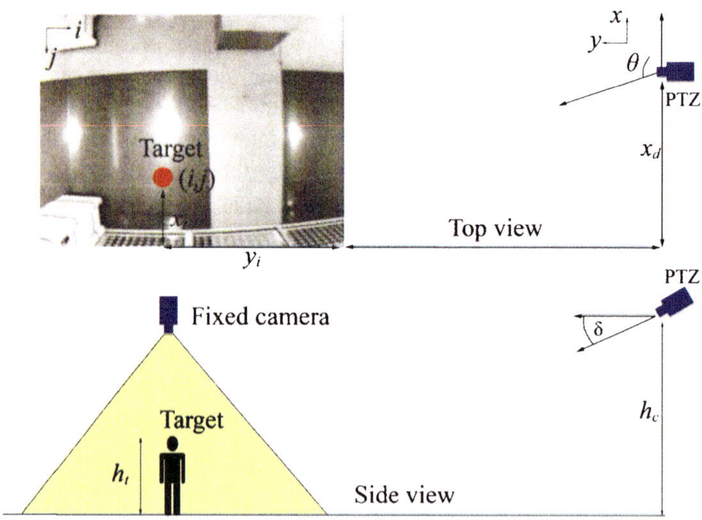

Figure 11.11. *Geometry of the dual-camera system.*

PTZ camera is located inside of the secure area and views the exit lane where the violation may occur, and (2) the height of the PTZ camera is much higher than the potential target height. For instance, if the target walks from left to right, the target appears to be moving from the top of the PTZ camera's view toward the bottom of the view. In a real situation, this method extracts not only the target to track, but also top-down motion caused by people walking out of the exit lane or shadows. Since we need to build a color model of the target, it is important to determine the correct target regions. If $B_t(i,j)$ is the result of thresholding, then $B_t(i,j)=1$ for moving regions and $B_t(i,j)=0$ for static pixels. We also define a mask M as an ellipse with semi-minor axis, x_r, and semi-major axis, y_r, at the center (0, 0). These values are determined in experiments. This mask is normalized, so the sum of pixel values in the mask is 1.

We define R_M as the convolution of B_t and M; then it holds that

$$R_M = B_t * M. \tag{11.5}$$

If a segment in B_t has a similar shape to the mask M, then $R_M(i,j)$ will be close to 1, so it can be determined whether segments in B_t are closer to the shape of the mask by

$$P(i,j) = \begin{cases} 1 & R_M(i,j) > t_n \\ 0 & otherwise. \end{cases} \tag{11.6}$$

An example illustrating this concept is shown in Fig. 11.12. One input image, out of two images for optical flow computation, is shown in Fig. 11.12a, and an extracted segment with a top-down motion is shown in Fig. 11.12b with noise present in the segment. R_M is shown in Fig. 11.12c, where the scale is 0 to 255 instead of 0 to 1 for display purposes. We can find the best location of the ellipse by calculating $P(i, j)$ and imposing that the center of existing segments belongs to P, as shown in Fig. 11.12d.

When an ellipse is superimposed on the detected segment, the corresponding color input segment is divided into achromatic and chromatic pixels using HSV color space. K-means segmentation is followed to divide chromatic pixels into multiple color pixels using hue components.

After color segmentation, the similarity between the divided segments and the surrounding pixels is determined by

$$S_{i,bg} = \sum_k \min\left(\frac{H_i(k)}{\sum_k H_i(k)}, \frac{H_{bg}(k)}{\sum_k H_{bg}(k)}\right), \tag{11.7}$$

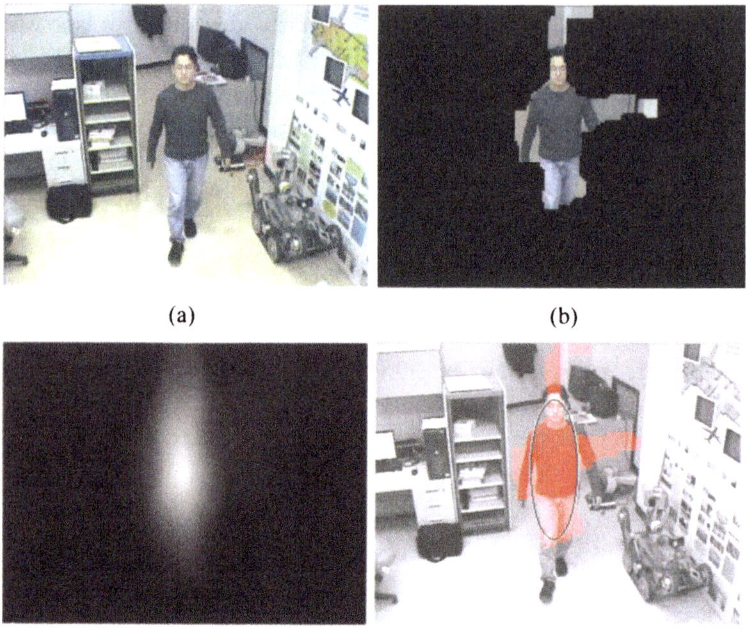

(a) (b) (c) (d)

Figure 11.12. *Finding a target from moving pixels. (a) One input image. (b) The extracted segment with a top-down motion. (c) The result of applying* $R_M = B_t * M$. *(d) The detected ellipse.*

for $k = 0, 1, \cdots, L-1$ and $i = 0, 1, \cdots, 18$, where H_i represents ith segments from the detected ellipse, H_{bg} is the histogram of the surrounding pixels, and L is the number of bins for each histogram model. This equation is a normalized version of the histogram intersection described in [SwaBal91]. For achromatic pixels, saturation and value channels are used for computing the similarity using Eq. (11.7), and hue and saturation channels are used for chromatic pixels. An example is shown in Fig. 11.13. A detected ellipse is shown in Fig. 11.13a and its segmentation result is shown in Fig. 11.13b. After applying Eq. (11.7) to each segment in Fig. 13(b) and the background rectangle defined in Figure 13(c), we can get major color that is different from the background. In this case, two segments were detected: One skin color, which is similar to the background color, and achromatic pixel, which is black. The similarity computed by Eq. (11.7) is 0.51 for the skin color and 0.07 for achromatic pixel. Finally, a chromatic color is chosen to represent the target's unique color for tracking.

The major color segmentation is then normalized by the maximum value of the histograms. For example, a histogram H can be normalized by

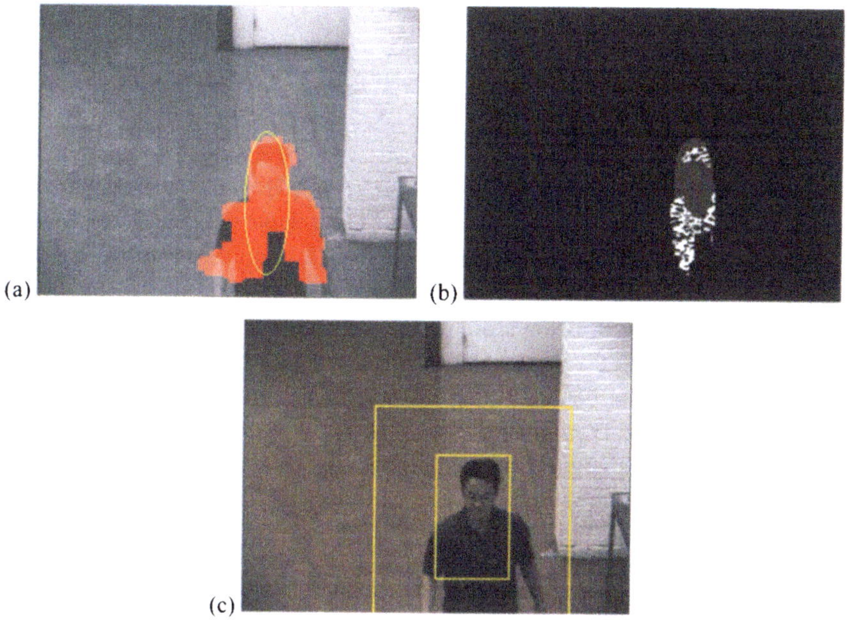

Figure 11.13. *An example of computing similarity between each segment and surrounding pixels. (a) Detected ellipse on an input frame. (b) Classification result to divide achromatic and chromatic pixels. White pixels represent achromatic pixels. (c) Definition of surrounding pixels that excludes the rectangular region.*

$$H_n = \frac{H(k)}{\max(H)}. \tag{11.8}$$

After normalization, the similarity between the model histogram and input image E can be formulated by

$$E(i, j) = H_n(E(i, j)). \tag{11.9}$$

The mean shift algorithm [Che95] is used to find the center of a distribution in a search window. This method is iterative and can be formulated as

$$i_c(t+1) = \frac{\sum\limits_{i \in S_t} \sum\limits_{j \in S_t} i \cdot P_c(i, j)}{\sum\limits_{i \in S_t} \sum\limits_{j \in S_t} P_c(i, j)}, \tag{11.10a}$$

$$j_C(t+1) = \frac{\sum_{i \in S_t} \sum_{j \in S_t} j \cdot P_C(i,j)}{\sum_{i \in S_t} \sum_{j \in S_t} P_C(i,j)}, \tag{11.10b}$$

where S_t is a search window placed at $(i_C(t), j_C(t))$ and t is the number of iterations. At each iteration, the search window will move to the center; where the center is usually the location of the target and the PTZ camera is controlled to keep the search window in the center of the frames.

Moreover, the size of the search window is changed by ± 2 pixels at each iteration by calculating the mean of P_C in the search window, M_S, and comparing to the mean of the four boundaries to deal with the size variation of the target. If the mean of a boundary is smaller than $0.2M_S$, then the searching window is shrunk. If the mean is larger than $0.8M_S$, then the search window is enlarged in the direction of the boundary.

11.4 COLOR ACTIVE SHAPE MODELS

Special considerations for digital image processing are required when tracking objects whose forms (and/or their silhouettes) change between consecutive frames. For example, cyclists in a road scene and people in an airport terminal belong to this class of objects denoted as *nonrigid objects*. ASMs can be applied to the tracking of nonrigid objects in a video sequence. Most existing ASMs do not consider color information [ParSay01]. We present several extensions of the ASM for color images using different color-adapted objective functions.

Detecting the shape and position of the target is a fundamental task for tracking a nonrigid target in a video sequence. Two-dimensional deformable models typically use a boundary representation (deformable contour) to describe an object in the image. Within the class of deformable models, the ASM is one of the best-suited approaches in the sense of both accuracy and efficiency for applications where a priori information about the object (or more precisely about the shape of the object) in the image is available. The basic concept of ASMs consists of modeling the contour of the silhouette of an object in the image by parameters in order to align the changing contours in the image frames to each other. More specifically, our ASM-based tracking algorithm consists of five steps:

1. Assignment of landmark points
2. Principal component analysis (PCA)
3. Model fitting
4. Local structure modeling
5. In this approach, an additional color component analysis

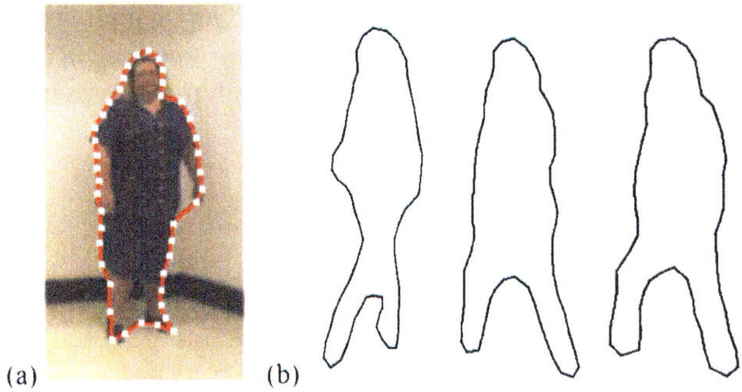

Figure 11.14. *(a) A human object with 42 landmark points (n = 42) and (b) three examples of different ASM alignments to the contour of a moving person in three different frames of a video sequence (reprinted from [Kos et al. 03] with permission from Elsevier).*

As an example of a target application, we tentatively set up the goal to track either people or suitcases in an airport. Figure 11.14 shows a person with 42 manually selected landmark points on the initial contour and three successful alignments of silhouettes to the contour of the object in the frames. The transformations needed for the alignments are determined in an iterative process.

11.4.1 Landmark Points

Given a frame of input video, suitable landmark points should be assigned on the contour of the object. Good landmark points should be consistently located from one image to another. In a two-dimensional image, we represent n landmark points by a $2n$-dimensional vector as

$$\mathbf{x} = [x_1, \ldots, x_n, y_1, \ldots, y_n]^T . \tag{11.11}$$

A typical setup in our system consists of 42 manually assigned landmark points ($n = 42$). Various automatic and systematic ways of obtaining landmark points were discussed by Tian et al. [Tia et al. 01]. The role of landmark points is controlling the shape of model contours. More specifically, the initially assigned landmark points are updated by minimizing the deviation from the original profile, which is normal to the boundary at each landmark point. More rigorous quantification of the deviation is given in section 11.4.5.

11.4.2 Principal Component Analysis

A set of n landmark points represents the shape of the object. Figure 11.15 shows a set of 56 different shapes, called a training set. Although each shape in the training set is in the $2n$-dimensional space, we can model the shape with a reduced number of parameters using the *principle component analysis* (*PCA*) technique. Suppose we have m shapes in the training set, presented by $\mathbf{x}_i$, for $i = 1, \ldots, m$. The PCA algorithm is as follows.

PCA algorithm

1. Compute the mean of the m sample shapes in the training set

$$\bar{\mathbf{x}} = \frac{1}{m}\sum_{i=1}^{m} \mathbf{x}_i . \tag{11.12}$$

2. Compute the covariance matrix of the training set

$$\mathbf{S} = \frac{1}{m}\sum_{i=1}^{m} (\mathbf{x}_i - \bar{\mathbf{x}})(\mathbf{x}_i - \bar{\mathbf{x}})^T . \tag{11.13}$$

3. Construct the matrix

$$\mathbf{\Phi} = [\phi_1 \mid \phi_2 \mid \cdots \mid \phi_q] , \tag{11.14}$$

   where $\phi_j, j = 1, \ldots, q$ represent eigenvectors of $\mathbf{S}$ corresponding to the q largest eigenvalues.

4. Given $\mathbf{\Phi}$ and $\bar{\mathbf{x}}$, each shape can be approximated as

$$\mathbf{x}_i \approx \bar{\mathbf{x}} + \mathbf{\Phi}\mathbf{b}_i , \tag{11.15}$$

where

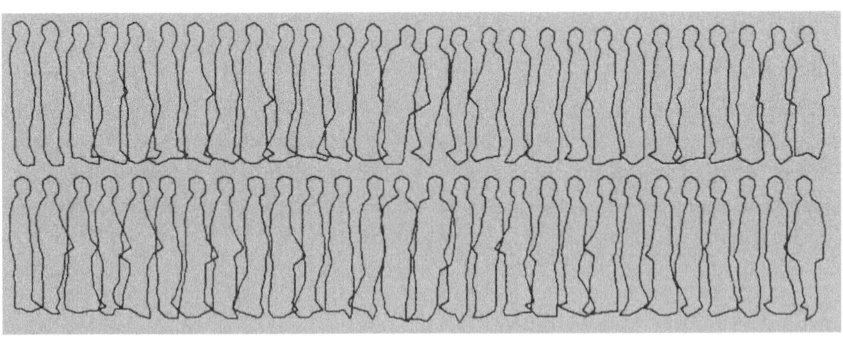

Figure 11.15. *Training set of 56 shapes (m = 56, reprinted from [Kos et al. 03] with permission from Elsevier).*

$$\mathbf{b}_i = \mathbf{\Phi}^T (\mathbf{x}_i - \bar{\mathbf{x}}). \tag{11.16}$$

In step 3 of the PCA algorithm, q is determined so that the sum of the q largest eigenvalues is greater than 98% of the sum of all eigenvalues.

In order to generate plausible shapes, we need to evaluate the distribution of $\mathbf{b}$. To constrain $\mathbf{b}$ to plausible values, we can either apply hard conditions to each element b_i or constrain $\mathbf{b}$ to be in a hyperellipsoid. The nonlinear version of this constraint is discussed in [Soz et al. 95].

11.4.3 Model Fitting

The best pose and shape parameters to match a shape in the model coordinate frame, $\mathbf{x}$, to a new shape in the image coordinate frame, $\mathbf{y}$, can be found by minimizing the following error function

$$E = (\mathbf{y} - \mathbf{M}\mathbf{x})^T \mathbf{W} \; (\mathbf{y} - \mathbf{M}\mathbf{x}), \tag{11.17}$$

where $\mathbf{W}$ is a diagonal matrix whose elements are weighting factors for each landmark point and $\mathbf{M}$ represents the geometric transformation of rotation θ, translation $\mathbf{t}$, and scaling s. The weighting factors are set in relation to the displacement between the computed positions of the old and the new landmark points along the profile. If the displacement is large, then the corresponding weighting factor in the matrix is set low; if the displacement is small, then the weighting is set high. Given a single point, denoted by $[x_0, y_0]^T$, the geometric transformation is defined as

$$\mathbf{M}\begin{bmatrix} x_0 \\ y_0 \end{bmatrix} = s\begin{bmatrix} \cos\theta & \sin\theta \\ -\sin\theta & \cos\theta \end{bmatrix}\begin{bmatrix} x_0 \\ y_0 \end{bmatrix} + \begin{bmatrix} t_x \\ t_y \end{bmatrix}. \tag{11.18}$$

After the set of pose parameters, $\{\theta, \mathbf{t}, s\}$, is obtained, the projection of $\mathbf{y}$ into the model coordinate frame is given as

$$\mathbf{x}_p = \mathbf{M}^{-1}\mathbf{y}. \tag{11.19}$$

Finally, the model parameters are updated as

$$\mathbf{b} = \mathbf{\Phi}^T (\mathbf{x}_p - \bar{\mathbf{x}}). \tag{11.20}$$

As the result of the searching procedure along profiles, the optimal displacement of a landmark point is obtained. The combination of optimally updated landmark points generates a new shape in the image coordinate frame $\mathbf{y}$.

This new shape is now used to find the nearest shape using Eq. (11.17). After computing the best pose, denoted by $\mathbf{M}$, this new shape is projected into Φ, which contains principal components of the given training set. This process updates the model parameter $\mathbf{b}$. As a result, only similar variation corresponding to the principal components can affect the model parameters. After computing the model parameters, the new shape, denoted by $\mathbf{x}$, can be generated by Eq. (11.15), and this new shape is used for the following iterations as in Eq. (11.17). After a suitable number of iterations, the final shape is obtained as $\mathbf{x}$.

11.4.4 Modeling a Local Structure

A statistical, deformable shape model can be built by assignment of landmark points, PCA, and model fitting steps. In order to interpret a given shape in the input image based on the shape model, we must find the set of parameters that best match the model to the image. Assuming that the shape model represents strong edges and boundaries of the object, a profile across each landmark point has an edge-like local structure.

Let $\mathbf{g}_j$, $j = 1,\ldots,n$, be the normalized derivative of a local profile of length K across the jth landmark point, and $\bar{\mathbf{g}}_j$ and $\mathbf{S}_j$ the corresponding mean and covariance, respectively. The nearest profile can be obtained by minimizing the following Mahalanobis distance between the sample and the mean of the model as

$$f(\mathbf{g}_{j,m}) = (\mathbf{g}_{j,m} - \bar{\mathbf{g}}_j)^T \mathbf{S}_j^{-1} (\mathbf{g}_{j,m} - \bar{\mathbf{g}}_j), \tag{11.21}$$

where $\mathbf{g}_{j,m}$ represents $\mathbf{g}_j$ shifted by m samples along the normal direction of the corresponding boundary. In practice, we use a hierarchical ASM technique because it provides a wider range for the nearest profile search.

Active shape models can be applied to the tracking of people. The shape of a human body has a unique combination of head, torso, and legs, which can be modeled with only a few parameters of the ASM. ASM-based video tracking can be performed in the following order: (1) shape variation modeling, (2) model fitting, and (3) local structure modeling,

$$E = (\mathbf{y} - \mathbf{Mx})^T \mathbf{W}(\mathbf{y} - \mathbf{Mx}), \tag{11.22}$$

where $\mathbf{M}$ represents the geometric transformation of rotation θ, translation t, and scale s. After the set of pose parameters, $\{\theta, \mathbf{t}, s\}$, is obtained, the projection of $\mathbf{y}$ into the model coordinate frame is given as $\mathbf{x}_p = \mathbf{M}^{-1}\mathbf{y}$. The model parameters are updated as $\mathbf{b} = \mathbf{\Phi}^T (\mathbf{x}_p - \bar{\mathbf{x}})$.

11.4.5 Hierarchical Approach for Multiresolution ASM

Video tracking systems inherently have variously shaped and sized input objects, which often results in a poor match of the initial model with an actual input shape. The hierarchical approach to multiresolution ASM is essential for video tracking systems to deal with such large deviation of initial fitting from the original shape. The idea of using pyramid models in image analysis was introduced by Tanimoto and Pavlidis [TanPav75] as a solution to edge detection. One important property of the pyramid model is that it is computationally efficient with comparable or better performance than with nonpyramidal approaches [Kro96]. Experiments with color stereo images have shown that matching is in general more accurate when using a hierarchical correspondence analysis instead of a nonhierarchical one. In addition, the computation time can be significantly reduced with a hierarchical approach [KosRod97].

Baumberg [Bau98] suggested a hierarchical implementation of snakes in intensity images. He discusses how a Kalman filter can be used with a snake model approach to improve shape-fitting robustness. He varies the number of landmark points in a coarse-to-fine sampling. The approach presented in this section differs from this in that (1) ASMs are used instead of snakes, (b) the same number of landmark points is used in every level of the image pyramid, and (3) a sequence of color image pyramids (one pyramid for every frame) instead of a sequence of intensity images is used for tracking. Furthermore, we will show that our approach applying an image pyramid can significantly improve the shape-fitting accuracy while Baumberg [Bau98] states that his hierarchical approach "does not appear to reduce the accuracy of image fitting" (p. 333).

The proposed hierarchical algorithm employs a quad pyramid of color images. In the calculation of a quad pyramid, each level is determined by a reduction of the resolution by a factor of four from the nearest lower level. A level L image represents an image that has been reduced by a factor 2^{2L} from the original image (level 0). The color values of the pixel are determined by calculating the mean values in each color component. It is noted that a color distortion appears when calculating the mean values in the color component [Zhe et al. 93]. This is, however, not important for our tracking algorithm, since in the upper levels of the pyramid only estimated values for the model fitting are determined. The final fitting values for the original color images are calculated at the lowest level (here, level 0). The example in Fig. 11.16 shows an image data pyramid with three resolutions (three levels, $L = 3$) of 320×240 pixels, 160×120 pixels, and 80×60 pixels.

The proposed hierarchical algorithm first reduces the size of the input image by a factor of 2^{2L}, and performs model fitting on the reduced image, which we denote "level L image." The result from the level L image is used as the initial model shape for the level $L-1$ image, and this hierarchical process continues until the result of the level 0 image is obtained.

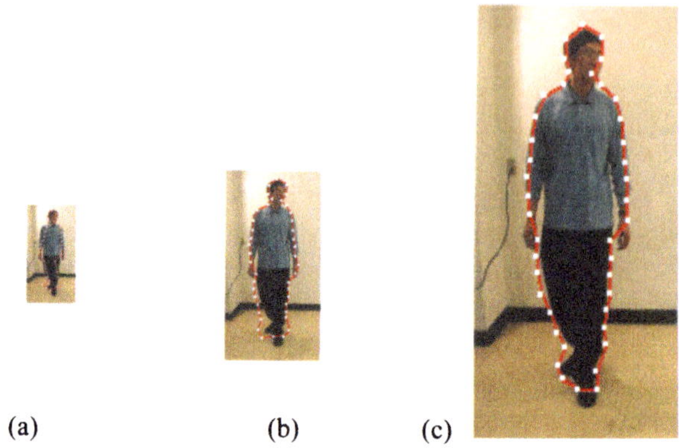

Figure 11.16. *Three different resolutions used in the hierarchical approach: (a) level 2, (b) level 1, and (c) level 0.*

In order to determine the optimal length of the local profiles and the corresponding number of hierarchies, denoted by K and L, respectively, different sets of these parameters are tested. Experimental results and discussions pertaining to the multiresolution ASM will be given in the next sections.

11.4.6 Extending ASMs to Color Image Sequences

In gray-level image processing, the objective functions for model fitting are determined along the normals for a representative point in the gray-value distribution. When selecting a vector-valued technique for extending ASMs to color image sequences, derivatives of vector fields can be incorporated into the objective functions for model fitting. However, the use of derivatives of vector fields in color image processing is based on classical Riemannian geometry, which makes it difficult to apply them to color spaces other than *RGB*. Our motivation for incorporating color information into ASM-based video tracking is to have the capability to distinguish between objects (or persons) of similar shape but with different colors.

Here, we present a simpler way to deal with color information by applying a monochromatic-based technique to the objective functions for model fitting. This can be done by first computing objective functions separately for each component of the color vectors. Afterward, a "common" minimum has to be determined by analyzing the resulting minima that are computed for each single color component. One method for doing this consists of selecting the smallest minimum in the three color components as a candidate. The common minimum becomes

$$\min\left\{\arg\min_{m} f_A(\mathbf{g}_{j,m}), \arg\min_{k} f_B(\mathbf{g}_{j,k}), \arg\min_{l} f_C(\mathbf{g}_{j,l})\right\}, \quad (11.23)$$

where f_A, f_B, and f_C are defined as in Eq. (11.21) for the three components in a tristimulus color space *ABC* (e.g., *RGB*). Consider the following example in the *RGB* space. We find the best fit (based on the minimization of Eq. (11.21)) for landmark point r between frame i and frame $i+1$ of the image sequence by a displacement (along the normal) of 4 pixels in the *R*-component, a displacement of 3 pixels in the *G*-component, and a displacement of 5 pixels in the *B*-component. The new updated position of landmark point r in frame $i+1$ is its old position in frame i shifted by 3 pixels along the normal. However, if one of the three color components contains an outlier (as in the example in Fig. 11.17), this outlier might be selected as a minimum.

Another procedure consists of selecting the mean value of the absolute minima in all three color components. The mean value becomes

$$\frac{1}{3}\left(\arg\min_{m} f_A(\mathbf{g}_{j,m}) + \arg\min_{k} f_B(\mathbf{g}_{j,k}) + \arg\min_{l} f_C(\mathbf{g}_{j,l})\right), \quad (11.24)$$

where all parameters are previously defined. However, outliers in one color component also lead in this case to a wrong result. Furthermore, the mean value may represent a value that does not correspond to any of the results of the energy functions' optimization. One way to overcome this problem is to use the median of the absolute minima in the three color components as a candidate.

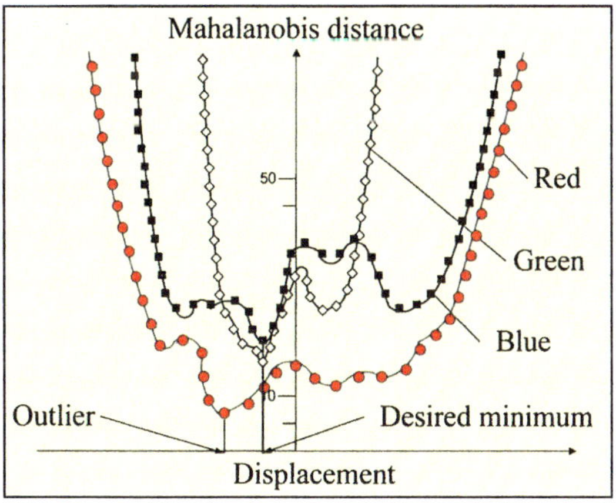

Figure 11.17. *Example of objective functions for the three color components in the RGB color space with an outlier in the red components.*

Thereby the influence of outliers in the minima of the objective functions is minimized. The median becomes

$$\text{median}\left\{\arg\min_m f_A(\mathbf{g}_{j,m}), \arg\min_k f_B(\mathbf{g}_{j,k}), \arg\min_l f_C(\mathbf{g}_{j,l})\right\}. \quad (11.25)$$

However, further false values may arise during the alignment of the contours.

Moreover, we will further address the question whether a contrast-adaptive optimization might improve the ASM performance. This approach is motivated by the observation that in general ASMs fit better to the object contour in high-contrast areas than in low-contrast areas. For every single landmark point we will select the color channel with the highest contrast and minimize the corresponding objective function. Based on the local contrast, we use, for example, the minimum of the objective function for the red channel for landmark point 1 and the minimum of the objective function for the blue channel for landmark point 2 to compute the fitting ASM.

We studied the performance of the ASM when employing the color spaces *RGB*, *YUV*, and *HSI*. So far, the same procedure has been applied to all color spaces. In these experiments, the best results were obtained when using the median in the *RGB* space (see Fig. 11.18 and Table 11.1). In addition, we applied a hierarchical implementation using image pyramids to speed up the process and decrease the error [Kan et al. 02].

The initial landmark points were manually placed as shown in Fig. 11.18. However, Hill et al. [Hil et al. 94] suggested a genetic algorithm that determines the "best" form parameters from a randomly specified set of initial values. Here a manual definition of the form parameters is suitable since the initial form has only to be determined once for a class of similar-shaped objects. Our goal in this example is to track persons and to ignore other moving objects.

Moreover, a maximum shift between two image frames is defined for an object to be tracked. This limitation is due to a reduction of the computing time and does not restrict the algorithm in general. The maximum shift parameter depends on the size of the object, the distance between the camera and the object,

Color Space	**Minimum**	**Median**	**Mean**
Intensity		164.32	
RGB	208.27	142.29	142.29
YUV	406.78	353.77	196.88
HSI	251.68	343.74	207.36

Table 11.1. *Error between the manually assigned points and the estimated points using three different minimum selection methods in different color spaces for a selected frame.*

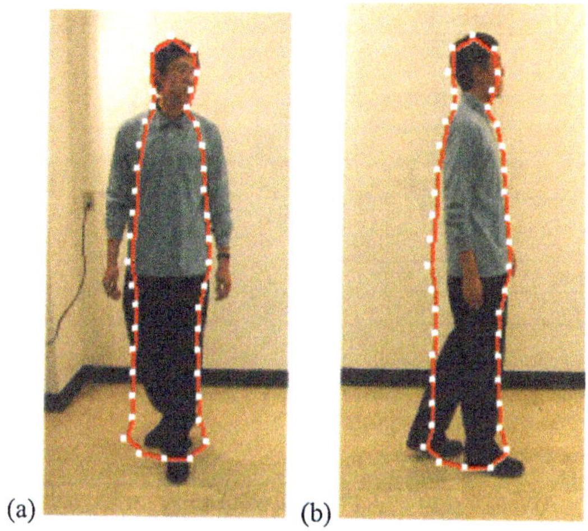

Figure 11.18. *Test images with initial points for (a) the 57th image and (b) the 7th image (reprinted from [Kos et al. 03] with permission from Elsevier).*

the velocity of the object, and the moving direction of the object. For example, for tracking a person in an airport we can predict the maximum size of a person, the maximum velocity of a walking or running person, and the minimum distance between the camera and a person. To limit the moving direction of a person, we can further assume that only a few persons might move toward a camera that is mounted on a wall. In this investigation the maximum shift is limited to 15 pixels for the hierarchical approach.

Both hierarchical (H) and nonhierarchical (NH) methods were tested for the image shown in Fig. 11.18a because its initial contour was set smaller than the real object. On the other hand, only the nonhierarchical method was tested in Figure 11.18b. In the hierarchical approach, level 0 represents the original given resolution, level 1 the half-sized resolution, and level 2 the quarter-sized resolution. Three different levels were shown in Fig. 11.16. We performed 5 iterations in level 2, another 5 iterations in level 1, and finally 10 iterations in level 0. For the nonhierarchical approach we performed 10 iterations. The hierarchical approach helps to enlarge the search regions and shows a better search result than the nonhierarchical approach. The model-fitting error for each experiment is summarized in Table 11.2.

The result of the hierarchical approach to Fig. 11.18a is shown in Fig. 11.19. The result of the nonhierarchical approach is shown in Fig. 11.20. The median method gives the best results in the sense of both the visual and the objective error measurements. Results using the *R*, *G*, and *B* color components show worse fitting than those methods using intensity. Table 11.2 summarizes error measurements of the different methods given in Table 11.3.

	Intensity	*R*	*G*	*B*
NH (57th)	196.25	241.33	157.51	190.40
H (57th)	146.66	182.84	159.18	164.37
NH (7th)	161.33	171.09	171.19	225.47
	Minimum	**Median**	**Mean**	**Adaptive**
NH (57th)	192.57	178.81	185.56	185.54
H (57th)	177.14	122.72	169.45	156.27
NH (7th)	158.94	152.08	170.46	162.96

Table 11.2. *The sum of distance between the estimated points by the different searching methods and the manually assigned points.*

Intensity	The result using the intensity image with the intensity profile.
R	The result using the color image with the Red profile.
G	The result with the Green profile.
B	The result with the Blue profile.
Minimum	The result using the color image after selecting the minimum of the minima of the Mahalanobis distance in the three color channels.
Median	The result with the median of the minima.
Mean	The result with the mean of the minima.
Adaptive	The result using the intensity image with the adaptive profile model that is modeled with the strongest edge among three color channels for each point.

Table 11.3. *Terminologies.*

The fitting results are also shown in Fig. 11.21. Based on Table 11.4 and Fig. 11.21, the result with 42 landmark points gives the best fitting in the sense of both quantitative and qualitative criteria. However, results with a reduced number of landmark points also give the correct location and size of the object.

Sequence (Frame)	$N = 10$	$N = 14$	$N = 21$	$N = 42$
Man_6 (19)	4.01	4.29	2.74	2.42
Man_9 (4)	8.06	7.83	7.22	3.01

Table 11.4. *Normalized error between the manually assigned and estimated points using different numbers of landmark points. The hierarchical approach with the median selection mode was used in the RGB color space (reprinted from [Kos et al. 03] with permission from Elsevier).*

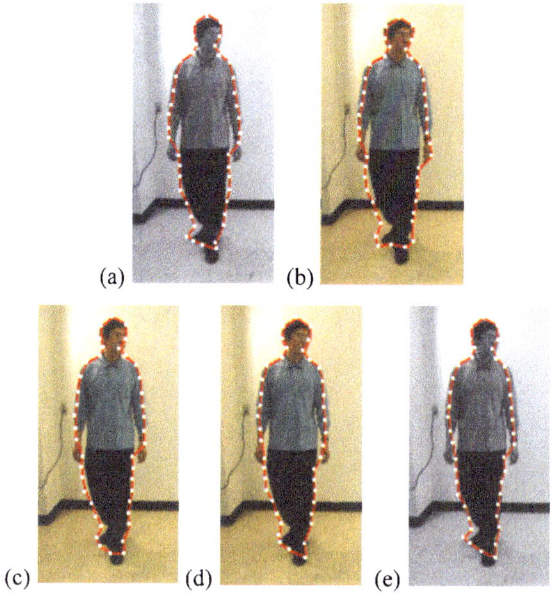

Figure 11.19. *Hierarchical search results of five different methods for the 57th image: (a) intensity, (b) minimum, (c) median, (d) mean, and (e) adaptive.*

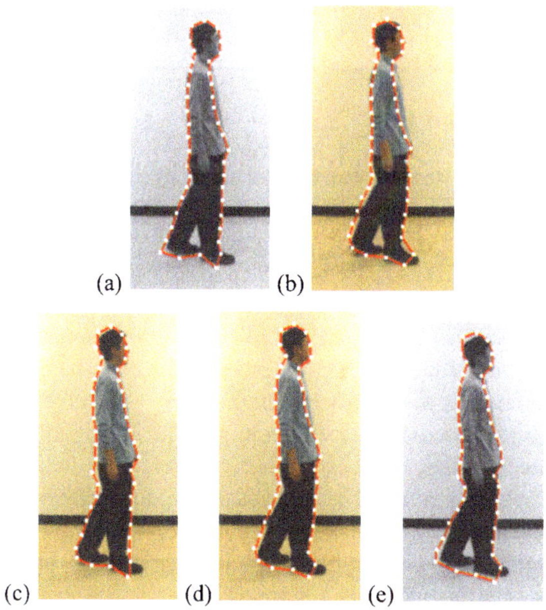

Figure 11.20. *Nonhierarchical search results of five different methods for the 7th image: (a) intensity, (b) minimum, (c) median, (d) mean, and (e) adaptive.*

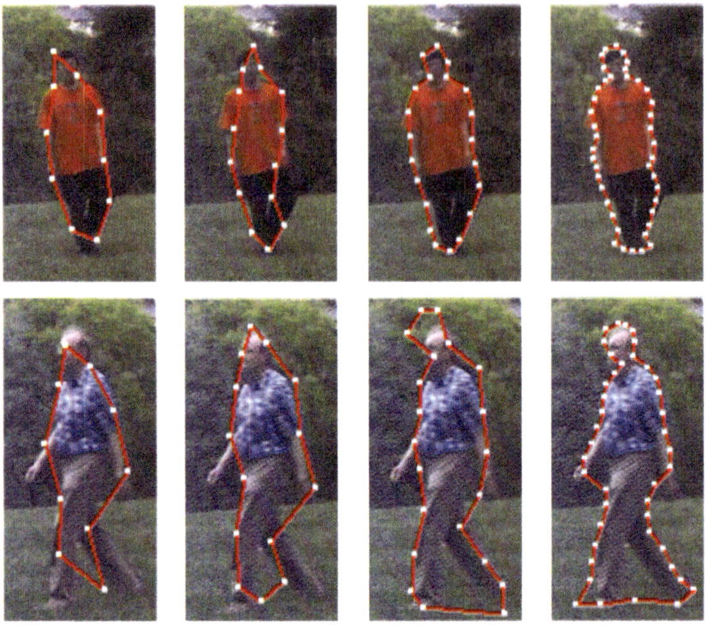

(a) (b) (c) (d)

Figure 11.21. *Fitting results using the 19th frame of the sequence Man_6 (top) and the 4th frame of the sequence Man_9 (bottom) with different numbers of landmark points: (a) 10, (b) 14, (c) 21, and (d) 42 (reprinted from [Kos et al. 03] with permission from Elsevier).*

11.4.7 Partial Occlusions

One advantage of ASM-based tracking is its ability to follow the shape of an occluded object. We studied outdoor sequences in the *RGB* color space, where individuals are partially occluded by different objects. Results obtained when applying the hierarchical method with the median selection mode to the sequence Man_11 are shown in Fig. 11.22.

In a second experiment using an additional outdoor sequence, the ASM was applied to each of the outdoor image frames and the mean, the minimum, and the median of the minima in the objective functions were selected for searching. The results for selecting the median of the minima are shown in Fig. 11.23. Note that the image sequence is slightly out of focus, which does not have a significant impact on the tracking results. The proposed tracking scheme provided good results in our experiments, even though the subjects are partially occluded by a bench. One property of the ASM-based tracking scheme is that the ASM can easily adjust to reappearing parts of the tracked object in an image sequence.

Figure 11.22. *Fitting results in four frames of a video sequence with a partially occluded person. The hierarchical method with the median selection mode in the RGB color space was used (reprinted from [Kos et al. 03] with permission from Elsevier).*

Tracking of a person becomes rather difficult if the image sequence contains several, similarly shaped moving people. In this case, a technique based exclusively on the contour of a person will have difficulties in tracking a selected individual. On the other hand, a technique exclusively evaluating the colors of a moving person (or object) may also fail. Any color-based tracker can lose the object it is tracking due, for example, to occlusion or changing lighting conditions. To overcome the sensitivity of a color-based tracker to changing lighting conditions, the color constancy problem has to be solved at least in part, which is a nontrivial and a computationally costly task.

A possible solution to this problem might consist of a weighted combination of ASM form-based and color-indexing tracking techniques. By applying such a combination technique to image sequences we might be able to distinguish between: (1) objects of similar colors but with different forms, and (2) objects of different colors but with similar forms. One drawback of such a combination approach is its high computational cost. Here a hardware implementation can be considered later for real-time applications.

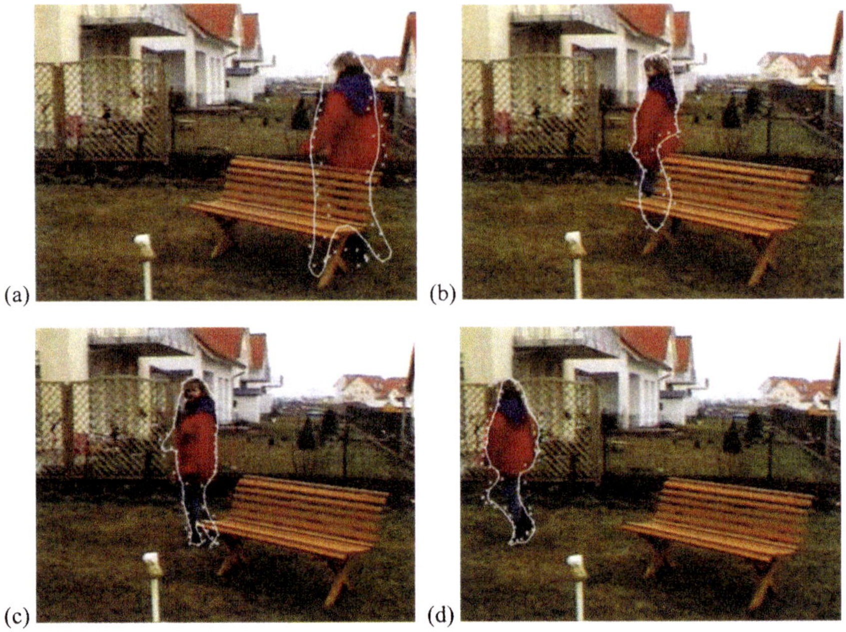

Figure 11.23. *Search results for an outdoor sequence using the nonhierarchical approach for (a) the 1st frame, (b) the 19th frame, (c) the 27th frame, and (d) the 33rd frame.*

11.4.8 Summary

A technique has been presented for recognizing and tracking a moving nonrigid object or person in a video sequence. The objective function for active shape models has been extended to color images. We have evaluated several different approaches for defining an objective function considering the information from the single components of the color image vectors. This tracking technique does not require a static camera (except to initialize the landmark points for the object to be recognized). Thus, it can be applied when using a PTZ (pan-tilt-zoom) camera for video tracking. However, the profile length has to be adapted to the pan, tilt, and zoom parameters of the PTZ camera.

In both our indoor and outdoor experiments, the median computation of the minima in the energy functions proved favorable. In general, the error in fitting an ASM to the real contour of an object was lower when using color information than when just using intensity information. Furthermore, we have shown that the fitting error is further reduced when applying a hierarchical instead of a nonhierarchical approach to the images. We showed that a small number of landmark points is sufficient for tracking if only a rough approximation of the object to be tracked is needed. When studying the *RGB*, *HSI*, and *YUV* color spaces, the method performed best in the *RGB* space. This was predominantly caused by a nonweighted analysis of the color components in the other spaces. Further investigations are necessary for a more detailed analysis of these color spaces.

The performance of the algorithm was rather robust regarding partial object occlusions. The problem of outliers in the objective functions could be partly solved by the evaluation of color information. One way to further enhance these results might be a refined analysis of the objective functions, where the neighbors of one point are also considered. Thereby the number of outliers can be further reduced.

The hierarchical, color active shape modeling algorithm took approximately 4 seconds for processing one frame using a Pentium 4, 1.3 GHz personal computer. This algorithm consists of: 5 iterations for level 2, 5 iterations for level 1, and 10 iterations for level 0. The processing time was measured without code optimization. This time can be significantly reduced if: (1) fewer landmark points are used, (2) the profile length becomes smaller, or (3) code optimization is performed. This may reduce the quality of shape fitting (as shown in Section 11.4.7) but will still allow a rough tracking of objects.

The tracking of a person becomes rather difficult if the image sequence contains several, similarly shaped moving people. In this case, a technique based exclusively on the contour of a person will have difficulties in tracking a selected person and the task may fail if the person is partially occluded. On the other hand, a technique exclusively evaluating the colors of a moving person (or object) may also fail. Any color-based tracker can lose the object it is tracking due, for example, to occlusion or changing lighting conditions. To overcome the sensitivity of a color-based tracker to changing lighting conditions, the color constancy problem has to be solved at least in part. This is a nontrivial and computationally costly problem that in general cannot be solved in video real-time. Another solution to the problem mentioned above could consist of a weighted combination of a form-based tracking technique using (for example) ASMs and a color-based tracking technique using (for example) color indexing. By applying such a combination technique to image sequences we might be able to distinguish between (1) objects of similar colors but with different forms and (2) objects of different colors but with similar forms.

11.5 REFERENCES

[Bau98] A. Baumberg. Hierarchical shape fitting using an iterated linear filter. *Image and Vison Computing* **16**, 1998, pp. 329-335.

[BlaIsa98] A. Blake, M. Isard. *Active Contours*. Springer, London, England, 1998.

[Bro et al. 94] S.A. Brock-Gunn, G.R. Dowling, T.J. Ellis. Tracking using colour information. *Proc. Int. Conference on Automation, Robotic and Computer Vision*, 1994, pp. 686-690.

[Che95] Y. Cheng. Mean shift, model seeking, and clustering. *IEEE Transactions on Pattern Analysis and Machine Intelligence* **17** (1995), pp. 790-799.

[ComRam00] D. Comaniciu, V. Ramesh. Robust detection and tracking of human faces with an active camera. *Proc. Visual Surveillance*, 2000, pp. 11-18.

[Coo et al. 95] T.F. Cootes, D.H. Cooper, C.J. Taylor, J. Graham. Active shape models - Their training and application. *Computer Vision and Image Understanding* **61** (1995), pp. 38-59.

[DelColl99] F. Dellaert, R. Collins. Fast image-based tracking by selective pixel integration. *Proc. ICCV 99 Workshop on Frame-Rate Vision*, September 1999.

[Dom et al. 06] S.M. Dominguez, T. Keaton, A.H. Sayed. A robust finger tracking method for multimodal wearable computer interfacing. *IEEE Transactions on Multimedia* **8** (2006), pp. 956-972.

[Har et al. 00] I. Haritaoglu, D. Harwood, L.S. Davis. W4: Real-time surveillance of people and their activities. *IEEE Transactions on Pattern Analysis and Machine Intelligence* **22** (2000), pp. 809-830.

[Hil et al. 94] A. Hill, C.J. Taylor, T.F. Cootes. A generic system for image interpretation using flexible templates. *Proc. European Conference on Computer Vision*, 1994, pp. 276-285.

[Hor et al. 00] T. Horprasert, D. Harwood, L.S. Davis. A robust background subtraction and shadow detection. *Proc. Asian Conference on Computer Vision*, Taipei, Taiwan, 2000.

[Kak et al. 07] P. Kakumanu, S. Makrogiannis, N. Bourbakis. A survey of skin-color modeling and detection methods. *Pattern Recognition* **40** (2007), pp. 1106-1122.

[Kan et al. 02] S.K. Kang, H.S. Zhang, J.K. Paik, A. Koschan, B. Abidi, M.A. Abidi. Hierarchical approach to enhanced active shape model for color video tracking. *Proc. Int. Conference on Image Processing*, Rochester, New York, 2002, Vol. I, pp. 888-891.

[Kim et al. 03] Y.-O. Kim, J. Paik, J. Heo, A. Koschan, B. Abidi, M. Abidi. Automatic face region tracking for highly accurate face recognition in unconstrained environments. *Proc. IEEE Int. Conference on Advanced Video and Signal Based Surveillance*, Miami, Forida, 2003, pp. 29-36.

[Kos et al. 03] A. Koschan, S. Kang, J. Paik, B. Abidi, M. Abidi. Color active shape models for tracking nonrigid objects. *Pattern Recognition Letters* **24** (2003), pp. 1751-1765.

[KosRod97] A. Koschan, V. Rodehorst. Dense depths maps by active color illumination and image pyramids. In: F. Solina et al., eds., *Advances in Computer Vision*, Springer, Vienna, Austria, 1997, pp. 137-148.

[Kro96] W.G. Kropatsch. Properties of pyramidal representations. *Computing Suppl. 11* (1996), pp. 99-111.

[Lee et al. 00] L. Lee, R. Romano, G. Stein. Monitoring activities from multiple video streams: establishing a common coordinate frame. *IEEE Transactions on Pattern Analysis and Machine Intelligence* **22** (2000), pp. 758-767.

[Lev97] H. Levkowitz. Color Theory and Modeling for Computer Graphics, Visualization and Multimedia Applications. Kluwer, 1997.

[Li et al. 02] J. Li , C.S. Chua, Y.K. Ho. Color based multiple people tracking. *Proc. 7th Int. Conference on Control, Automation, Robotics and Vision*, Singapore, 2002, pp. 309-314.

[Li et al. 00] Y. Li, A. Goshtasby, O. Garcia. Detecting and tracking human faces in videos. *Proc. Int. Conference on Pattern Recognition*, 2000, vol. 1, pp. 807-810.

[LuTan01] W. Lu, Y.-P. Tan. A color histogram based people tracking system. *Proc. ISCAS*, 2001, vol. 2, pp. 137-140.

[MarVil02] F. Marqués, V. Vilaplana. Face segmentation and tracking based on connected operators and partition projection. *Pattern Recognition* **35** (2002), pp. 601-614.

[McK et al. 99] S.J. McKenna, Y. Raja, S. Gong. Tracking colour objects using adaptive mixture models. *Image and Vision Computing* **17** (1999), pp. 225-231.

[Nic et al. .00] M. Nicolescu, G. Medioni, M. Lee. Segmentation, tracking and interpretation using panoramic video. *Proc. IEEE Workshop on Omnidirectional Vision*, pp. 169-174, 2000.

[ParSay01] M. Pardàs, E. Sayrol. Motion estimation based tracking of active contours. *Pattern Recognition Letters* **22** (2001), pp 1447-1456.

[PläFua01] R. Plänkers, P. Fua. Tracking and modeling people in video sequences. *Computer Vision and Image Understanding* **81** (2001), pp. 285-302.

[Rob et al. 06] T.J. Roberts, S.J. McKenna, I.W. Ricketts. Human tracking using 3D surface colour distributions. *Image and Vision Computing* **24** (2006), pp. 1332-1342.

[Roh et al. 00] H. Roh, S. Kang, S.-W. Lee. Multiple people tracking using an appearance model based on temporal color. *Proc. Int. Conference on Pattern Recognition*, vol. 4, pp. 643 -646, 2000.

[Son et al. 99] M. Sonka, V. Hlavac, R. Boyle. *Image Processing, Analysis, and Machine Vision*. Brooks/Cole, 1999.

[Soz et al. 95] P. Sozou, T.F. Cootes, C.J. Taylor, E.D. Mauro. A nonlinear generalization of point distribution models using polynomial regression. *Image and Vision Computing* **12** (1995), pp. 451-457.

[SwaBal91] M.J. Swain, D.H. Ballard. Color indexing. *Int. J. of Computer Vision* **7** (1991), pp. 11-32.

[TanPav75] S. Tanimoto, T. Pavlidis. A hierarchical data structure for picture processing. *Computer Graphics and Image Processing* **4** (1975), pp.104-119.

[Tia et al. 01] Q. Tian, N. Sebe, E. Loupias, T.S. Huang. Image retrieval using wavelet-based salient points. *J. Electronic Imaging* **10** (2001), pp. 935-849.

[WuHua02] Y. Wu, T.S. Huang. Non-stationary color tracking for vision-based human computer interaction. *IEEE Transactions on Neural Networks* **13** (2002), pp. 948-960.

[WuYu06] Y. Wu, T. Yu. A field model for human detection and tracking. *IEEE Transactions on Pattern Analysis and Machine Intelligence* **28** (2006), pp.753-765.

[XioDeb04] T. Xiong, C. Debrunner. Stochastic car tracking with line- and color-based features. *IEEE Transactions on Intelligent Transportation Systems* **5** (2004), pp. 324-328.

[Zhe etal93] J. Zheng, K.P. Valavanis, J.M. Gauch. Noise removal from color images. *J. Intelligent and Robotic Systems* **7** (1993), pp. 257-285.

12 MULTISPECTRAL IMAGING FOR BIOMETRICS

Multispectral imaging has been widely used in various applications, such as remote sensing for resource monitoring, astronomy, medical imaging, analysis of museological objects, agriculture, manufacturing, forensics, and high-fidelity color printing. In this technology, information is collected over contiguous narrow-wavelength intervals across the visible, near-infrared, or infrared regions and can generate precise optical spectra at every pixel. Multispectral images carry information about a number of spectral bands: from three components per pixel for *RGB* color images to over a hundred bands for hyperspectral images. This can allow for extracting additional information that the human eye fails to capture.

In the following sections, the term *multispectral image* is defined and several techniques for acquiring multispectral images are discussed. Then the relatively new application of multispectral imaging in biometrics is addressed. Several techniques for multispectral image fusion are presented for improved face recognition. The first group of techniques, presented in Section 12.3, fuses conventional three-channel color images (*RGB*) with IR images to overcome sensitivity to changing illumination. The second group of techniques, introduced in Section 12.4, uses multiple bands in the visible spectrum to analyze face image data. Results are presented that are obtained in face recognition when applying different data-fusion techniques to multispectral images.

12.1 WHAT IS A MULTISPECTRAL IMAGE?

A *multispectral image* is a collection of several monochrome images of the same scene, each of them taken with additional receptors sensitive to other frequencies of the visible light, or to frequencies beyond visible light, like the infrared region of electromagnetic continuum. Each image is referred to as a *band* or a *channel*.

Regarding to Eq. (1.2) in Chapter 1, a *multispectral* or *multiband image* can be represented as

$$\mathbf{C}(x, y) = \left(u_1(x, y), u_2(x, y), \ldots, u_n(x, y)\right)^T = \left(u_1, u_2, \ldots, u_n\right)^T, \qquad (12.1)$$

whose special case for $n = 3$ can be, for example, a three-channel color image in the *RGB* space. While a monochrome image, $n = 1$, has only one band, which is represented as a gray-value image, a multispectral image consists of at least three bands, $n \geq 3$. Thus, the image value of a pixel in a multispectral image is represented by vectors with n components, as opposed to scalar image values representing pixels in a monochrome image. Although a color image with three bands constitutes in theory the simplest form of a multispectral image, the term is more commonly used for images with more than three bands. One example would be a four-band image using the three *RGB* bands and an additional band beyond the visible spectrum, like in the infrared (IR). Satellites usually take several images from frequency bands in the visible and nonvisible range. *Landsat 5*, for example, produces seven-band images, $n = 7$, with the wavelength of the bands being between 450 and 1250 nm.

There is no common agreement yet on the definition of the term *hyperspectral image*. However, the term is commonly used for images with more than a hundred bands, $n > 100$. While *multi* in multispectral means *many* spectral bands, the *hyper* in hyperspectral means *over* as in *more than many* and refers to the large number of measured wavelength bands.

12.2 MULTISPECTRAL IMAGE ACQUISITION

Multispectral images (MSIs) are typically acquired by specialized systems. In the late nineteenth century, several optical engineers developed platforms for capturing and reproducing the spectra of original scenes. Lancaster in 1895, for example, described a microdispersion process [Hunt95] that used a grating and a prism. The grating divided the original scene into small strips and the prism spectrally dispersed the strips onto a silver-halide negative. A dimensionally identical positive was made for viewing and it was placed into the apparatus where the negative had been. Nevertheless, the system did not make it beyond the laboratory because of the complexity for both capture and viewing.

In recent years, modern spectral image capture systems consider combinations of CCD cameras with various types of narrow- or broadband filters. The images are then processed using conventional computers with software developed to properly process the spectral data. Ohta et al. [Oht et al. 81] used a film-based system for multispectral image capture. They used a mechanical rotating filter wheel with eight gelatin filters and imaged rigid objects. Tominaga proposed a six-color camera system with six color filters [Tom96], which had the six spectral channels of the color filters' fixed-wavelength bands. The VASARI imaging system (Visual Art System for Archiving and Retrieval of Images) developed at the National Gallery in London employs a seven-channel multispectral camera to capture paintings [SauCup93]. It aims to show that high-resolution colorimetric imaging can be done directly from paintings. However, it is not portable and limits the image acquisition to predefined illumination and object sizes. In general, fixed-filter systems have three essential restrictions:

1. The selection of color filters with regard to their central wavelength and their bandwith is limited.
2. Filters with a narrow passband are difficult to manufacture.
3. Misalignments between image bands may occur due to mechanical vibrations of the imaging system when the filter wheel is spinning.

With the advances in filter technology, *electronically tunable filters* (*ETFs*) were used in conjunction with a monochrome camera to produce a stack of images at a sequence of wavelengths, forming the MSIs. A wide selection of different ETFs is commercially obtainable. The majority of them can be classified under three categories [PogAng01]:

1. Acousto-optical devices based on diffraction
2. Interferometer-type filters
3. Liquid crystal tunable filters based on birefringence (see also [Gat00] for further details)

The operation of the *acousto-optic tunable filter* (*AOTF*) is based on the interaction of electromagnetic and acoustic waves. The main module of an AOTF is an optically transparent crystal that possesses a certain combination of optical and acoustic properties. While the incoming light falls on the crystal, a radio-frequency acoustic wave is sent to the crystal simultaneously. It is used in creating a refractive index wave within the crystal. The incident beam when passing through the refractive index wave breaks into its component wavelengths. In the end, a single wavelength of light is selected for transmission [PogAng01]. Proper design makes one of these wavelengths much more prominent and that becomes the output color of the filter. The wavelength of the filtered light is selected by changing the frequency of the acoustic wave. AOTFs are lightweight and very fast spectral filtering devices. One disadvantage of such devices is the requirement that the incident light be collimated [Got94].

Another category of electronically tunable filters applies the principle of *optic interference.* A Fabry-Perot cavity is the basic component consisting of two parallel planar surfaces, whose inner face is coated with partially transparent films of high reflectivity, enclosing a rectangular volume of air or some dielectric material [PogAng01]. Light enters through one of the partially transparent mirrors and is multiply reflected within the cavity. The multiply transmitted rays interact with each other, creating optical interference effects, which result in the transmission, through the opposite semitransparent mirror, of only one particular wavelength and its harmonics. To block the unwanted harmonics, often two cavities in a row are employed, constituting a *dual tunable Fabry-Perot* (*DTFP*) device [PogAng01]. *Electro-optic Fabry-Perot* (*EOFP*) devices adjust the bandpass spectrum by varying the refractive index of the cavity through the application of electric potential. Recently, *liquid crystals* (LCFP) are employed as cavity medium. On average, single-cavity ETFs can select the output wavelength out of an input range that is no larger than 100 nm wide. Thus, a cascade of Fabry-Perot cavities is needed in order to have an EOFP that can analyze the entire visible spectrum. Such designs are more costly and have a lower transmission rate (20 - 50% instead of 90% for a single cavity [PogAng01]).

Recently, the Applied Spectral Imaging SpectraCube has been introduced, which is an interferometry-based portable digital camera. This camera is based on the idea that if interference of the color signal is created and measured, the spectrum of the original signal can be recovered applying the *inverse Fourier transform* [Fin et al. 04]. With this device, a full 2D array of spectra is captured at once and, unlike filter-based systems, a single exposure is acquired. The spectral resolution of this device can be set higher than most filter-based systems (e.g., about 4 nm), but it also comes at a high expense. Moreover, the single-image acquisition time ranges from 30 to 150 seconds (depending on spatial and spectral resolution and aperture) [Fin et al. 04].

The third and most commonly used category of filter devices is *liquid crystal tunable filters* (*LCTFs*), which use electrically controlled liquid crystal elements to select a specific visible wavelength of light for transmission through the filter at the rejection of all others. A typical wavelength-selective liquid crystal tunable filter is build from a stack of fixed filters consisting of interwoven birefringent crystal/liquid crystal combinations and linear polarizers. The spectral region passed by LCTFs depends on the choice of polarizers, optical coatings, and the liquid crystal characteristics. In general, visible-wavelength devices of this type usually perform quite well in the 400 - 700 nm range.

The LCTF is polarization sensitive. The minimum time needed to change between two successive bands is restricted by relaxation time of the crystal and is typically ~50 msec. Special devices can be designed for fast switching (~5 msec) through a short sequence of wavelengths. The spectral resolution, or bandpass, of an LCTF is on average of the order of several nm, although a narrower bandpass can be constructed. The physical limitations of the spectral resolution result also from the small amount of light that passes through a very narrow bandwidth filter, which is difficult to measure by a sensor.

LCTF-based multispectral imaging systems have been employed by several research groups [ImaBer98], [Ima et al. 01], [Pan et al. 03] [Har et al. 02], [Tom96] for different applications. Contrary to traditional filter wheels, liquid crystal tunable filters have no moving parts and are lightweight, which makes them suitable for mobile sensor platforms. The tunable filter provides the capability of finer spectral sampling with narrow bandwidths. In addition, the large aperture and imaging capability of these devices represent a distinct advantage over conventional dispersive spectral analysis techniques. Figure 12.1 shows a multispectral imaging system consisting of a monochrome camera and an LCTF.

An example of three bands from a multispectral image and a corresponding color image are depicted in Fig. 12.2. The multispectral bands are acquired with a width of 7 nm each where the center of the interval defines the band's name (e.g., band 590 refers to an image taken over the interval 586.5 – 593.5 nm).

Although each of the three types of ETFs relies on different principles of optics, all of them are successful in selecting individual bandpasses over a continuum of spectral ranges with high speed and accuracy. A large variety of LCTFs, AOTFs, and EOFPs is commercially available these days. Most of them have comparable performance characteristics. Table 12.1 lists typical characteristics of ETFs (after [PogAng01]).

Spectroradiometers are a precise alternative to filter-based systems. After light passes through the shutter, it is directed to a concave diffraction grating that breaks up the signal into a photosensitive array and focuses the diffracted signal onto a photosensitive array. These devices have a very high spectral resolution, precision, and stability [Fin et al. 04]. Nevertheless, one disadvantage of spectroradiometers is that they measure only single points. Therefore, it is nearly impossible to use them to capture a full scene.

Only a very few attempts have been made so far to employ multispectral imaging for face recognition. The Munsell Color Science Laboratory initiated efforts with multispectral images using a LCTF over the visible spectrum, especially for high resolution art portrait reconstruction (see [ImaBer98] and [Ima et al. 01]). They also acquired the Lippmann2000 database [RosJia99] that contains spectral images of several objects, including faces from four Caucasians and three East-Asians.

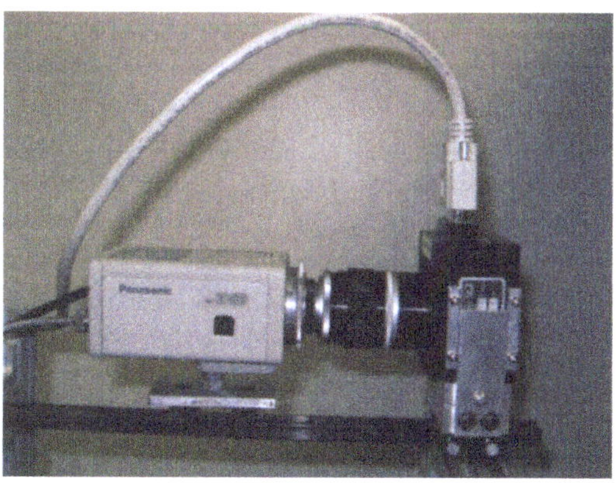

Figure 12.1. *Multispectral imaging system consisting of a monochrome camera and a liquid crystal tunable filter.*

Attributes	LCTF	AOTF	EOFP
Tunability time	~50 msec	~15 - 30 msec	~40 msec (LCFP) ~4 msec (DTFP)
Operating spectral range	400 -1 800 nm	200-5000 nm	400 - 1550 nm
Max width of tunable range	450 nm (vis + NIR) 950 nm (MIR)	700 nm (vis + NIR) 3900 nm (MIR)	100 nm
Min. output bandwidth	5 nm	0.4 nm	0.05 nm
Max. output bandwidth	30 nm	50 nm	10 nm
Mean error in central wavelength	0.5 nm	1 nm (varies with λ)	1 nm (varies with λ)
Average transmission rate	20 - 50 %	98%	20 - 50%
Transmission rate over wavelength	increases with wavelength	constant	increases with wavelength
Out of band transmission	0.01 - 0.05%	0.05 - 0.1%	0.5 -1%
Incident light limitations	none	requires collimated light	none

Table 12.1. *Typical tunable filter characteristics (after [PogAng01]).*

This data was acquired by a film camera with approximately 15 – 25 sec lapses between exposures and 16 exposures for each person, under flash lighting. Pan et al. [Pan et al. 03], [Pan et al. 04], [Pan et al. 05] acquired spectral images over the near-infrared spectrum (700 - 1000 nm) and demonstrated that spectral images of faces acquired in the near-infrared range can be used to recognize an individual under different poses and expressions. So far not much research has been done using multispectral imaging in the visible domain to address the problem of face recognition, especially with respect to changes in illumination conditions. The multispectral databases mentioned above either have very few data records or are not in the visible spectrum. Moreover, these datasets have not yet been used in conjunction with conventional face images in recognition engines. The following sections present some techniques and results for face recognition (FR) with multispectral images.

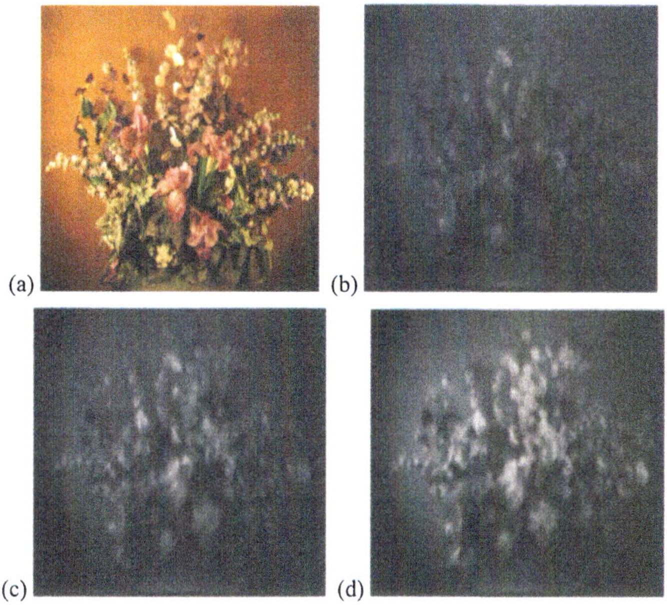

Figure 12.2. *Example of a multispectral image. (a) Color image "Flower" in RGB, (b) band at 590 nm, (c) band at 670 nm, and (d) band at 710 nm.*

12.3 FUSION OF VISIBLE AND INFRARED IMAGES FOR FACE RECOGNITION

The fusion of imaging modalities toward enrichment of the interpretive information from the image data is called *image fusion*. It is the capacity of being able to produce a single fused image from a set of input images. The fused image should have enhanced information that is more comprehensible and decipherable for human perception, machine learning, and computer vision. The automatic recognition of faces has become an important need in recent years. FR based only on the visible spectrum has shown difficulties in performing consistently in uncontrolled operating conditions. The accuracy of face recognition degrades significantly when the lighting is dim or when the face is not uniformly illuminated. Since the face is essentially a three-dimensional (3D) object, lighting sources from different directions may significantly change visual appearances. Light reflected from human faces also varies depending on the skin color of people from different ethnic groups. This variability, coupled with changing lighting conditions, may cause great difficulties in recognizing the face in applications such as outdoor surveillance tasks.

Face recognition using different imaging modalities, particularly infrared (IR) imaging sensors, has become an area of growing interest [Kon et al. 05]. The use of thermal IR images can improve the performance of face recognition in uncontrolled illumination conditions. The fusion of visible and infrared images reduces the effects of poor illumination, as shown in Fig. 12.3, since infrared images are invariant to changing illumination. In the fused image in Fig. 12.3f, the left side of the subject's face, the eyeballs, and edges of the subject's face are clearer than in the visible image in Fig. 12.3a.

The fusion technique used to obtain Figure 12.3 (f) [Har et al. 06] is based on *empirical mode decomposition* (*EMD*) and will be detailed later. Visible and infrared face images are fused to increase the FR rate. The visible and infrared images in Fig. 12.3 are taken from the NIST/Equinox database [Equ06], which consists of 1622 pairs of visible and calibrated thermal IR face images from 90 individuals. The visible and thermal IR image pairs are coregistered within 1/3 pixel with a hardware setting having a spatial resolution of 320 × 240 pixels, and a grayscale resolution of 8 bits (visible) and 12 bits (IR). Image fusion of multiple imaging modalities can be performed in various ways. Singh et al. [Sin et al. 04] use *principal components analysis* (PCA) to decompose input images into components for fusion using a genetic algorithm. Fusion is performed by selecting the maximum of the visible and infrared image wavelet coefficients by Li et al. [Li et al. 95].

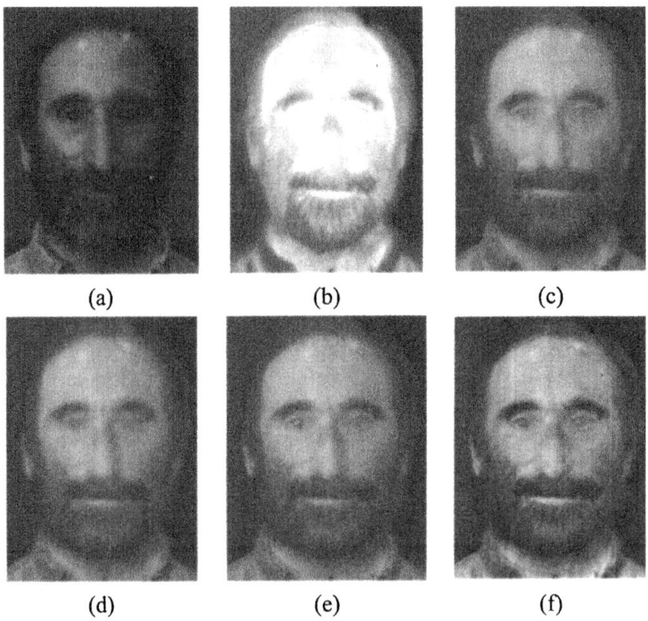

Figure 12.3. *An example where fusion of visible and infrared images reduces the effects of poor illumination. (a) Visible image (from [Equ06]), (b) infrared image (from [Equ06]), (c) averaged image, (d) PCA-fused image, (e) image fused using wavelet-based fusion [4], and (f) image fused using EMD.*

Here several major techniques are addressed that are widely used as image-fusion tools, namely the wavelet-fusion technique used by Kong et al. [Kon et al. 07], the PCA fusion scheme used in [RocFle98], and pixel-by-pixel averaging for comparisons. Besides the selection of the image-fusion technique, the selection of the matching algorithm for FR is equally important. Discussing different face recognition techniques is beyond the scope of this chapter. The interested reader is referred, for example, to Li and Jain [LiJa05] for further details. Face-it®, a renowned FR engine as indicated by Phillips et al. [Phi et al. 02] is applied in the tests presented here. Image data fusion requires the pixelwise alignment or registration of the images to be fused. In the next section, a technique for the registration of visible and thermal images is introduced. Then, empirical mode decomposition and its use in image fusion are detailed in the following sections.

12.3.1 Registration of Visible and Thermal Face Images

Bringing multiple images of the same scene into spatial correspondence is an important step in typical image-fusion procedures. A special-purpose imaging sensor assembly that produces coregistered image pairs in the visible and thermal IR domains may not be always practical in FR scenarios due to high cost and low availability. Software registration of images, which is applied as a first processing step to the acquired image pairs, enables the use of off-the-shelf visible and IR cameras for large-scale deployment at a reasonable cost. Although the accuracy of the pixel-to-pixel alignment can sometimes be less than the one that can be obtained with a special-purpose imaging sensor alignment for hardware registration, the salient image features can be matched sufficiently well in a software registration process. In general, registration methods vary depending on the similarity measures between the images. Popular similarity measures used in image registration include cross-correlation, correlation ratio, and mutual information.

Most area-based registration criteria generally assume global statistical dependence between the images to be aligned. This condition may not be satisfied in the case of multisensor imagery such as visible and thermal IR images. For images containing prominent features, or sharp changes in intensity in certain regions, feature maps such as edge or frequency maps enhance registration accuracies. To align visible and IR images, Irani and Anandan [IraAna98] designed a criterion that employed directional energy maps computed from the visible and IR images, which were used to reduce the visual differences between the two modalities and to highlight the common features. The technique constrains local statistical dependence between the two modalities and the criterion was obtained by summing local cross-correlation measures of small image patches. This method was applied to the alignment of manmade structures such as airport runways and buildings.

A Gaussian-fields technique was introduced for the registration of 3D point-sets in the context of scene modeling [Bou et al. 04]. This method relies on a differentiable criterion and employs a standard gradient-based optimization scheme. The inputs to the Gaussian-fields algorithm are the coordinates of the points that can be augmented by adding attributes such as color and local shape descriptors. The framework can also be applied to the two-dimensional (2D) registration of visible and thermal IR images. The criterion can be customized and applied to the task of binarized edge-map matching, which was addressed previously by Huttenlocher et al. [Hut et al. 93] through the use of the Hausdorff distance. The registration criterion maximizes the overlap between features that are present in both visible and thermal images. As a point-sets registration technique, the method is also closely related to the popular class of methods known as the *iterative closest point* (*ICP*) techniques [Bes92], [Dal02], although ICP was mostly employed for rigid registration. In the Gaussian-fields framework, a smooth point-set and a shape-matching criterion substitutes for conventional nondifferentiable registration algorithms such as ICP [Fit03] and Hausdorff distances [Cha et al. 03]. Moreover, an additional advantage of the Gaussian-fields approach is its computational efficiency, which leads to fast linear implementation [Elg et al. 03].

A point X of either datasets to be registered can be augmented with an associated 1D attribute vector $S(X)$ whose components can be local shape descriptors as well as intensity or color information. Let $S(X)$ and $S(Y)$ denote the attribute vectors of X and Y in the binarized feature-map computed from thermal IR and visible images. The binary feature-maps can be described as point-sets $M = \{X, S(X)\}$ of N_M points and $D = \{Y, S(Y)\}$ containing N_D points of the visible and IR data, respectively. The basic registration function measures the spatial proximity and local feature similarity of the two points X and Y in terms of a Gaussian function F defined by

$$F(X,Y) = \exp\left(-\frac{d^2(X,Y)}{\sigma^2} - \left(S(X) - S(Y)\right)^T \Sigma^{-1} \left(S(X) - S(Y)\right) \right), \quad (12.2)$$

where $d(X,Y)$ denotes the Euclidean distance between points X and Y. The Gaussian function can be interpreted as a force field decaying with Euclidean distance between the image points and the Mahalanobis distance between the attribute vectors. The parameter σ controls the decay with Euclidean distance, while a diagonal matrix $\mathbf{\Sigma}$ with small components penalizes the difference in the attributes. The matrix $\mathbf{\Sigma}$ is used to decorrelate the feature descriptors and is obtained by computing the statistical dependence of the different descriptors in an approach similar to that proposed by Sharp et al. [Sha et al. 02]. The attribute vectors correspond to the edge maps extracted from visible and thermal IR face images. Commonly used local feature descriptors such as curvature are not useful in visible-thermal image registration due to wide intensity variations between the image pair and large noise in thermal data. The criterion that measures the registration of the two point-sets is defined as the integration over all the Gaussian forces exerted by one point-set over the other:

$$E(T) = \sum_{\substack{X \in M \\ Y \in D}} \exp\left(-\frac{d^2(X,T(Y))}{\sigma^2} - (S(X) - S(T(Y)))^T \Sigma^{-1} (S(X) - S(T(Y))) \right), \quad (12.3)$$

where $T(\cdot)$ denotes an affine transformation for the registration of the two point-sets:

$$\begin{pmatrix} x' \\ y' \end{pmatrix} = T\begin{pmatrix} x \\ y \end{pmatrix} = \begin{pmatrix} a_{11} & a_{12} & a_{13} \\ a_{21} & a_{22} & a_{23} \end{pmatrix}\begin{pmatrix} x \\ y \end{pmatrix}. \quad (12.4)$$

Moments that are invariant to affine transformations [Mur93] can be used as visual attributes. For a small value of the parameter σ, the criterion will consist of Boolean operators that count the number of overlapping points between the two binary images for a given transformation. Equation (12.3) can be derived from a simple combinatorial function that determines point-to-point overlap by analytical mollification [ZitFlu03], or by the Gaussian expression and relaxation. The registration method uses a continuously differentiable function that maximizes both overlap and local shape similarity between the feature maps. To avoid spurious results (mainly due to local maxima in the criterion), a regularizing term is introduced to restrict the 6-parameter affine deformation $T(\cdot)$. The resulting registration criterion E_{reg} can be defined as

$$E_{reg}(T) = E(T) + \lambda \sum_X \|Y - T(Y)\|^2, \quad (12.5)$$

where λ denotes a Lagrange multiplier associated with the constraint. The parameters of the affine transformation are computed by minimizing the criterion function in Eq. (12.5) using a standard quasi-Newton algorithm [Pre et al. 92]. Once the images are aligned, fusion techniques may be applied.

12.3.2 Empirical Mode Decomposition

Empirical mode decomposition (EMD) is a nonparametric data-driven analysis tool that decomposes nonlinear nonstationary signals into *intrinsic mode functions* (IMFs). In this method, images from different imaging modalities are decomposed into their IMFs. Fusion is performed at the decomposition level and the fused IMFs are reconstructed to form the fused image. The effect of fusion on face recognition can be measured by obtaining the *cumulative match characteristics* (*CMCs*) between training sets (galleries) and testing sets (probes). For comparison, CMCs are obtained for the visible and infrared images, images fused using averaging, principal component (PCA) fusion, wavelet-based fusion, and EMD fusion. The FR rate due to EMD-fused images is higher than the FR rates due to raw visible, raw infrared, and other fused images, which will be shown by examples for fused images and illustrative CMC comparison charts.

EMD is intuitive and direct, with the basis functions derived from and based on the data. Huang et al. [Hua et al. 98] introduced EMD for time series data. The assumptions for EMD are:

1. The signal has at least one pair of extrema.
2. The characteristic time scale is defined by the time between the successive extrema.
3. If there exist no extrema and only inflection points are present, then the signal can be differentiated to realize the extrema, whose IMFs can be extracted. Integration may be employed for reconstruction.

As per IMF definition, the decomposition method can simply employ the envelopes defined by the local maxima and minima of the data individually. The image can, for example, be converted into a one-dimensional vector, $X(t)$, using lexicographical rearrangement, and the extrema of the same are identified. All local maxima are interpolated by a cubic spline to form the upper envelope. This process is repeated for the local minima and the lower envelope is constructed. The pointwise mean of the envelopes is called m_1, and is subtracted from the data r_0 to obtain the first component h_1. For the first iteration, $r_0 = X(t)$ and, thereby, $h_1 = r_0 - m_1$. In the second sifting process, h_1 is considered as the data and h_{11} is calculated by

$$h_{11} = h_1 - m_{11}, \tag{12.6}$$

where m_{11} is the mean of the h_1 extrema envelopes. The sifting is continued k times till the first IMF,

$$h_{1k} = h_{1(k-1)} - m_{1k}, \tag{12.7}$$

is obtained. The first IMF, c_1, is defined as

$$c_1 = h_{1k}. \tag{12.8}$$

In sifting, the finest oscillatory modes are separated from the data, analogous to separating fine particles through a set of fine-to-coarse sieves. To retain the physical meanings of the IMF, a standard deviation based-stopping criterion is used. Sifting is stopped if the standard deviation, SD,

$$SD = \sum_{t=0}^{l} \left[\frac{\left| \left(h_{1(k-1)}(t) - h_{1(k)}(t) \right) \right|^2}{h_{1(k-1)}^2(t)} \right], \quad (12.9)$$

calculated from two consecutive sifting results, falls below a threshold. The computation is performed over the entire length l of the vector, which in this case is equivalent to the total number of pixels in the image. The isolated intrinsic mode function, c_1, contains the finest scale of the signal. Then, c_1 is separated from the data to obtain

$$r_1 = r_0 - c_1. \quad (12.10)$$

The new signal, r_1, called the *residue* still holds lower-frequency information. In the next iteration, the residue r_1 is treated as the new data in place of r_0 and subjected to the sifting process. This procedure is repeated on all the subsequent residues (r_j's),

$$r_1 - c_2 = r_2, \ldots, r_{n-1} - c_n = r_n, \quad (12.11)$$

to establish a set of IMFs. The sifting through residuals is stopped when the residue becomes a monotonic function containing no IMFs. Reconstruction of the signal is performed using the relation,

$$\hat{X}(t) = \sum_{i=1}^{n} c_i + r_n. \quad (12.12)$$

12.3.3 Image Fusion Using EMD

The EMD theory was originally proposed for one-dimensional data. It has been extended for two-dimensional data toward image compression [Lin04] and texture analysis [Yan et al. 04]. The results shown in this section are obtained by vectorizing the input images in lexicographical order to imitate time-series data. EMD is performed on these vectors to obtain the visible and infrared IMFs. These components were reconstructed as images to understand the visual significance of the components. The lower IMFs pertain to edges and the higher IMFs recede into details and illumination, as shown in the example image in Fig. 12.4.

Based on the nature of the IMFs, experiments were conducted to utilize this knowledge toward image fusion. In general, image fusion is more meaningful if the components being fused are distinct from each other, and the reduction of mutual information between visible and infrared IMFs usually increases the interpretive information of the fused image. At this prereconstruction stage, the visible and infrared IMFs are multiplied by a set of weights that decrease the mutual information between them. Features from both modalities are emphasized well, when the mutual information is reduced between the IMF pairs, making the resultant image richer in infrared and visible features [Har et al. 06b]. The compact form of our fusion method is given by,

$$F(x,y) = \sum_{j=1}^{k} \left(\alpha_j V_j + \beta_j T_j\right), \quad (12.13)$$

where $F(x,y)$ is the fused image, α_j is the weight by which the jth visible IMF is multiplied, V_j is the jth visible IMF, β_j is the weight by which the jth infrared IMF is multiplied, T_j is the jth infrared IMF, and k is the number of IMFs.

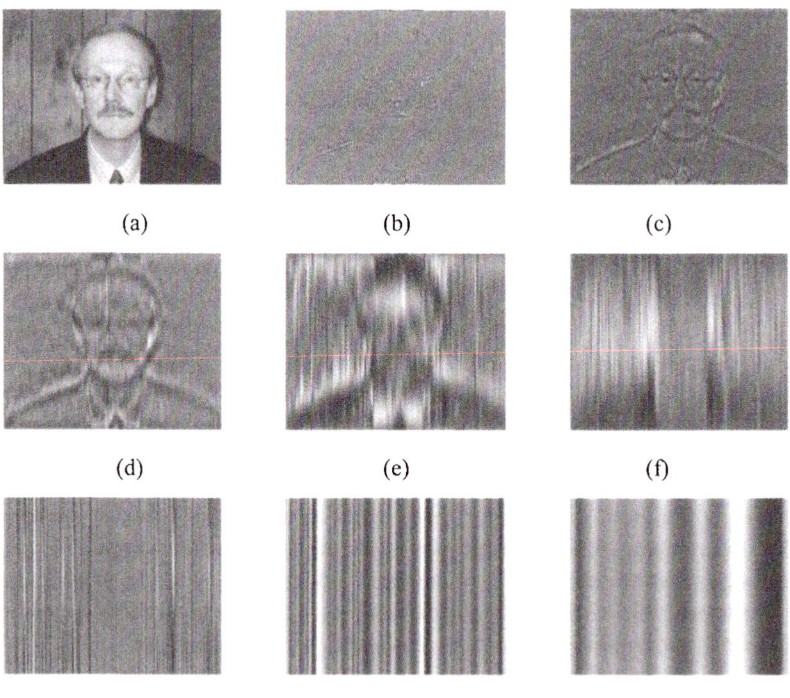

(a) (b) (c) (d) (e) (f) (g) (h) (i)

Figure 12.4. *An example of the decomposition of an image into IMFs that emphasize edges, details, and illumination (a) Original, (b) IMF1, (c) IMF2, (d) IMF3, (e) IMF4, (f) IMF5, (g) IMF6, (h) IMF7, and (i) IMF8.*

12.3.4 Experimental Results

The basic experimental construct in this work is the comparison of face recognition efficiencies using raw and fused images as inputs to a face-recognition engine, namely Face-it®. Face recognition efficiency of a recognition system, for a given training set, is measured by the cumulative match characteristic (CMC) obtained by authenticating a set of testing images (probe images) against a predefined training set (gallery images). The gallery images are enrolled in the recognition engine, and the rank of the first correct match is observed for each image in the probe set. Each cumulative match is calculated by

$$M_i \frac{1}{\Omega}\left[\sum_{j=1}^{i} R_j\right], \qquad (12.14)$$

where R_j is the rank of the first correct match for probe j, and M_i is the sum of instances of all $R_{j<i}$'s . The total number of probe images is denoted by Ω , with i being the index of correct matches. The data items used for our tests are subsets of the Equinox database [Equ06].

Two sets of experiments, namely the comparison of face recognition efficiencies of raw and fused images, imaged under (1) frontal illumination while the subjects maintained a neutral expression and (2)lateral illumination with changing subject expressions, were performed. The gallery is common to both experiments to test the effects of illumination on the face recognition efficiencies. The experiments are conducted for raw visible and infrared images also. We fuse the gallery and probe images using pixel-by-pixel averaging, PCA fusion, wavelet-based fusion, and EMD fusion. In Fig. 12.5a,b, an example of a visible-infrared image pair is depicted. In Fig. 12.5c, the averaged image is shown. Since averaging is a low pass operation, there is a loss of finer detail in the averaged images. In PCA fusion, a finite set of the most principal components are retained for fusion. The loss of information due to the rejection of minor components results in a blurry PCA fused image as illustrated in Fig. 12.5d.

In Fig. 12.5e, the image fused using wavelet fusion is depicted, and coarser details are more visible than in PCA-fused or averaged images. In the EMD fused image shown in Fig. 12.5f, a synergy of features from both the modalities is seen. The edges on the face are in higher contrast, augmented by the infrared signature, to show the ear lobes. The eyeballs are distinct than all of the other visible and fused images. The EMD-fused image shows the thermal signature of the face showing the colder nose and warmer eye sockets. The thermal information is not as visible in the other fusion outputs. The nature of our fusion scheme augments distinct features from both modalities, which contribute to a more discernable fused image.

Results of the first experiment (CMC between same illumination conditions) are shown in Fig. 12.6. The infrared image alone does not perform well because eye locations are difficult to detect, which is a very important factor in many FR systems including Face-it®. The wavelet-fused image has lower face recognition rates as compared to PCA fusion. The low pass action of averaging and the loss of minor components in PCA fusion both lead to loss of detail. The PCA-fused images and averaged images have very similar face recognition performance. The EMD-fused images show a better recognition rate compared to other raw or fused images, as it increases the synergy between the components of the visible and infrared images.

In Fig. 12.7, results for the second test conducted in different illumination conditions are depicted. The results show a performance consistent with the experiments conducted under same illumination conditions, supporting the inclusion of infrared images for robustness against illumination changes. The EMD-fused images have better performance than the raw input and other fused images.

The methods presented above constitute multimodal fusion schemes to improve face recognition rates. First, the input images from the different modalities were decomposed using EMD. Then, fusion was performed at the decomposition level by minimizing the mutual information between the component IMFs. The synergistic confluence of information from the visible and infrared spectrum increases the feature reliability. The fusion of visible and thermal information increases feature reliability and thereby increases recognition rate. EMD fusion exhibits improved face recognition, under changing illumination and expressions.

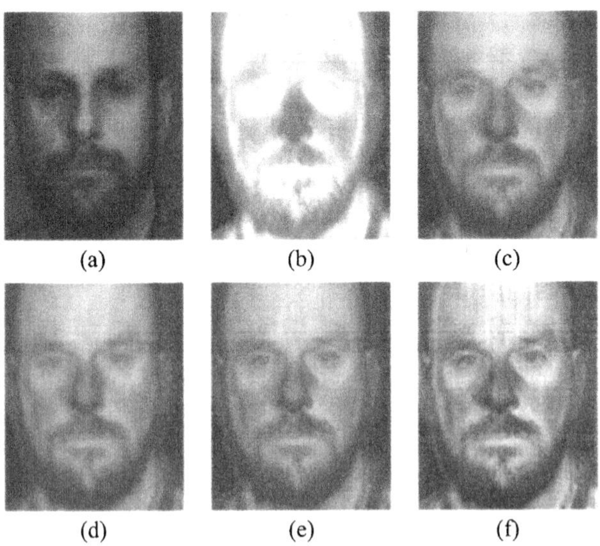

Figure 12.5. *(a) Visible image (from [Equ06]), (b) infrared image (from [Equ06]), (c) averaged image, (d) image fused using PCA fusion, (e) image fused using wavelet-based fusion, and (f) image fused EMD.*

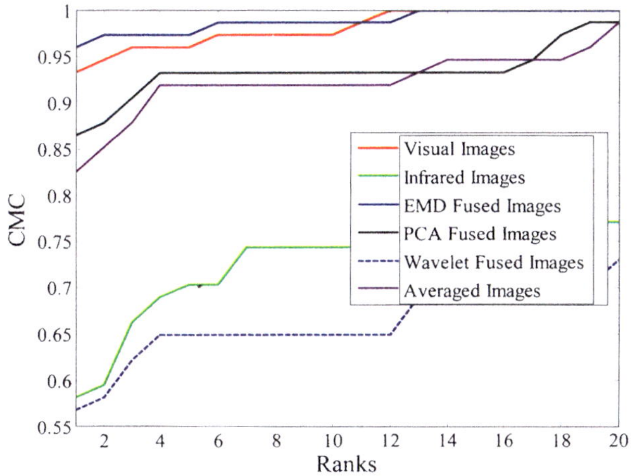

Figure 12.6. *First experiment: A comparison of face recognition rates (CMCs) of raw and fused images, using gallery and probe images acquired with frontal lighting while the subject maintained a neutral expression (Gallery: 62 images; Probe 1: 214 images, from [Har et al. 06]).*

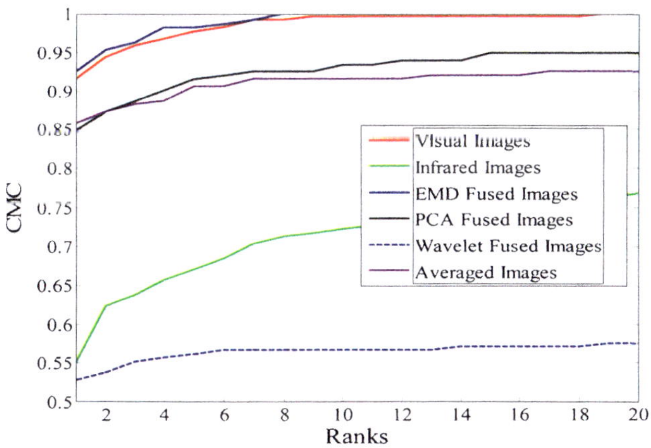

Figure 12.7. *Second experiment: A comparison of face recognition rates (CMCs) of raw and fused images, using the same gallery as in the first experiment and probe images acquired under lateral lighting while subject changed expressions (Gallery: 62 images; Probe 2: 174 images, from [Har et al. 06]).*

12.4 MULTISPECTRAL IMAGE FUSION IN THE VISIBLE SPECTRUM FOR FACE RECOGNITION

Several algorithms for the fusion of multiple bands are addressed in the following to investigate the effects of using and fusing multispectral images on face recognition. Note that the fusing techniques may be used in several different applications but their performance is evaluated here in the context of FR. Physics-based weighted fusion and illumination adjustment are addressed in Sections 12.4.1 and 12.4.2. Fusion employing wavelet-based techniques is discussed in Section 12.4.3. A measure to evaluate FR performance is introduced in Section 12.4.4, and an insight into a multispectral, multimodal, and multiilluminant database is given in Section 12.4.5. Experimental results are shown in Section 12.4.6.

12.4.1 Physics-Based Weighted Fusion

Physics-based weighted fusion of MSIs for FR was introduced in [Cha et al. 06]. The signal strength $u_k(x,y)$ of a camera sensor in a certain wavelength range, $\lambda_{\min}$ to $\lambda_{\max}$, can be represented as

$$u_k(x,y) = \int_{\lambda_{\min}}^{\lambda_{\max}} R(x,y,\lambda)L(x,y,\lambda)S_k(x,y,\lambda)d\lambda \qquad (12.15)$$

with $k = 1,\ldots,n$, where $n = 1$ for monochromatic images and $n = 3$ for color images. The parameters (x,y) indicate the pixel location in the image. $R(x,y,\lambda)$ is the spectral surface reflectance of the object, $L(x,y,\lambda)$ is the spectral distribution of the illumination, and $S_k(x,y,\lambda)$ is the spectral sensitivity of the camera corresponding to channel k. The entire possible integration wavelength range can be in the visible spectrum, 400 - 720 nm, or in addition may include the infrared spectrum depending on the camera design. To simplify the analysis, it is assumed that the illumination is uniform, and, therefore, (x,y) can be omitted. Using Eq. (12.15), and the knowledge of three factors mentioned above, the image intensity values can be predicted. In theory, the intensity of one image acquired with one light source can be adjusted to look like a corresponding image acquired with a (spectrally) different light source, when the spectral power distributions of the two different light sources are known. However, because spectral power distributions of most light sources have high-dimensional information, the information obtained from conventional color or monochromatic cameras is usually not sufficient to accomplish the adjustment.

Contrary, multispectral imaging systems can make use of this high-dimensional information. For a multispectral imaging system employing a LCTF, the camera response u_{λ_i} corresponding to band i centered at wavelength λ_i can be represented as

$$u_{\lambda_i} = \int_{\lambda_{i,\min}}^{\lambda_{i,\max}} R_{\lambda_i}(\lambda) L_{\lambda_i}(\lambda) S_{\lambda_i}(\lambda) T_{\lambda_i}(\lambda) d\lambda \,, \; i = 1,2,\ldots n, \qquad (12.16)$$

where i indicates the ith spectral band, n is the total number of bands, and T_{λ_i} is the spectral transmittance of the LCTF. Figure 12.8 illustrates the spectral distribution of the LCTF used in the IRIS lab. With the bandwidth of 7 nm for each filter, for example, the values of $\lambda_{i,\min}$ and $\lambda_{i,\max}$ for $\lambda_i = 700$ nm are equal to 696.5 nm and 703.5 nm respectively. The imaging process is illustrated in Fig. 12.9. The incident light is first reflected by the surface of an object and then passes through the LCTF and lens. The photons finally reach the CCD chip and, hence, the spectral sensor response is created.

As pointed out in Eq. (12.15), the camera response is the result of an integration process that can be also calculated in a discrete form as the summation of samples. Since each spectral image is acquired within a very narrow band, it is sufficient to use only one sample of each factor per band. Thus, the sensor output at wavelength λ_i can be represented as

$$u_{\lambda_i} = R_{\lambda_i} L_{\lambda_i} S_{\lambda_i} T_{\lambda_i} \,. \qquad (12.17)$$

Theoretically, any designed intensity values can be obtained from each spectral image. However, in practice it is not easy to achieve this due to physical limitations, such as the lower transmittance at the lower wavelength in the LCTF, which is depicted in Fig. 12.8.

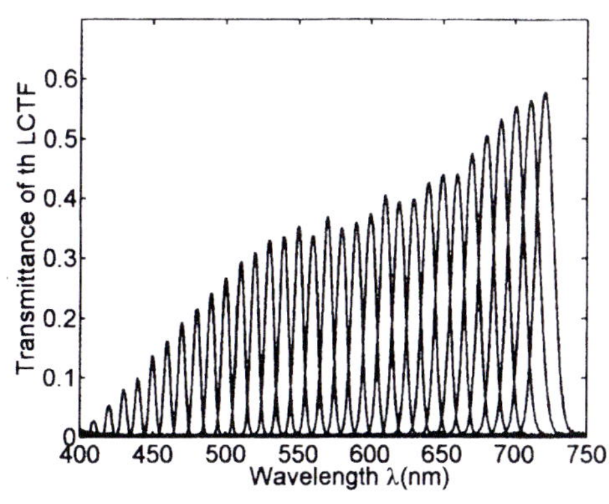

Figure 12.8. *Spectral distribution of the LCTF used in the IRIS imaging system.*

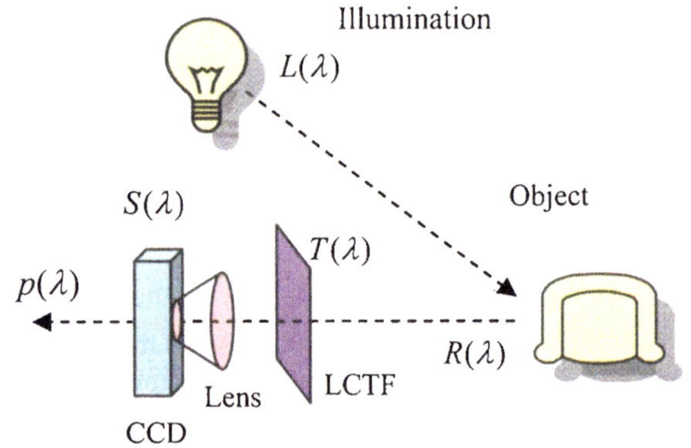

Figure 12.9. *The camera response is the result of integration of all the factors involved, including the spectral distribution of illumination, reflectance of the object, the transmittance of the filter, and the spectral response of the camera.*

Assuming that the rest of the three factors in Eq. (12.17) are ideally uniformly distributed in the spectrum, it can be expected that the spectral images at the shorter wavelengths would appear darker than the images at the longer wavelengths because of the LCTF's transmittance differences. In reality, each of the factors has a nonuniform distribution across the spectrum. For example, the normalized human skin reflectance in the visible spectrum is shown in Fig. 12.10a. The spectral transmittance of the IRIS imaging system, including the LCTF and spectral response of the monochromatic camera, is shown in Fig. 12.10b. Some band image examples are shown in Section 12.4.5.

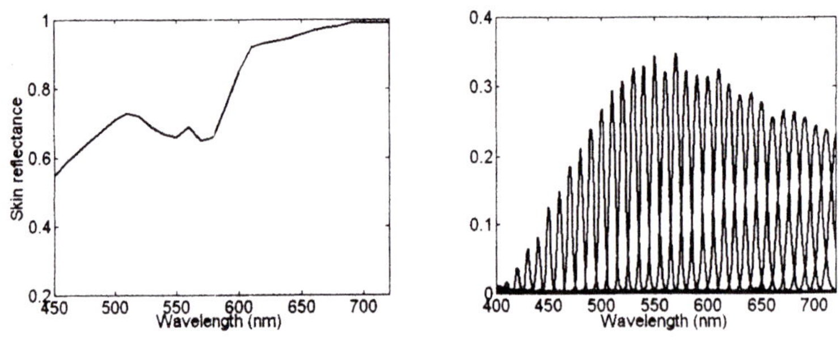

(a) (b)

Figure 12.10. *(a) Normalized skin reflectance (after [Ang01]). (b) Spectral response function of the multispectral imaging components, including the transmittance of the LCTF and the spectral response of the monochromatic camera.*

An "ideal" multispectral imaging system would have a uniform transmittance along the wavelength axis, as the dashed line shows in Fig. 12.11. However, the transmittance of an acquisition system in practice is always nonuniformly distributed. If skin reflectance is considered together with the spectral transmittance of the imaging system, a human-skin-oriented spectral response can be obtained, which is plotted in Fig. 12.11 in a solid line for the IRIS equipment. Furthermore, the spectral distribution of the illumination also influences the output images as discussed in the next subsection.

From the transmittance curve of the IRIS imaging system, we can see that there is a global intensity difference between the spectral images. The dataset acquired for this research proved these differences, which can negatively affect feature extraction and bias the fusion. Therefore, compensating for the intensity differences is considered by adding weights, w_{λ_i}, to each band image for the purpose of achieving a uniform transmittance. The pixel values of the weighted fusion, u_W, can be represented as

$$u_W = \frac{1}{C}\sum_{i=1}^{n} w_{\lambda_i} \cdot p_{\lambda_i}, \text{ with } C = \sum_{i=1}^{n} w_{\lambda_i}. \tag{12.18}$$

For example, if the weights are the reciprocal values of the transmittance of the LCTF at each wavelength, the intensity difference caused by the transmittance of the LCTF is compensated. The same rules can be applied on the other factors in Eq. (12.17). Physics-based weighted fusion is designed to improve images by incorporating physics information into the fusion process. Physics information is utilized as weights for the multispectral image fusion, and the information includes the transmittance of the LCTF, the spectral power distribution of the illuminations, the sensor's spectral response, and the skin reflectance.

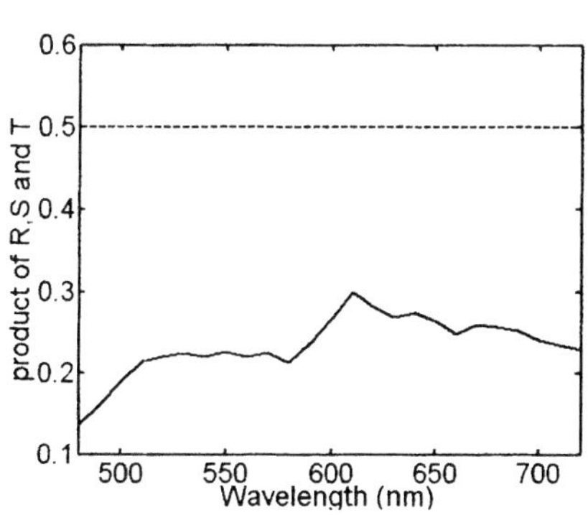

Figure 12.11. *Spectral response of a skin-oriented multispectral imaging module (solid line) and the ideal uniform transmittance (dashed line).*

12.4.2 Illumination Adjustment via Data Fusion

In physics-based weighted fusion, the weights can be set as the ratio of the spectral distribution of two different light sources. This is called *illumination adjustment* (*IA*), since by using this ratio, the image acquired under one illumination can be adjusted to appear like the image acquired under a different illumination. The details are explained in the following.

Different light sources have different spectral properties. For example, halogen light has a very smooth spectral distribution and has more high-wavelength components than low wavelength components. One example of the spectral distribution of a halogen light is plotted in Fig. 12.12a. Another type of commonly used light source, fluorescent light, shows a spiky distribution where the spike locations are decided by the chemical elements in the bulb. In other words, different fluorescent light tubes may show different spectral properties, which is clearly shown between Fig. 12.12b and d. The spectral distribution of one sample of daylight is depicted in Fig. 12.12c. In general, daylight is smoother than fluorescent light, but spikier than halogen light. Moreover, it is comprised of more components in the middle- and short-wave range than the halogen light source.

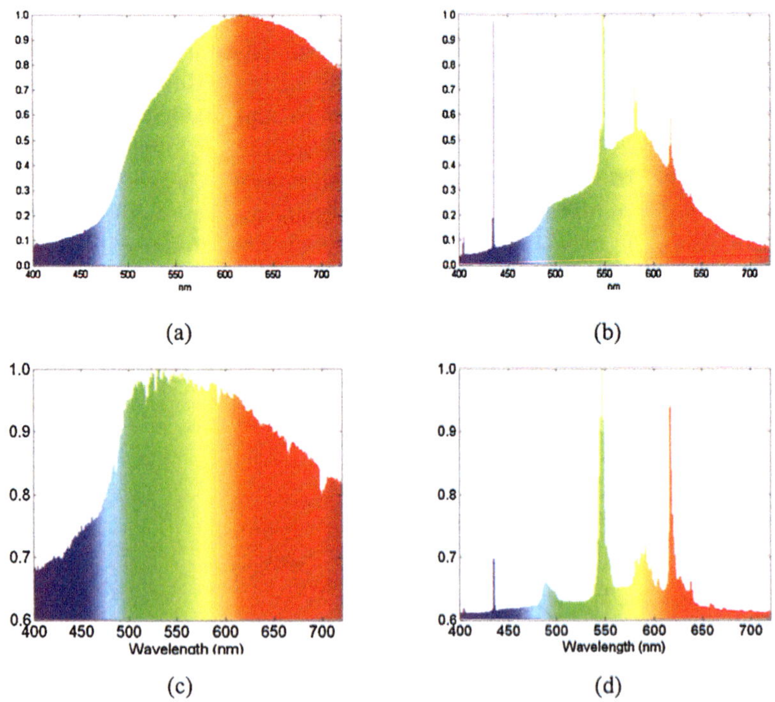

Figure 12.12. *Normalized spectral power distributions of (a) halogen light, (b) fluorescent light, (c) daylight, and (d) a second type of fluorescent light.*

Here, the spectral power distributions of different light sources are used to improve face recognition performance. Given a particular camera, a filter, and an object, the product $F_{\lambda_i} = R_{\lambda_i} S_{\lambda_i} T_{\lambda_i}$ remains the same. In this case, the camera response has a direct relationship to the incident illumination. The camera response, u_{1,λ_i}, at band λ_i acquired using a particular illumination (denoted by L_1), can be represented as

$$u_{1,\lambda_i} = F_{\lambda_i} L_{1,\lambda_i}, \tag{12.19}$$

where L_{1,λ_i} is the spectral power distribution of L_1 at band λ_i. The camera response, u_{2,λ_i}, acquired under another illumination source, L_2, can be represented as

$$u_{2,\lambda_i} = F_{\lambda_i} L_{2,\lambda_i}. \tag{12.20}$$

Considering Eqs. (12.19) and (12.20), the spectral image acquired at λ_i under L_1 can be transformed to the corresponding image acquired under L_2 by applying the weight

$$w_{\lambda_i} = L_{2,\lambda_i} / L_{1,\lambda_i}, \tag{12.21}$$

and the corresponding image can be represented as

$$u'_{2,\lambda_i} = u_{1,\lambda_i} w_{\lambda_i}. \tag{12.22}$$

The intensity of the fused image can be represented as

$$u'_2 = \frac{1}{C} \sum_{i=1}^{n} w_{\lambda_i} u_{1,\lambda_i} \quad \text{with} \quad C = \sum_{i=1}^{n} w_{\lambda_i}. \tag{12.23}$$

Here, the probe images acquired under one particular illumination are transformed to appear as acquired under a different illumination via applying specific weights on each band image and averaging the weighted images.

12.4.3 Wavelet Fusion

Wavelet-based methods are commonly used toward image fusion. The wavelet transform is a data analysis tool that provides a multiresolution decomposition of an image. The input image is decomposed into a set of wavelet decomposition levels. The basis functions are generated from one single basis function popularly referred to as the *mother wavelet*. The mother wavelet is shifted and scaled to obtain the basis functions. Wavelet decomposition can be applied on an image in several ways. Here, the wavelet-based pixel-level data fusion is used on two sets of probes. Given two registered images E_1 and E_2 of the same participant in these two sets of probes, two-dimensional discrete wavelet decomposition is performed on E_1 and E_2, to obtain the wavelet approximation coefficients a_1, a_2 and the detail coefficients d_1, d_2. Wavelet approximation and detail coefficients of the fused image, a_f and d_f, are calculated as follows:

$$a_f = W_{a_1} \times a_1 + W_{a_2} \times a_2, \tag{12.24}$$

$$d_f = W_{d_1} \times d_1 + W_{d_2} \times d_2, \tag{12.25}$$

where W_{a_1}, W_{a_2} and W_{d_1}, W_{d_2} are weights determined empirically. The weights are chosen such that $W_{a_1} + W_{a_2} = 1$, $W_{a_1} = W_{d_1}$, and $W_{a_2} = W_{d_2}$. The two-dimensional discrete wavelet inverse transform is then performed to obtain the fused image.

12.4.4 CMC Measure

In face recognition experiments, where many probe sets resulting from different fusion techniques are compared with a single gallery, often very similar CMC curves are obtained, with the identification rates of different ranks intersecting and crossing over other CMC curves. In this case, it is difficult to distinguish and choose the better probe sets according to the CMC results visually and numerically. A mapping operation projecting the multi-index CMC curve to a single number can aid the comparison of various CMC curves. The many-to-one mapping Q_{CMC}, named *CMC measure* (*CMCM*), is expressed by

$$Q_{CMC} = \sum_{k=1}^{M} C_k \cdot \frac{1}{k}, \tag{12.26}$$

where M is the number of gallery images and k represents the rank number. $1/k$ can be viewed as weight, which decreases monotonously as k increases. C_k is equal to P_k/N, where P_k is the number of probe images that can be correctly identified at (not at and below) rank k. N is the number of probe images. Note that Q_{CMC} is a normalized value between 0 and 1 and the proof is represented as

$$0 < Q_{CMC} = \sum_{k=1}^{M} C_k \cdot \frac{1}{k} = \sum_{k=1}^{M} \frac{P_k}{N} \cdot \frac{1}{k} \\ \leq \sum_{k=1}^{M} \frac{P_k}{N} = \frac{1}{N}\left(\sum_{k=1}^{M} p_k \right) = \frac{N}{N} = 1. \tag{12.27}$$

12.4.5 Multispectral, Multimodal, and Multi-illuminant IRIS-M^3 Database

The *multispectral, multimodal, and multi-illuminant IRIS-M^3 database* was acquired under controlled and uncontrolled illumination situations in indoor and outdoor environments to support research in multispectral image fusion for face recognition [Cha et al. 06b]. While images produced by conventional cameras contain one or three channels, multispectral images provide more spectral information for each pixel over a selected wavelength range. In the database, each spectral band image is recorded for each different wavelength via a LCTF in front of a monochromatic camera. The IRIS multispectral imaging system, shown in Fig. 12.1, was integrated into a mobile data acquisition platform to acquire well-aligned face images in a short duration. This allows participants to maintain their expression and pose. The imaging system consists of a multispectral imaging module, a digital *RGB* camera, a thermal camera, a frame grabber, and an onboard computer. In addition, the illumination was also characterized using a light meter and a spectrometer during the database acquisition. These tools and the IRIS multimodal image acquisition system are shown in Fig. 12.13.

Several datasets were acquired with three different illumination scenarios: halogen light, fluorescent light, and daylight. The three illumination setups are shown in Fig. 12.14. The quadruple halogen lights with a pair on each side of the participant are shown in Fig. 12.14a. The second illumination setup was a pair of fluorescent light panels (denoted as "fluorescent-1"), which is shown in Fig. 12.14b. For the indoor illuminations, halogen light and fluorescent light, a uniform illumination distribution on the face is assumed. Furthermore, face data was acquired in daylight with side illumination. This was due to the fact that many participants were unable to maintain pose or expression with bright sunlight and wind streaming directly into their eyes. The outdoor data acquisition is grouped into eight different sessions according to weather conditions and acquisition time. The weather conditions reach from sunny to cloudy and the passing clouds and wind cause the rapid changes in lighting conditions. Note that illumination conditions during indoor data acquisition are stable over time for different participants, while as opposed to this illumination conditions during outdoor data acquisition can change in a few seconds and can as a result be rather different for different participants. An outdoor data acquisition setup with side illumination is shown in Fig. 12.14c.

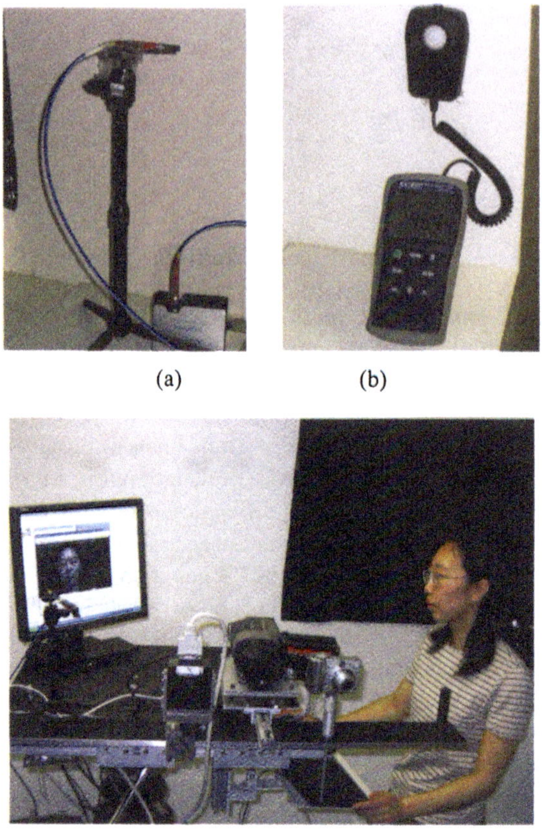

(a) (b)

(c)

Figure 12.13. *(a) Spectrometer and (b) light meter used in (c) the IRIS multimodal image acquisition system.*

An additional set of images was acquired with a digital color camera under a different type of fluorescent light, which we refer to as "fluorescent-2." The image acquisition setup for "fluorescent-2" is illustrated in Fig. 12.14d and the normalized spectral power distribution of "fluorescent-2" is shown in Fig. 12.12d. There are a total of 2624 face images from 82 participants of different ethnicities, ages, and facial, and hair characteristics, and both genders in the database. The corresponding illumination information for each image is recorded and included as well. The image resolution is 640×480 pixels and the eye-to-eye distance is about 120 pixels. The database was collected in 11 sessions between August 2005 and May 2006 with some participants being photographed multiple times. Figure 12.15 shows samples from a single data record in the IRIS-M^3 database, collected by the IRIS data-acquisition equipment, with changes in lighting conditions and elapsed time.

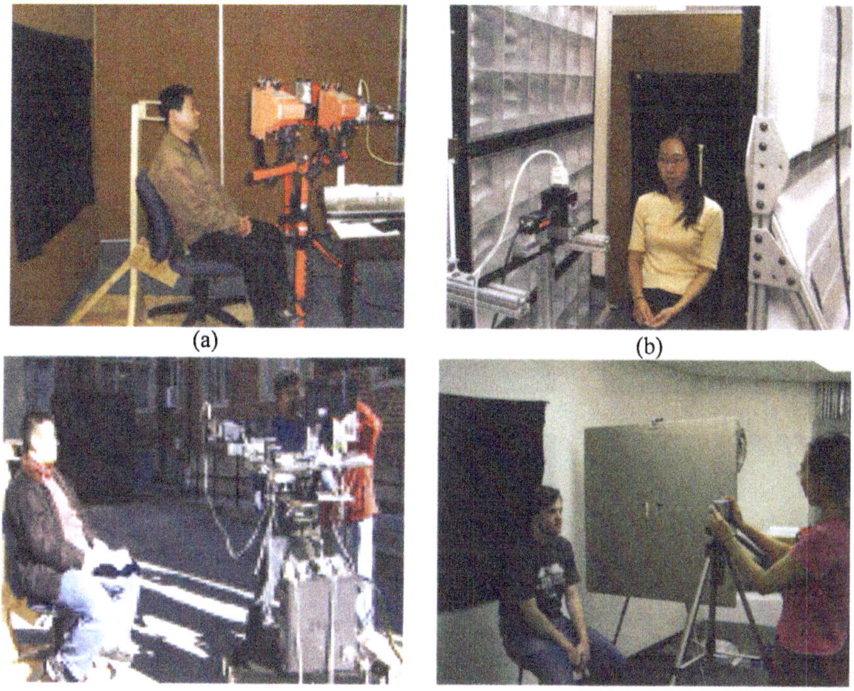

Figure 12.14. *Image-acquisition setup: (a) Halogen lighting setup, (b) "fluorescent-1" lighting setup, (c) outdoor daylight acquisition, and (d) "fluorescent-2" lighting setup.*

The data records in the database were taken from 76% male and 24% female participants. The ethnic diversity is defined as a collection of 57% Caucasian, 23% Asian (Chinese, Japanese, Korean, and similar ethnicity), 12% Asian Indian, and 8% of African descent. Figure 12.16 illustrates example images of eight participants where the images were taken indoors under fluorescent illumination and outdoors under daylight conditions.

The face images were acquired over an academic period of two semesters, including indoor and outdoor lighting conditions. In Table 12.2, the face images are categorized and abbreviated according to the acquisition time and light conditions. There are three parts to the abbreviations: The first part shows the data acquisition time. For example, "1stS" indicates the data was acquired in the first semester, and the "2ndS" in the second semester. The second part specifies the type of illumination. The third part of the abbreviation describes the type of the images, either monochromatic images (BW) or multispectral images (MSI).

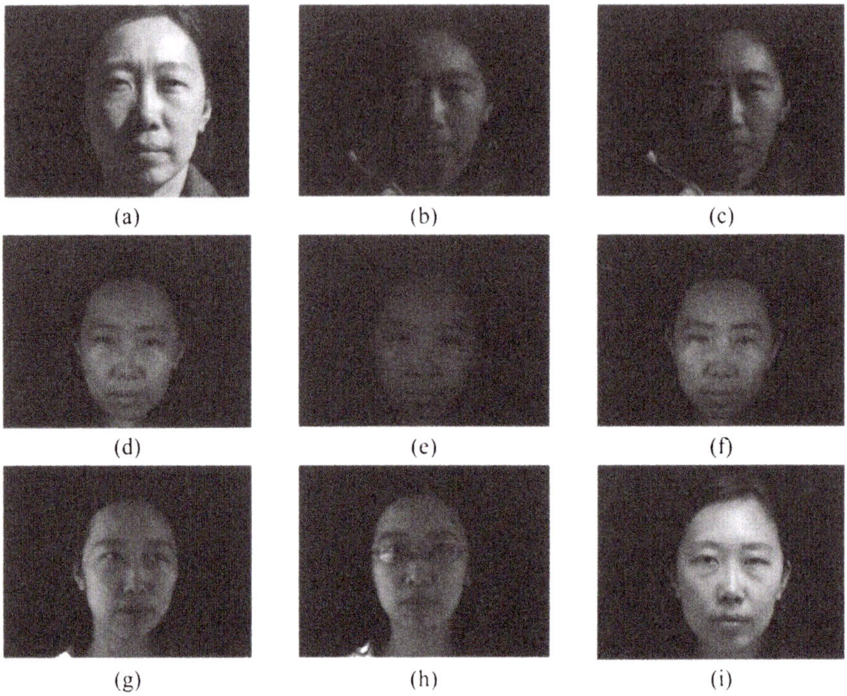

Figure 12.15. *Sample images in a data record in the IRIS-M^3 database: (a) under daylight, side illumination, (b) band 640 nm multispectral image under daylight, (c) band 720 nm multispectral image under daylight, (d) under indoor halogen light, (e) band 640 nm multispectral image under indoor halogen light, (f) band 720 nm multispectral image under indoor halogen light, (g) under "fluorescent-1" with slightly facing left, (h) under "fluorescent-1" with glasses, and (i) under indoor "fluorescent-2."*

Session	Illumination	Imaging Modality	Full Description	Abbreviation
First Semester				
1	Halogen	Mono-chromatic	1^{st}SHalogen_BW	1HBW
1	Halogen	MSI	1^{st}SHalogen_MSI	1HMSI
2	Fluorescent	Mono-chromatic	1^{st}SFluorescent_BW	1FBW
2	Fluorescent	MSI	1^{st}SFluorescent_MSI	1FMSI
Second Semester				
3	Fluorescent	Mono-chromatic	2^{nd}SFluoresecent_BW	2FBW
4-11	Daylight	Mono-chromatic	2^{nd}SDaylight_BW	2DBW
4-11	Daylight	MSI	2^{nd}SDaylight_MSI	2DMSI

Table 12.2. *Different sessions, modalities, illuminations, and their corresponding abbreviations in the IRIS-M^3 Face database.*

Figure 12.16. *Example images of eight participants. Images taken indoors under fluorescent illumination: (a) male Caucasian, (b) female Asian, (c) male of African descent, (d) female of African descent, (i) male Asian Indian, (j) male Asian, (k) female Caucasian, (l) female Caucasian. Images taken outdoors under daylight: (e) male Caucasian, (f) female Asian, (g) male of African descent, (h) female of African descent, (m) male Asian Indian, (n) male Asian, (o) female Caucasian, and (p) female Caucasian.*

12.4.6 Experimental Results

The goal of the following experiments in this section is to prove that fused MSIs provide better recognition performance than conventional images, especially when the gallery and probes are acquired under different illuminations. To support this hypothesis, fusion by averaging, PCA fusion, wavelet fusion, physics-based weighted fusion, illumination adjustment, and decision-level fusion are evaluated by using MSIs from the IRIS-M^3 database.

The results obtained with the multispectral image fusion algorithms for three sets of experiments are discussed in this section. Each experiment has different time lapses and illumination conditions between the gallery and probe images. There are 82 persons with a total of 2624 face images involved in the following experiments.

Experiment-1: The objective of this experiment is to compare the identification performance of differently fused images to conventional images where the same gallery images are used in all the experiments and all images are acquired within 3 months.

Experiment-2: The second set of experimental tests is conducted with the same probes but with a different fluorescent gallery that was acquired 5 months later than the probe images.

Experiment-3: Here, different outdoor probe sets are tested against indoor gallery images.

In each experimental set, physics-based weighted fusion, illumination adjustment (IA), wavelet fusion, PCA fusion, and averaging fusion are conducted and the identification characteristics and CMCMs of different experiments are compared. The fused images are compared with conventional images using the commercial face recognition engine Face-It® (similar to the experiments discussed above in Section 12.3.4). All fusion algorithms, with the exception of wavelet fusion, were implemented on the entire set of spectral band images. Wavelet fusion was investigated in the fusion of single-band images with visible images considering calculation complexity.

Before fusion, all band images need to be well registered. To assure accurate registration, a headrest was used in the chair design to arrest erratic and involuntary motion of the subject's heads. This assures that images acquired within 5 seconds are well aligned, including the corresponding conventional images acquired without the filter. Conventional images are acquired as probes for comparison purposes and are also used as gallery images. The images in the IRIS-M^3 database are well aligned except for a few minor exclusions in scale and translation. Typically, a simple affine transform registration is sufficient to reduce the effects of these minor exclusions. In cases where larger misalignments between the image bands occur, the registration technique introduced in Section 12.3.1 can be applied.

Experiment-1 within Semester: Gallery-1FBW vs. Probes-1HMSI/1HBW

In this set of experiments, the probes and gallery images were acquired in the same semester. The gallery images were taken under fluorescent light, and the probe sets under halogen light. In other words, the monochromatic images acquired under fluorescent light (1FBW) are compared against raw (1HBW) and fused probe sets (1HMSI) acquired under halogen illumination. The images used in "Experiment-1" and the CMCM values are presented in Table 12.3.

Data Set	Method	CMCM
Gallery	1FBW	
Probe set 0	1HBW	95.92%
Probe set 1	Physics-based weighted fusion	95.99%
Probe set 2	IA	95.99%
Probe set 3	Wavelet	96.38%
Probe set 4	Averaging	95.92%
Probe set 5	PCA	93.79%

Table 12.3. *Images used in gallery and probes and the corresponding CMCM in "Experiment-1."*

The calculated CMCMs show that images fused by physics-based weighted fusion (Probe set 1) and illumination adjustment (Probe set 2), using the ratio of the spectral distributions of fluorescent and halogen lights, slightly outperformed conventional images (Probe set 0) in this set of experiments. Furthermore, fused images (Probe set 3) by wavelet fusion via Probe set 0 and Probe set 3 had the highest CMCM value of 96.38%. Averaging (Probe set 4) and PCA fusion (Probe set 5) did not outperform conventional images. Contrary, CMCMs have shown that physics-based weighted fusion (Probe set 1), IA (Probe set 2), and wavelet fusion (Probe set 3) outperform conventional images (Probe set 0).

Experiment-2 between Semesters: Gallery-2FBW vs. Probes-1HMSI/1HBW

In the second set of experiments, visible monochromatic images acquired in the second semester under fluorescent light (2FBW) constitute gallery images. Different raw (1HBW) and fused probe sets from multispectral images (1HMSI) acquired under halogen illumination are the probe sets. The difference between the first and the second experiments is the time lapse between the gallery and probe images. The time lapses in "Experiment-2" are longer than those of "Experiment-1." The probe sets and gallery used in "Experiment-2" and the corresponding CMCMs are given in Table 12.4.

The results in Table 12.4 illustrate that wavelet fusion between 1HBW and 1HMSI640 gave the second-best FR performance, 97.1%, of these two probe sets. Fusion by illumination adjustment slightly improved the recognition performance, 93.8%, from conventional images, 93.7%. In addition, averaging and PCA fusion of MSIs did not provide a higher recognition rate than conventional images. Some examples of probe images of one subject are shown in Fig. 12.17. In this set of experiments, wavelet fusion and illumination adjustment performed more robustly to longer time lapses between image acquisitions than physics-based weighted fusion. Moreover, they always provided better performance than conventional images.

Data Set	Method	CMCM
Gallery	2FBW	
Probe set 0	1HBW	93.7 %
Probe set 1	Physics-based weighted fusion	92.4 %
Probe set 2	IA	93.8 %
Probe set 3	Wavelet	97.1 %
Probe set 4	Averaging	91.7 %
Probe set 5	PCA	90.6 %

Table 12.4. *Images and the corresponding CMCM values of "Experiment-2" (Gallery 2FBW).*

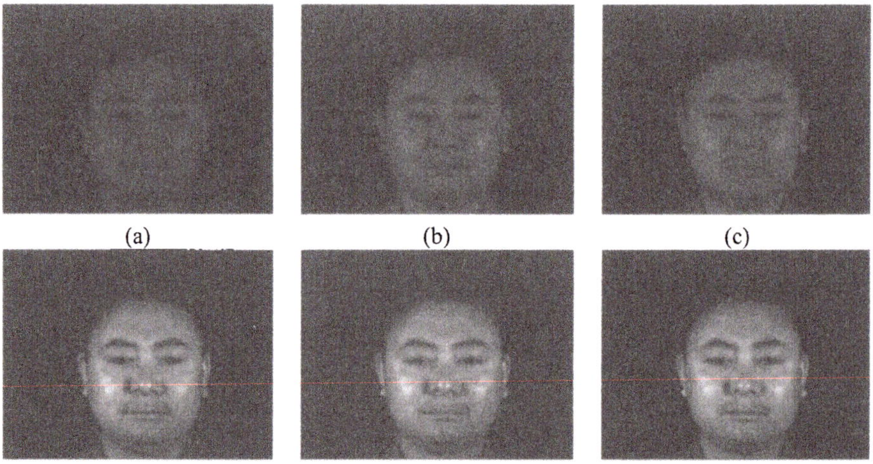

(a) (b) (c)

(d) (e) (f)

Figure 12.17. *Examples of images used for one subject in "Experiment-2:" (a) Gallery, (b) Probe 0, (c) Probe 1, (d) Probe 2, (e) Probe 3, and (f) Probe 4.*

Experiment-3. Indoor Gallery vs. Outdoor Probes

Every experiment set involves a time separation between gallery and probe images in this study. In the third set of experiments, a time separation of 6 months or longer, which is often the case in face recognition situations, is applied. Different raw (2DBW) and fused probe sets from the multispectral images (2DMSI) acquired under side sunny daylight are compared against monochromatic images acquired in the first semester under halogen illumination (1HBW). The data used in these tests and the corresponding CMCM values are shown in Table 12.4.

The CMCMs in Table 12.5 demonstrate that physics-based weighted fusion (Probe set 1), IA (Probe set 2), and wavelet fusion (Probe set 3) all provided higher FR rates than monochromatic images. Wavelet fusion was implemented between spectral band images at 700 nm (1DMSI700) and the corresponding conventional images 2DBW. The weights used for weighted fusion are obtained by multiplying the spectral power distribution of the light sources and the transmittance of the LCTF. A few examples of fused images are shown in Fig. 12.18.

So far, FR using narrow-band MSIs in the visible spectrum has not received much attention. Here the usage of MSIs was introduced for face recognition not simply because MSIs carry more information than conventional images, but also for the reason that the spectral information of illumination can be separated from other spectral information using MSIs. To our knowledge, this is the first effort that uses and fuses MSIs in the visible spectrum (400 - 720 nm) for improving face recognition and that compares fused images with conventional images by a commercial face recognition engine.

"Experiment-1" demonstrated that physics-based weighted fusion, IA, and wavelet fusion can provide better recognition rates than conventional images. In "Experiment-2," a longer time lapse between gallery and probe image acquisition, six months, took place as opposed to a time lapse of one month in "Experiment-1." Still, wavelet fusion and IA outperformed conventional images in the experiments.

From "Experiment-1" and "Experiment-2," the conclusion can be drawn that with large illumination differences and time lapses, IA and wavelet fusion are very promising fusion approaches for multispectral face images that provide a stable and better performance than conventional monochromatic images. Last but not least, outdoor probes with a six-month time lapse between image acquisition for gallery and probes, in which situation usually a huge drop of recognition rate occurs, were studied in "Experiment-3." The most promising improvement in recognition rate was obtained with wavelet fusion. In addition, physics-based weighted fusion and IA also provided higher FR rates than monochromatic images.

Data Set	Method	CMCM
Gallery	1HBW	
Probe set 0	2DBW	53.21%
Probe set 1	Physics-based weighted fusion	58.49%
Probe set 2	IA	59.70%
Probe set 3	Wavelet	64.35%

Table 12.5. *Images used and the CMC measurement used in "Experiment-3."*

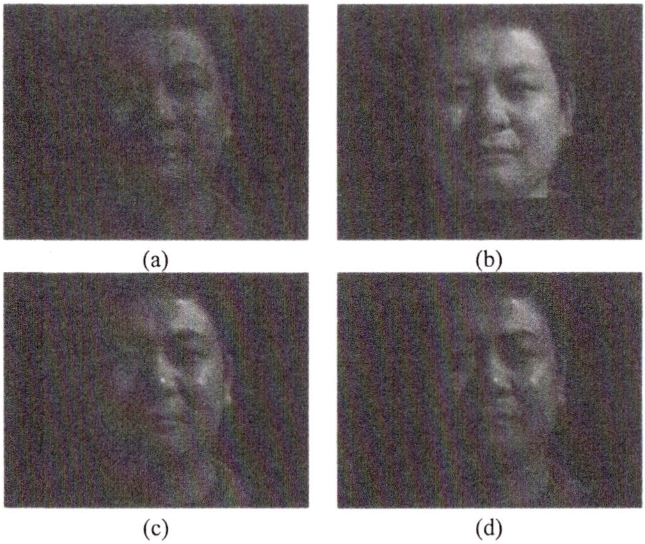

Figure 12.18. *Examples of images used for one subject in "Experiment-3." (a) Probe 0, (b) Probe 1, (c) Probe 2, and (d) Probe 3.*

In summary, multispectral imaging systems are rapidly developing and have been proven extremely useful in numerous imaging applications. FR is one possible application where MSI can aid biometrics. There still remain several questions that need to be answered from "How many bands are useful for a special application?" to "How do we improve multispectral image acquisition for cost-effective, fast, and reliable data collection?" Nevertheless, multispectral image processing will gain much more attention in the future.

12.5 REFERENCES

[Ang01] E. Angelopoulou. Understanding the color of human skin. *Proc. SPIE* **4299**, pp. 243-251, 2001.

[Bes92] P.J. Besl, N.D. McKay. A method for registration of 3-D shapes. *IEEE Transactions on Pattern Analysis and Machine Intelligence* **14** (1992), pp.239-256.

[Bou et al. 04] F. Boughorbel, A. Koschan, B. Abidi, M. Abidi. Gaussian fields: A new criterion for 3D rigid registration. *Pattern Recognition* **37** (2004), pp. 1567-1571.

[Cha et al. 03] G. Charpiat, O. Faugeras, R. Keriven. Shape metrics, warping and statistics. *Proc. Int. Conference on Image Processing*, Vol. 2, pp. 627-630, Barcelona, 2003.

[Cha et al. 06] H. Chang, A. Koschan, B. Abidi, M. Abidi. Physics-based fusion of multispectral data for improved face recognition. *Proc. Int. Conference on Pattern Recognition*, Hong Kong, pp. 1083-1086, 2006.

[Cha et al. 06b] H. Chang, H. Hariharan, M. Yi, A. Koschan, B. Abidi, M. Abidi. An indoor and outdoor, multimodal, multispectral and multi-illuminant database for face recognition. *Proc. IEEE Conference on Computer Vision and Pattern Recognition, Workshop on Biometrics*, New York, pp. 54-61, 2006.

[Dal02] G. Dalley, P. Flynn. Pair-wise range image registration: A study in outlier classification. *Computer Vision and Image Understanding* **87** (2002), pp. 104-115.

[Elg et al. 03] A. Elgammal, R. Duraiswami, L. Davis. Efficient kernel density estimation using the fast Gauss transform with applications to color modeling and tracking. *IEEE Transactions on Pattern Analysis and Machine Intelligence* **25** (2003), pp.1499-1504.

[Equ06] http://www.equinoxsensors.com/products/HID.html

[Fin et al. 04] G.D. Finlayson, P.M. Morovic, S.D. Hordley. Using the SpectraCube for mulitspectral imaging. *Proc. 2nd Conference on Color in Imaging, Vision and Graphics*, pp. 268-274, 2004.

[Fit03] A.W. Fitzgibbon. Robust registration of 2D and 3D point sets. *Image and Vision Computing* **21** (2003), pp.1145-1153.

[Gat00] N. Gat. Imaging spectroscopy using tunable filters: A review. *Proc. SPIE* **4056**, Wavelet Applications VII, pp. 50-64, 2000.

[Got94] M.S. Gottlieb. Acousto-optic tunable filters. In: *Design and Fabrication of Acousto-Optic Devices,* A.P. Goutzoulis, D.R. Pape, (eds.), Marcel Dekker, New York, 1994. pp. 197-284.

[Har et al. 02] J.Y. Hardeberg, F. Schmitt, H. Brettel. Multispectral image capture using a liquid crystal tunable filters. *Optical Engineering* **41** (2002), pp. 2532-2548.

[Har et al. 06] H. Hariharan, A. Koschan, B. Abidi, A. Gribok, M.A. Abidi. Fusion of visible and infrared images using empirical mode decomposition to improve face recognition. *Proc. IEEE Int. Conference on Image Processing*, Atlanta, Georgia, pp. 2049-2052, 2006.

[Har et al. 06b] H. Hariharan, A. Gribok, M. Abidi, A. Koschan. Image fusion and enhancement via empirical mode decomposition. *J. Pattern Recognition Research* **1** (2006), pp. 16-32.

[Hua et al. 98] N.E. Huang, Z. Shen, S.R. Long, M.C. Wu, H.H. Shih, Q. Zheng, N-C. Yen, C.C. Tung, H.H. Liu. The empirical mode decomposition and the Hilbert spectrum for nonlinear and non-stationary time series analysis. *Proc. R. Soc. London A.* **454** (1998), pp. 903-995.

[Hunt95] R.W.G. Hunt. *The Reproduction of Colour.* 5th ed., Fountain, England, 1995.

[Hut et al. 93] D.P. Huttenlocher, G.A. Klanderman, W.J. Rucklidge. Comparing images using the Hausdorff distance. *IEEE Transaction on Pattern Analysis and Machine Intelligence* **15** (1993), pp.850-863.

[ImaBer98] F.H. Imai, R.S. Berns. High-resolution multispectral image archives: A hybrid approach. *Proc. IS&T/SID Sixth Color Imaging Conference*, pp. 224-227, 1998.

[Ima et al. 01] F.H. Imai, M.R. Rosen, R.S. Berns. Multi-spectral imaging of a van Gogh's self-portrait at the National Gallery of Art, Washington, D.C. *Proc. of IS&T PICS*, pp. 185-189, 2001.

[IraAna98] M. Irani, P. Anandan. Robust multi-sensor image alignment. *Proc. 6th Int. Conference. on Computer Vision*, pp. 959-965, 1998.

[Kon et al. 05] S.G. Kong, J. Heo, B. Abidi, J. Paik, M. Abidi. Recent advances in visual and infrared face recognition - A review. *Computer Vision and Image Understanding* **97** (2005), pp. 103-135.

[Kon et al. 07] S.G. Kong, J. Heo, F. Boughorbel, Y. Zheng, B.R. Abidi, A. Koschan, M. Yi, M.A. Abidi. Adaptive fusion of visual and thermal IR images for illumination-invariant face recognition. *Int. J. of Computer Vision* **71** (2007), pp. 215-233.

[LiJa05] S.Z. Li, A.K. Jain (eds.), *Handbook of Face Recognition*, Springer, New York, 2005.

[Li et al. 95] H. Li, B.S. Manjunath, S.K. Mitra. Multisensor image fusion using the wavelet transform. *Graphical Models and Image Processing* **57** (1995), pp. 235-245.

[Lin04] A. Linderhed. Image compression based on empirical mode decomposition. *Proc. of SSAB Symposium Image Analysis,* Uppsala, pp. 110-113, 2004.

[Mur93] D.A. Murio. The Mollification Method and the Numerical Solution of Ill-Posed Problems. Wiley, New York, 1993.

[Oht et al. 81] N. Ohta, K. Takahashi, H. Urabe, T. Miyagawa. Image simulation by use of a laser color printer. *O plus E* **22**, (1981) pp. 57-64.

[Pan et al. 03] Z. Pan, G. Healey, M. Prasad, B. Tromberg. Face recognition in hyperspectral images. *IEEE Transaction on Pattern Analysis and Machine Intelligence* **25** (2003), pp. 1552-1560.

[Pan et al. 04] Z. Pan, G. Healey, M. Prasad, B. Tromberg. Hyperspectral face recognition under variable outdoor illumination. *Proc. SPIE* **5425**, pp. 520-529, 2004.

[Pan et al. 05] Z. Pan, G. Healey, M. Prasad, B. Tromberg. Multiband and spectral eigenfaces for face recognition in hyperspectral images. *Proc. SPIE* **5779**, pp. 144-151, 2005.

[Phi et al. 02] P.J. Phillips, P. Grother, R.J. Michaels, D.M. Blackburn, E. Tabassi, M. Bone. Face recognition vendor test 2002, evaluation report. www.frvt.org/DLs/FRVT_2002_Evaluation_Report.pdf.

[PogAng01] S. Poger, E. Angelopoulou. Selecting components for building multispectral sensors. *IEEE CVPR Technical Sketches* 2001.

[Pre et al. 92] W.H. Press, S.A. Teukolsky, W.T. Vetterling, B.P. Flannery. *Numerical Recipes in C: The Art of Scientific Computing*. Cambridge University Press, 2nd ed., 1992.

[RocFle98] O. Rockinger, T. Fletcher. Pixel-level image fusion: The case of image sequences. *Proc. SPIE* **3374**, pp. 378-388, 1998.

[RosJia99] M. Rosen, X. Jiang. Lippmann2000: A spectral image database under construction. *Proc. Int. Symposium on Multispectral Imaging and Color Reproduction for Digital Archives*, pp. 117-122, 1999.

[SauCup93] D. Saunders, J. Cupitt. Image processing at the national gallery: The Vasari project. Technical Report 14:72, National Gallery, 1993.

[Sha et al. 02] G.C. Sharp, S.W. Lee, D.K. Wehe. ICP registration using invariant features. *IEEE Transactions on Pattern Analysis and Machine Intelligence* **24** (2002), pp.90-102.

[Sin et al. 04] S. Singh, A.V. Gyaourova, G. Bebis, I. Pavlidis. Face recognition using fusion. *SPIE Defense and Security Symposium*, Orlando, Florida, 2004.

[Tom96] S. Tominaga. A multi-channel vision system for estimating surface and illuminant functions. *J. Optical Society of America A* **13** (1996), pp. 2163-2173.

[ZitFlu03] B. Zitova, J. Flusser. Image registration methods: a survey. *Image and Vision Computing* **21** (2003), pp. 977-1000.

[Yan et al. 04] Z. Yang, D. Qi, L. Yang. Signal period analysis based on Hilbert-Huang transform and its application to texture analysis. *Proc. Int. Conference on Image and Graphics*, pp. 430-433, 2004.

13 PSEUDOCOLORING IN SINGLE-ENERGY X-RAY IMAGES

Based on the psychological and physiological processing involved in the human perception of color, a series of linear and nonlinear pseudocoloring techniques can be designed and applied to single-energy x-ray luggage scans, in an effort to assist airport screeners in identifying and detecting threat items, particularly hard-to-see low-density weapons in luggage. The *RGB* and *HSI* color models were both explored. Original grayscale data, various enhanced images, as well as entropy-based segmented scenes were used as input to the various colormapping schemes designed. A highly interactive, user-friendly graphical user interface and a portable test were developed and used in a performance evaluation study involving a population of actual federal airport screeners. The study proved the advantages of using color over gray-level data and also allowed the ranking of colormaps and selection of the best-performing coloring scheme. Rate improvements in weapon detection of up to 97% were achieved through the use of color.

13.1 PROBLEM STATEMENT

Achieving higher detection rates of threat objects during inspection of x-ray luggage scans is a very pressing and desirable goal for airports and airplane security personnel. Because of the complexities present in knowing the content of each individual bag and the increasingly sophisticated methods adopted by terrorists in concealing threat objects, x-ray luggage scans obtained directly from major luggage inspection machines still do not reveal 100% of objects of interest [But02], especially potential low-density threat objects.

Low-density threat items, in reference to x-ray images, are items of composition, thickness, and color such that their absorption coefficients are too low and consequently register at very low gray-level values (close to 0) in the output image. Materials such as glass, plexiglass, various grades of wood, ceramic, aluminum, carbon/epoxy, and plastic can all be used to make lethal weapons that do not show up in an x-ray image in the same way conventional metallic weapons (high-density) appear. Most existing x-ray systems focus on objects of metallic composition (knives and guns), while unconventional weapons could easily go unchecked. Grayscale enhancement and segmentation techniques have been applied in an effort to increase the rate of detecting low-density threat items in x-ray scans [Abi04]. It was shown, through screener evaluation studies that an improvement of 57% in detection rates accrued after enhancement as compared to the inspection results from the raw data.

Since it is known that humans can discern only a couple dozen gray-level values while they can distinguish thousands of colors, the use of color for human interpretation can only improve the number of objects that can be distinguished. In addition to that, color adds vivacity to images, which in turn decreases boredom and improves the attention span of screeners. Pseudocoloring of grayscale images is a typical process used as a means of supplementing the information in various fields such as medicine, inspection, military, and several other data-visualization applications. This process can significantly improve the detectability of weak features, structures, and patterns in an image by providing image details that otherwise would not be noticed [Cze99]. The main purpose of colorcoding is to harness the perceptual capabilities of the human visual system to extract more information from the image. This will provide a better qualitative overview of complex data sets and will assist in identifying regions of interest for more focused quantitative analysis when similarly joined areas in the scene become more distinguishable [Dai96]. By helping in differentiating objects of various densities, colorcoding also minimizes the role of human operators in monitoring and detection, reduces the time required to perform inspection, and lessens the probability of errors due to fatigue.

Since the introduction of color into x-ray luggage inspection machines, mainly one coloring scheme has been used that requires two x-ray images, one at low energy and the second at high energy, resulting in the need for a costlier system. The atomic number of the material is determined using the two images and color is assigned based on the value of the atomic numbers. There was no perceptual or cognitive basis, psychological or physiological, on which color combinations were selected. Pseudocoloring has been used in other imaging applications to detect the boundary (for improved visibility) in ultrasound medical images [Cze99]. Each pixel is allocated a particular color, where the hue is related to the orientation of the most prominent line segment at that point. The idea of applying pseudocoloring to ultrasound was to display statistical properties of the backscatter at each point. The visibility of the boundaries is improved by assigning contrasting colors to the area inside the boundaries and outside the boundaries. Color tinting, a similar technique, has been also used for the colorcoding of ultrasound images [May85]. An *HSI* colorcoding scheme is applied with the hue and saturation kept as two independent variables. The primary image produced is a high-resolution intensity image with the color tinting shown at a considerably lower resolution. Different pseudocoloring schemes can be obtained by varying the values of the individual components of the *HSI* space.

Most visualization techniques generally contain a step in which data values are mapped to color to make the overall range of data visible. The interpretation of results produced by these visualization techniques depends crucially on this mapping, given that the human eye is more sensitive to some parts of the visible spectrum of light than to others and the brain may also interpret color patterns differently.

In an effort to address the relatively new problem of better visualizing low-density threat items in luggage scenes while incorporating considerations of the perceptual and cognitive characteristics of the human operator, a set of *RGB*-based and *HSI*-based color transforms was designed and applied to single-energy x-ray luggage scans. The effectiveness of using colorcoding schemes over grayscale representations is established by conducting a preliminary online survey followed by a comparative performance study using actual airport screeners. The results are tabulated and statistical analysis carried out.

Section 13.2 introduces the physiological and psychological processes associated with the human perception of color (compare Chapter 2). General recommendations for optimum color assignment are drawn based on these physiological and psychological considerations. In Section 13.3, theoretical backgrounds on color spaces and color transforms are presented, followed by our design and implementation of *RGB*-based color transforms and *HSI*-based color transforms in Sections 13.4 and 13.5. Experimental results and performance studies are shown in Sections 13.6 and 13.7.

13.2 ASPECTS OF THE HUMAN PERCEPTION OF COLOR

Human color perception can be devided into physiological and psychological aspects (compare Chapter 2). In the following sections, these different aspects are addressed regarding colorcoding of monochrome values for improved data visulization. After a short introduction, several recommendations and guidelines are presented.

13.2.1 Physiological Processing of Color

The eye lens, retina, and color-processing unit along the optic nerve play the main role in the *physiological processing* of light. The function of the lens is to focus the incoming light on the retina. Different wavelength lights require different focal lengths. Therefore, for pure hues the lens must change its shape so that the light is focused correctly. If pure blue (short wavelength) and pure red (long wavelength), for instance, are intermixed, the lens has to constantly change shape and the eye will become tired. A related effect, called *chromostereopsis*, is the appearance of pure colors located at the same distance from the eye as being at different distances (e.g., reds appear closer and blues more distant). Sometimes, pure blue focuses in front of the retina and appears unfocused. The lens also absorbs light about twice as much in the blue region as in the red region, meaning that people are more sensitive to longer wavelengths (yellows and oranges) than to shorter wavelengths (cyan and blue).

The optic nerve connects the retina to the brain and a color-processing unit along the optic nerve transforms cone signals into the composite signals $red + green$, $red - green$, and $blue - yellow$ ($yellow = (R+G)/2$) [Oue04], then transmits them to the brain using three corresponding channels, called *opponent processing channels*. Blue plays no part in lightness, therefore, colors differing only in the amount of blue do not produce sharp edges, because edges with lightness differences appear clear.

13.2.2 Psychological Processing of Color

Psychological processing is a vast field of study and not as well understood as physiological processing. In this field, simultaneous contrast, color constancy, and the effects of various backgrounds on color perception are the main aspects usually examined. *Simultaneous contrast* is the effect caused when the color of a patch is shifted perceptually by the color of adjacent patches [War88]. For example, colors tend to look darker and smaller against white, and lighter and larger against black as, shown in Fig. 13.1.

Color surroundings cause *chromatic induction* (a color region appears tinged with the complementary hue of the surround), which can make the same colors look different or different colors look the same, as shown in Fig. 13.2. Many other factors such as field size and viewing conditions also affect the appearance of the color perceived.

Figure 13.1. *Simultaneous contrast: Colors look darker and smaller against a white background and lighter and larger against a dark background [Mac99].*

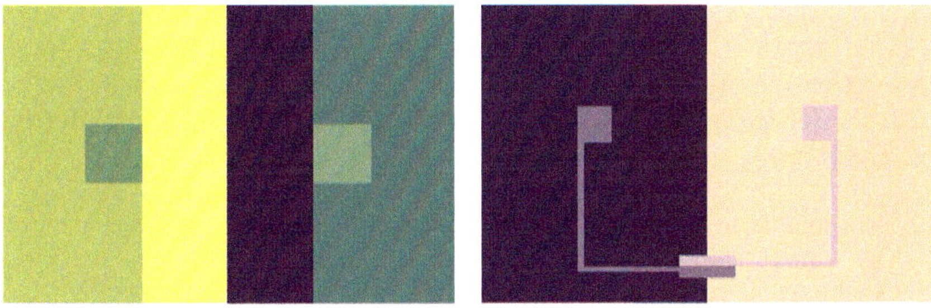

(a) (b)

Figure 13.2. *Simultaneous contrast can make (a) the same colors look different, or (b) different colors look the same [Tay86].*

13.2.3 General Recommendations for Optimum Color Assignment

According to Taylor and Murch [Tay86], the effective use of color can be a very powerful tool, while the ineffective use of color can degrade an application's performance and lessen user satisfaction [Wri97]. In order to create an application with optimum colors, the following questions [Hea96] have to be answered:

1. How effectively can colors be chosen to provide good differentiation between objects?
2. How many colors can be displayed?
3. Which color space should be used?
4. What factors determine target element color relative to nontarget elements?

Color is usually used in a qualitative rather than a quantitative fashion, that is, to show that one item is different from another and not to display a relationship of degree. In general, for a set of colors to work well in a design, some unifying attribute should tie them together, following the Gestalt law of closure (completeness). This could be a particular hue or range of saturations or lightnesses that appears throughout the composition to designate a given aspect or function, relieved by small areas of a contrasting accent color. Key issues to consider in finalizing a color scheme are clarity, comprehensibility, and how well the user will be able to pick out the desired information and understand its significance. Contributory factors include:

1. Discernibility: How easy is it to distinguish an item from its background?
2. Conspicuity: How obvious is the item relative to its neighbors?
3. Salience: How well does the item "pop out" from the display as a whole?

13.2.4 Physiologically Based Guidelines

Based on the physiology of the human visual system as described in Section 13.2.1, the following guidelines for the use of color were drawn from Murch's principles [Mur84]:

1. The simultaneous display of extreme-spectrum colors should be avoided. This causes the lens to rapidly change shape and thus tires the eyes. Desaturate the colors or use colors that are close together in the spectrum.
2. Pure blue should be avoided for fine details such as text, thin lines, and small shapes. Since there are no blue cones in the center of the retina, fine details are difficult to see. But blue makes an excellent background color.
3. Red and green should be avoided in the periphery of large displays.
4. Adjacent colors that differ only in the amount of blue should be avoided. Since blue does not contribute to brightness, this creates fuzzy edges.
5. Adjacent areas of strong blue and strong red should be avoided to prevent unwanted depth effects [Mac99].
6. Opponent colors go well together.
7. Older operators need higher brightness levels to distinguish colors.
8. The use of hue alone to encode information should be avoided in applications where serious consequences might ensue if a color-deficient user were to make an incorrect selection.

13.2.5 Psychologically Based Guidelines

The following recommendations are based on the considerations given in Section 13.2.2:

1. The number of colors used should be reasonable. If users are overwhelmed or confused by too many colors vying for their attention, they are unlikely to develop an effective mental model [Wri97].
2. Consistency is vital when meanings are assigned to colors. The intuitive ordering of colors helps establish intuitive consistency in a design. The spectral and perceptual order of red, green, yellow, and blue can guide the order of the concepts attached to color. For instance, red is first in the spectral order and focuses in the foreground, green and yellow focus in the middle, while blue focuses in the background.
3. If the color of a kind of item is known ahead of time, or if a color applies only to a particular type of item, the search time for finding an item decreases.
4. Where accurate visual judgment of a color is necessary, the surrounding should be a neutral mid-gray to avoid unwanted perceptual color changes.
5. Color can be described more meaningfully in terms of the perceptual dimensions of lightness, hue, and colorfulness than in terms of the measured dimensions.

13.3 THEORETICAL ASPECTS OF PSEUDOCOLORING

The two main aspects that influence the appearance of pseudocolored objects are the color space selected and the color transform applied within that space. A detailed presentation of different color spaces can be found in Chapter 3. The following is an analysis of how each of these can affect the perceived colored scene.

The main objective of pseudocoloring is to obtain an ergonomic color representation of the data that can be easily recognized by a human with normal vision. A variety of mapping schemes can be used to achieve this task. Pseudo-colored images are all obtained using the basic technique shown in Fig. 13.3. Pseudocolor mappings are nonunique and extensive interactive trials have to be performed to determine an optimum mapping for displaying a given set of data. Traditionally, color scales were designed by having the hue sequence range from violet, via indigo, blue, green, yellow, and orange, to red, following the color order of the visible spectrum. Since the human visual system has different sensitivities to different hues, researchers such as Clark and Leonard [Cla89] indicated that spectrum-based color scales were not perceived to possess a color order that corresponds to the natural order of the grayscale in the image.

Several studies focused on constructing a uniform color scale where adjacent colors are equally spaced in terms of just-noticeable differences (JNDs) and maintain a natural order along this color scale. Levkowitz and Herman's [Lev92] research resulted in a scale with maximum and uniform resolution. They were hoping that their optimal color scale would outperform the grayscale, but evaluation results did not confirm that, at least for their particular application. They presented several possible reasons for the unexpected results. For example, the CIELUV they used to adjust final colors might not be adequate to model perceived uniformity; in addition, the perceived change in color due to the surround of a color was not considered. Shi et al. [Shi02] designed a uniform color scale by visually setting up its color series to run from inherently dark colors to inherently light colors (i.e., from black through blue, magenta, red, yellow to white), then further adjusting the colors to make them equally spaced. Both color scales were evaluated by comparing them to the grayscale. Shi et al. [Shi02] then indicated that the contrast sensitivity had been improved after applying their uniform scale, but they failed to demonstrate any significant outcome. Some

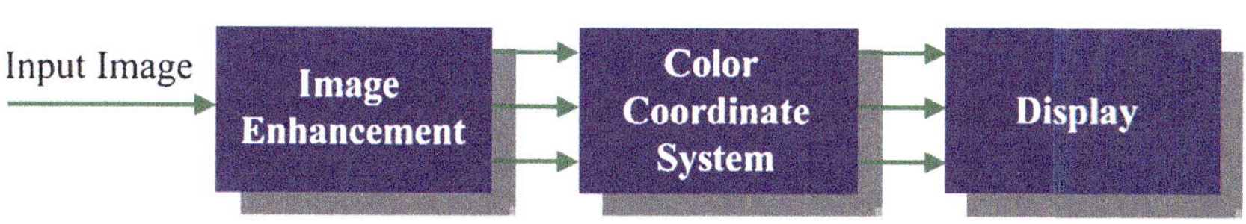

Figure 13.3. *General approach for pseudocoloring.*

researchers focused on decreasing the perceptual artifacts of the human visual system, such as the simultaneous contrast, to convey color information accurately and effectively. Ware [War88] divided the information available in images into (1) metric or value and (2) form information. They proposed theoretical principles to predict which color scale would be effective in conveying both the metric and the form information. Through a series of psychophysical experiments, Ware demonstrated that simultaneous contrast was a major source of error when reading metric information, but only partially verified his form hypothesis. General rules for designing color scales that can effectively convey metric and form information were provided.

Other research utilized common mathematical functions such as the sine function to construct desired mappings or color scales. Gonzalez and Wood [Gon02] described an approach where three independent transforms were performed on the gray-level data, and the three output images fed into the *R*, *G*, and *B* color channels to produce a specific colormapping. The natures of these mathematical functions would determine the characteristics of the color scale.

Through interactions with various types of color scales under different circumstances, most researchers agree that color is useful to reveal more information in images, but for certain applications some types of color scales are superior to others. Therefore, Levkowitz and Herman [Lev92] concluded that several easily interchangeable color scales could substantially increase the perception of information in images over the use of a single scale.

One of the basic methods of pseudocoloring is to directly apply a single color hue to replace a particular grayscale. For example, the color range from C_{0-255} can be used to code the grayscale range F_{0-255}. Another approach [Dai96] is based on the desired tristimulus value of the output color image, where analysis of the tristimulus value of the required output image is performed and a function $P[]$ is defined that maps the original grayscale data $E(x,y)$ to the primary color values $R(x,y)$, $G(x,y)$, and $B(x,y)$. This process can be represented as follows:

$$\begin{aligned} R(x,y) &= P_R[E(x,y)] \\ G(x,y) &= P_G[E(x,y)] \\ B(x,y) &= P_B[E(x,y)] \end{aligned} \tag{13.1}$$

where the transforms $P_R[], P_G[], P_B[]$ could be either linear or nonlinear functions, based on the desired output image. The complete color-coding process can be described as follows:

$$E(x,y) \Rightarrow \begin{Bmatrix} P_R[] \\ P_G[] \\ P_B[] \end{Bmatrix} \Rightarrow \begin{Bmatrix} R(x,y) \\ G(x,y) \\ B(x,y) \end{Bmatrix} \Rightarrow \mathbf{C}(x,y) \tag{13.2}$$

where $\mathbf{C}(x, y)$ is the final pseudocolored image. By varying the functions $P[]$, different color-coded images can be obtained.

To accurately represent gray values using color, the following properties are desirable in a color scale [Lev92]:

Given a sequence of numerical gray values $\{v_1 \le ... \le v_N\}$ represented by the colors $\{\mathbf{c}_1 \le ... \le \mathbf{c}_N\}$, respectively:

1. Color should perceivably preserve the order of gray values; the relationship among the colors should be c_1 perceived-as-preceding ... c_i perceived-as-preceding c_N.
2. Color should perceivably reflect the distances between the gray values, that is, for any $1 \le i, j, m, n \le N$, *if* $v_i - v_j = v_m - v_n$, we should also have $pd(c_i, c_j) = pd(c_m, c_n)$, where $pd(c_i, c_j)$ is the perceived distance between c_i and c_j.
3. Boundaries not existing in gray-level data should not be perceived after using color [Lev92].

The combination of the concepts of increasing the number of JNDs [Lev92] and conveying both value and form information effectively by reducing the effects of simultaneous contrast would result in a color scale sequence that increases monotonically in luminance while cycling through a range of hues [War88]. Combining those two major ideas with the general recommendations for optimum color assignment given in Section 13.2.3, the following rules for designing an optimum color scale can be used to produce a number of optimal transforms.

For transforms designed based on the *RGB* color model, the *R*, *G*, and *B* versus gray-level transform functions should not decrease simultaneously and the intensity, $I = (R + G + B)/3$, should increase monotonically.

$$\text{For } 1 \le n \le N-1 \text{ and } r_n \le r_{n+1}, g_n \le g_{n+1} \text{ and } b_n \le b_{n+1},$$

we should have

$$r_n + g_n + b_n \le r_{n+1} + g_{n+1} + b_{n+1}.$$

Colors should be triple mixtures (*RGB*), and not pairwise mixtures, to avoid exaggerated saturation.

For transforms designed based on the *HSI* color model, good properties for revealing both shape and value are as follows: the intensity *I* versus gray level should increase monotonically; the hue value *H* versus gray level should cycle through a particular range of hues such that for

$$1 \le m \le N, h_1 \le ... \le h_m \prec h_{m+1} + 360 ... \le h_N + 360 \quad \text{or}$$

$$h_1 \ge ... \ge h_m \succ h_{m+1} - 360 ... \ge h_N - 360 \, ;$$

and the saturation S versus gray level should be monotonic, that is,

$$s_1 \le s_2 \le s_3 \le ... \le s_{N-1} \text{ or } s_2 \ge s_3 \le ... \ge s_{N-1} \ge s_N \quad \text{[Lev92]}.$$

The hue vaues should be chosen in such a way that the color scale runs from inherently dark hues to inherently light hues.

Based on the characteristics of major color spaces and recommendations for color transforms, a number of colormaps were designed, implemented, and applied to x-ray luggage scans. The following sections describe the implementation, results, and performance of each of these colormapping approaches.

13.4 *RGB*-BASED COLORMAPS

RGB-based colormaps can be further classified into two subcategories: perceptually based colormaps and mathematically formulated approaches. Linear and nonlinear maps are also addressed.

13.4.1 Perceptually Based Colormaps

In these mappings, the color series are usually set up visually according to the color preferences of the user and then the transforms defined accordingly.

Linear Mapping

For initial trials, some colormaps were adopted from the Matlab image processing package. Those maps were also used in our final performance evaluation for the sake of performance comparison with the transforms that we designed. The "Hot" and "Jet" color scales were first applied to x-ray luggage scans. As Figs. 13.4a and b show, "Hot" changes smoothly from black, through shades of red, orange, and yellow, to white; "Jet" ranges from blue to red and passes through cyan, yellow, and orange. The two color scales can be produced using Eqs. (13.3) and (13.4) for "Hot" and "Jet" respectively:

$$R = \begin{cases} \dfrac{I+1}{n} & I \le n-1 \\ 1 & I > n-1 \end{cases} \tag{13.3a}$$

(a) (b)

Figure 13.4. *(a) The "Hot" color scale and (b) the "Jet" color scale [Abi06].*

$$G = \begin{cases} 0 & I \le n-1 \\ \dfrac{I+1-n}{n} & n-1 < I \le 2n-1\text{, and} \\ 1 & I > 2n-1 \end{cases} \tag{13.3b}$$

$$B = \begin{cases} 0 & I \le 2n-1 \\ \dfrac{I+1-2n}{m-2n} & I > 2n-1 \end{cases} \tag{13.3c}$$

where I represents the gray value, m is the number of colors of the "Hot" scale, and $n = f_1(3/8 \cdot m)$, in which $f_1(x)$ rounds x to the nearest integer toward zero. In Eq. (13.4), I represents the gray value, $n = f_2(m/4)$ and $\Delta = f_2(n/2) - (\text{mod}(m,4) == 1)$, in which m is the number of colors of the "Jet" scale and $f_2(x)$ rounds x to the nearest integer:

$$G = \begin{cases} 0 & I \le \Delta - 1 \\ \dfrac{I+1-\Delta}{n} & \Delta - 1 < I \le \Delta + n - 1 \\ 1 & \Delta + n - 1 < I \le \Delta + 2n - 2 \\ \dfrac{\Delta + 3n - I - 1}{n} & \Delta + 2n - 2 < I \le \Delta + 3n - 2 \\ 0 & I > \Delta + 3n - 2 \end{cases} \tag{13.4}$$

The Red and Blue component values can be obtained by shifting the Green component G to the right and left by $n/\max(\textit{grayvalue})$, respectively. Figures 13.5a and b illustrate the R, G, and B transforms of the "Hot" and "Jet" color scales for the 256-step case.

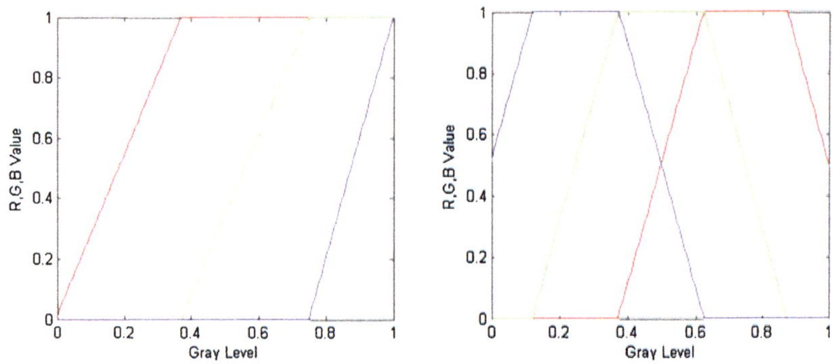

(a) (b)

Figure 13.5. *(a) R, G, and B values versus gray level for the "Hot" scale with 256 steps; (b) R, G, and B values versus gray level for the "Jet" scale with 256 steps [Abi06].*

Nonlinear Mapping

This scale, which we refer to as the "Warm" color scale, varies from dark blue, through magenta and orange, to light yellow. The distances of adjacent colors on this scale are perceivably equal. A 256-step scale as seen in Fig. 13.7b was developed and applied to x-ray luggage scans based on the 16-step color scale shown in Fig. 13.6 [Shi02]. The 16 colors on the 16-step scale were utilized as base colors and intermediate colors computed by linearly interpolating the red, green, and blue intensity values from each base color to the next. Let R_i, G_i, B_i and R_{i+1}, G_{i+1}, B_{i+1} represent any two adjacent base colors; I_i and I_{i+1} denote their corresponding gray levels. Thus, given a gray level I ($I_i < I < I_{i+1}$ for $1 \le i \le 15$), the associated intermediate color $\mathbf{C}$ between base colors $\mathbf{C}_i$ and $\mathbf{C}_{i+1}$ can be found using Eq. (13.5):

$$R = \left(R_{i+1} - R_i\right)\left(\frac{I - I_i}{I_{i+1} - I_i}\right) + R_i ,$$

$$G = \left(G_{i+1} - G_i\right)\left(\frac{I - I_i}{I_{i+1} - I_i}\right) + G_i , \qquad (13.5)$$

$$B = \left(B_{i+1} - B_i\right)\left(\frac{I - I_i}{I_{i+1} - I_i}\right) + B_i .$$

where R, G, and B are the intensity values in the three channels of color $\mathbf{C}$. Figure 13.7a illustrates the R, G, and B values versus the gray levels obtained by linearly fitting 256 colors to the 16-step data.

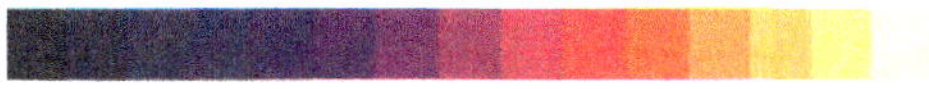

Figure 13.6. *The 16-step "Warm" color scale [Abi06].*

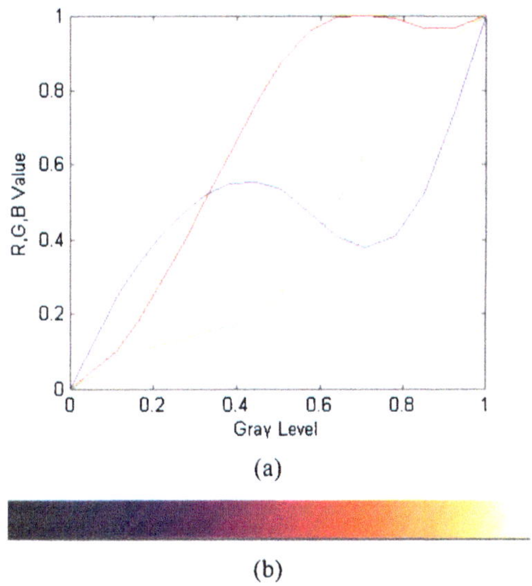

Figure 13.7. *(a) R, G, and B values versus gray-level transforms obtained by linearly fitting 256 steps to the16-step data; (b) the 256-step "Warm" color scale [Abi06].*

13.4.2 Mathematical Formulations

The mapping between colors and gray values in this case is obtained through the use of a mathematically formulated transform. The nature of the mathematical formula determines the color content of the output image. In this section, I denotes the value of a pixel before transform; R, G, and B denote the values in the three channels of the colored pixel. Various types of transforms can be used for colormapping.

Algebraic Transforms

In this mode, color is assigned through a mathematical algebraic formula applied to individual or group pixel values to create various combinations of these original pixels and obtain their resulting color counterparts. An example formula for such a function is given in Eq. (13.6), where the number of colors is a parameter (L) that can be supplied by the user and is intrinsically incorporated into the formula for color generation.

$$
\begin{aligned}
R &= N / L \\
G &= 1 - N / L \\
B &= \left| 2N / L - 1 \right| .
\end{aligned} \tag{13.6}
$$

N is an integer varying from 1 to L. The grayscale was divided into L equal intervals corresponding to the L colors; therefore, for all *I*s belonging to the same interval, one single color is assigned.

In this mode, the number of basic colors actually remains the same but variations occur in the color range. For example, when $L = 4$, a color image with only the "basic" colors is produced. The information in this color image is minimal, because only the higher pixel values from the grayscale image are coded. However, when $L = 16$ an image with the same basic colors is produced but the individual range for each color is expanded to obtain 16 colors overall. Most of the information is now retreived as even the darker (or lower) pixel values in the grayscale image are coded. Figure 13.8 illustrates the use of this method to color code a grayscale image and shows how the variation of the number of colors can achieve different effects in the colored image, such as a variable amount of detail and clutter.

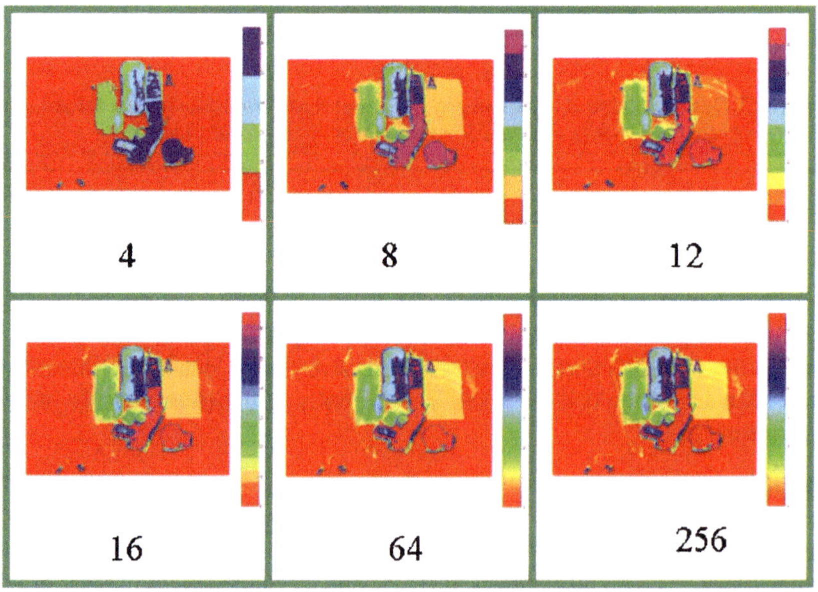

Figure 13.8. *Varying the number of colors using the mathematical expression of Eq. (13.6) for colormapping [Abi06].*

Sine/Cosine Transform

One of the characteristics of the sine function is that it contains regions of relatively constant value around the peaks as well as regions that change rapidly near the valleys. The advantage of a continuous color scale becomes evident when considering overlapping materials inside luggage. In systems with abrupt color switchover between, for example, organic and nonorganic materials, even thin layers of overlapping materials, such as steel, copper, or PVC, will lead to organic materials being classified as nonorganic materials, which is incorrect and might result in false negatives and/or false positives. The typical form of the sine transform can be expressed as:

$$\begin{aligned} R &= \left|\sin(\omega_R I + \theta_R)\right| \\ G &= \left|\sin(\omega_G I + \theta_G)\right| \\ B &= \left|\sin(\omega_B I + \theta_B)\right| \end{aligned} \tag{13.7}$$

where ω_R, ω_G, and ω_B are the radiant frequencies for the *R*, *G*, and *B* channels and θ_R, θ_G, and θ_B are their corresponding phase angles. Changing the frequency and phase of each sine function can emphasize (in color) certain ranges in the grayscale. The effect of mathematical manipulation of the sine-based color assignment algorithm on viewer perception of objects and presence of details in x-ray data can be seen in Fig. 13.9, where an increased perception of the level of details and a greater discrimination power between the various scene components are apparent between Figs. 13.9a, b, and c as periodicity and phase of the sine color functions are varied.

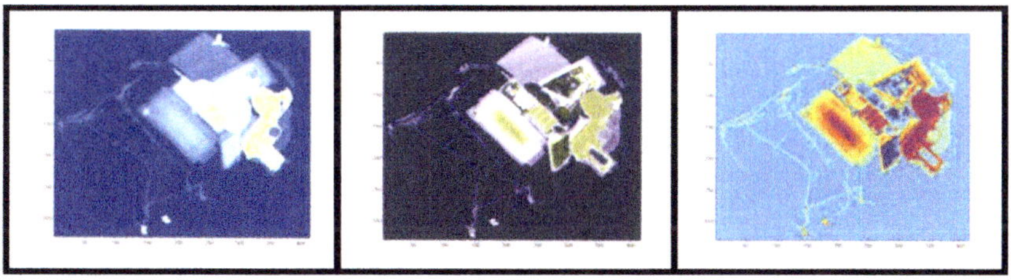

(a) (b) (c)

Figure 13.9. *Effects of mathematical manipulation of sine color function on visualization and interpretation capabilities of viewer. Note increased level of detail and power of discrimination from (a) to (c) [Abi06].*

Rainbow Transform

The rainbow transform can be seen as a special case of the sine/cosine transform. The three transform functions used for the rainbow map are given in Eq. (13.8).

$$
\begin{aligned}
R &= (1+\cos(\frac{4\pi}{3\cdot 255} I))/2 \\
G &= (1+\cos(\frac{4\pi}{3\cdot 255} I - \frac{2\pi}{3}))/2 \\
B &= (1+\cos(\frac{4\pi}{3\cdot 255} I - \frac{4\pi}{3}))/2
\end{aligned}
\tag{13.8}
$$

Specifically, the three periodic functions were used in such a way that one of them was at a peak in a selected color interval, as shown in Fig. 13.10.

13.5 *HSI*-BASED COLORMAPS

HSI-based color transforms are classified into two subcategories. Color transforms in the first subcategory provide a direct mapping between the gray values and their color counterparts while color transforms in the second subcategory were created according to the approach illustrated in Fig. 13.11, where a selected series of enhancement operations (e.g., some statistical properties of an x-ray luggage scan can help in choosing specific enhancement operations) are first performed on the x-ray luggage scan to extract or emphasize features of interest and then the results are fed into the *H*, *S*, and *I* components to create a composite chromatic image.

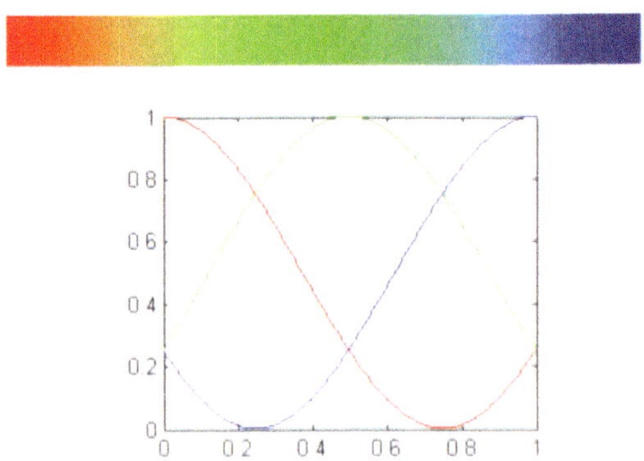

Figure 13.10. *Top: the rainbow color scale. Bottom: the three periodic functions used for the rainbow transformation. Red, green, and blue curves represent the R, G, and B vs. gray-level transform functions, respectively.*

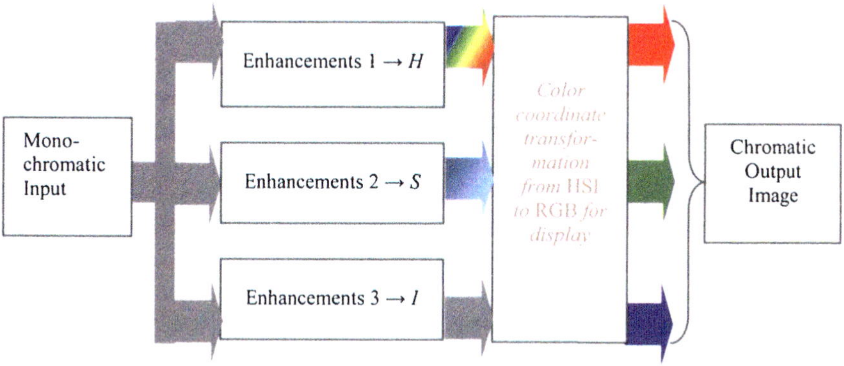

Figure 13.11. *General process producing the second-subcategory of HSI-based color transforms*

13.5.1 Mapping of Raw Grayscale Data

Two methods are investigated in this subcategory:

1. Histogram-based approach resulting from a visually selected scale
2. Nonlinear approach based on the recommendations of Section 13.3.2

Histogram-Based Colormapping

The colors of the various components in the scene are assigned based on the values of the raw image. Pixel ranges are selected from the data's histogram and automatically given certain colors. For instance, it is known that high-density (metallic) material has a low degree of transparency and consequently higher pixel intensities. Color components for such gray levels can then be set to result in certain values. Based on the results derived from the color study discussed earlier, this basic procedure can be conducted as follows:

1. Set threshold values.
2. Define the number of colors to be used.
3. Define the hues to be used for each group of pixels.
4. Set the saturation to one and the intensity to the gray value of the pixel.
5. Transform the *HSV* image to *RGB* space for display.

If, for example, four gray-level regions were created, the chances of the low-density threat being present would be greatest in the first two regions. Four colors will be picked based on the recommendations of Section 13.3.3. Blue will be used as background and other easy-to-remember basic colors like red and green applied to the other pixels in each bin. The output image would have four hues, which vary as a function of the gray intensity value of each pixel.

Function-Based Mapping

The following color scales were designed based on the criteria mentioned in section 13.3.2. The intensity was selected to be monotonically increasing, which could be a logarithm transform, an exponential, a linearly increasing, or any other nondecreasing transform. The optimum selection of the appropriate transform depends on the specific application (i.e., the low-intensity or high-intensity part of the image that needs to be emphasized). A key issue in the implementation is the design of the hue transform. A right/left semi-ellipse curve is selected for the hue transform and the saturation set to a constant. Considering the fact that the objective is to better visualize low-density threat items, the logarithm transform for intensity is used as a way of increasing the value of the low gray levels and improving their appearance and recognition by screeners.

Figure 13.12a shows the transformations of intensity, hue, and saturation and Fig.13.12b and c the resulting color bars for the right and left semi-ellipses. The color scale of Fig. 13.12 (b), called "Springtime", uses $h_s = 0, 0 \le h_e \le 180$ and $dir = up$, where h_s is the starting hue, h_e is the ending hue, and *up* means a counterclockwise progression.

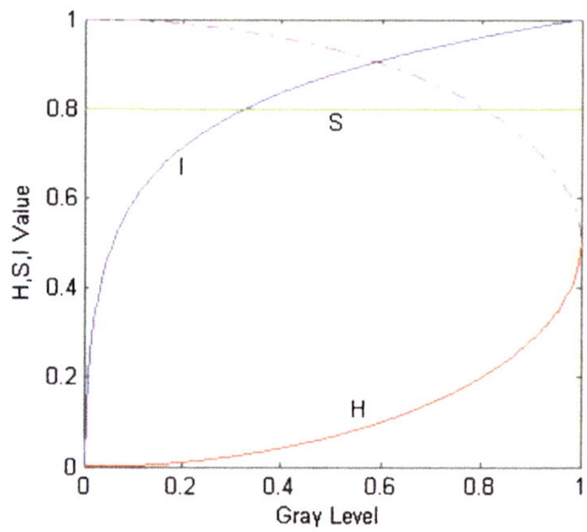

(a)

(b)

(c)

Figure 13.12. *(a) Intensity, hue, and saturation transforms; (b) color scale "Springtime" produced by using the concave part (solid curve) of the semi-ellipse for the hue transform; (c) color scale produced by using the convex part (dotted curve) of the semi-ellipse for the hue transform [Abi06].*

13.5.2 Color Applied to Preprocessed Grayscale Data

Based on the theoretical background addressed in Section 13.3.2 and the recommendations of Section 13.2.3, different coloring schemes were designed where preprocessed grayscale data is used as input to the hue, saturation, and intensity color channels. The first set of transforms uses a constant saturation and the second set uses a data-dependent saturation.

Constant Saturation

Let E_1 and E_2 be two enhanced images produced using some desired enhancement operations. In this first approach, E_1 is fed into the hue component and E_2 is fed into the intensity component. The saturation, S, is set to a constant within the interval $[0.6, 1]$, that is high enough to ease the distinction between colors but low enough to avoid eye fatigue due to refocusing [Mur84]. Using this scheme, zero-value pixels in E_2 would not appear in the colored image and the coloring scheme generally helps improve the appearance of image E_2.

"CS1" and "CS2", described below, are two sets of transformations using gray-level enhancement operations found, through preliminary tests, to be effective in revealing low-density threats in x-ray luggage scans [Abi04]. The notation "$Original + A + B + C + ...$" denotes a sequence of preprocessing using operations *A, B, C,* applied to the original x-ray luggage scan. The two sets are described as:

1. "CS1": $H = E_1 =$ original + histogram equalization + contrast stretching, $S =$ constant, and $I = E_2 =$ original + negative + H-domes [Abi04] + contrast stretching and
2. "CS2": $H = E_1 =$ original + negative, $S =$ constant, and $I = E_2 =$ original + histogram equalization + contrast stretching.

Another approach, based on the segmentation results of the original data, uses the segmented image as an input to the pseudocoloring process. *N* classes in the scene are clustered through segmentation and a single hue is assigned for each class. For any two adjacent classes complementary colors are used. Contrast stretching is applied to each class such that the intensities of objects in each class cover the full range of grayscale. Objects belonging to a given class will have the same hue value but with different intensities. This transform is similar to the histogram-based colormap described in Section 13.5.1, except that the thresholds in this case are selected automatically via the segmentation algorithm. For each class, the transforms for the *H, S,* and *I* channels are expressed as: $H =$ constant, $S =$ constant, and $I =$ image slice containing one class of objects + contrast stretching.

A single chromatic image containing the *N* classes of objects is obtained by replacing the image slice in the intensity transform by the segmented image and assigning the hue values accordingly.

Variable Saturation (VS)

This scheme E_1 feeds into both the hue and saturation components, while E_2 is fed into the intensity component. Thus, E_1 determines whether some areas of the output image are shown in color or in black and white. If the pixel values of an area in E_1 are zero, the area will appear black in the output image. Therefore, colors are assigned only to the areas where pixel values of both E_1 and E_2 are nonzero, that is, the areas of $E_1 \cap E_2 \neq 0$. The constants used for the S component in sets "CS1" and "CS2" of Section 13.5.2 were substituted for by the same transformed data used for the H component. The two sets obtained are therefore:

1. "VS1": $H = E_1 =$ original + histogram equalization + contrast stretching, $S =$ original + histogram equalization + contrast stretching, and $I = E_2 =$ original + negative + H-domes [Abi04] + contrast stretching,
2. "VS2": $H = E_1 =$ original + negative, $S =$ original + negative, and $I = E_2 =$ original + histogram equalization + contrast stretching.

13.6 EXPERIMENTAL RESULTS

The *RGB* and *HSI*-based color transforms described in Sections 13.4 and 13.5 were applied to x-ray luggage scans containing various potential threat objects made of low-density materials. Three x-ray luggage scans with three different low-density knives in order of increasing density are shown in Figs. 13.13a, b, and c. The three knives are made of soft wood, light-purple glass, and aluminum, respectively. Color-coded resulting images generated by applying the various aforementioned pseudocoloring approaches are shown as follows.

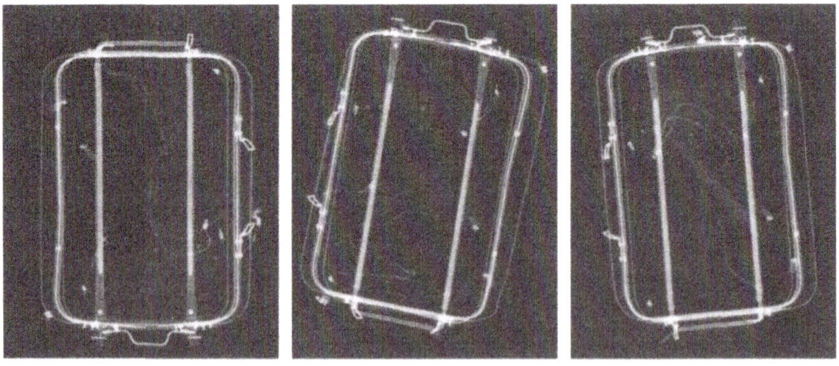

(a) (b) (c)

Figure 13.13. *Original x-ray luggage scans containing knives made of (a) soft wood, (b) light-purple glass, and (c) aluminum [Abi06].*

13.6.1 Color-Coded Images Generated by *RGB*-Based Transforms

Colored versions of figures were generated by applying the "Hot" and "Jet" scales to the enhanced images of Fig. 13.14 produced by using *"original + negative + h-domes + contrast stretching."* Figures 13.15 and 13.16 illustrate these two mappings, respectively. Figure 13.17 was obtained by applying the designed color scale "Warm" to the same preprocessed images of Fig. 13.14.

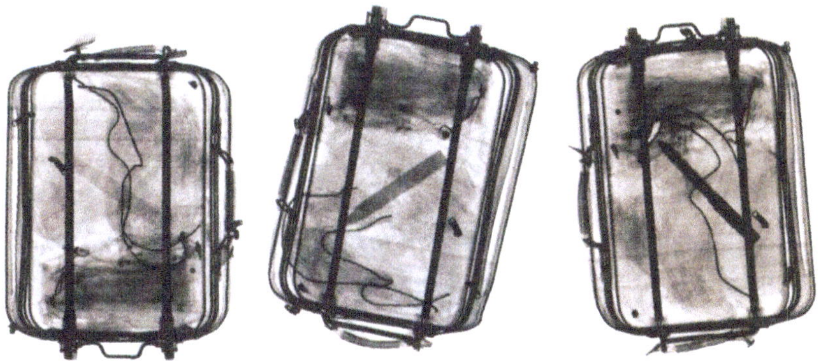

(a) (b) (c)

Figure 13.14. *Enhanced version of Figure 13.13 obtained by using original + negative + h-dome + contrast stretching [Abi06].*

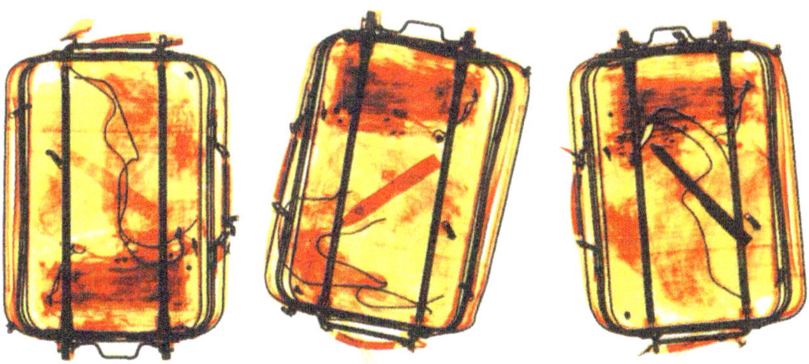

(a) (b) (c)

Figure 13.15. *Colored version of Fig. 13.14 generated by applying the color scale "Hot" [Abi06].*

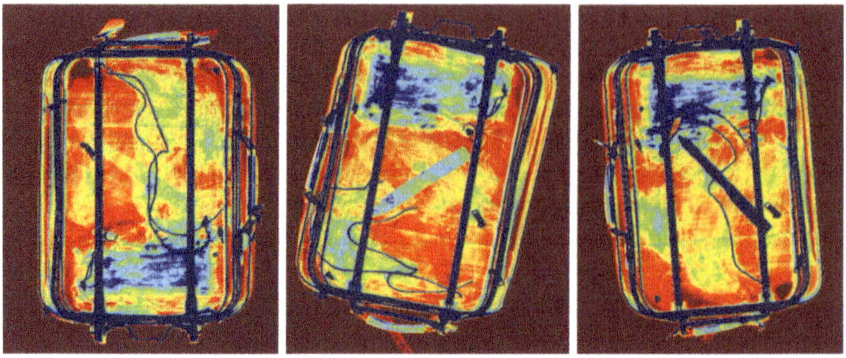

(a) (b) (c)

Figure 13.16. *Colored version of Fig. 13.13 generated by applying color scale "Jet" [Abi06].*

(a) (b) (c)

Figure 13.17. *Colored version of Fig. 13.13 generated by applying the designed color scale "Warm" shown in Fig. 13.6 [Abi06].*

Figure 13.18 shows a different example illustrating improvement in distinguishability between various scene elements after using the "Jet" colormap. The luggage image contains a set of three knives of various low-density materials. The colored version of Fig.13.13 generated by using the algebraic transform of Eq. (13.6) is shown in Fig. 13.19, while Fig. 13.20 illustrates the colored version of Fig. 13.13 produced by using the sine transform. Another example illustrating the result of applying a sine-based colormapping is shown in Fig. 13.21. The two knives are made of two different grades of wood (soft and hard).

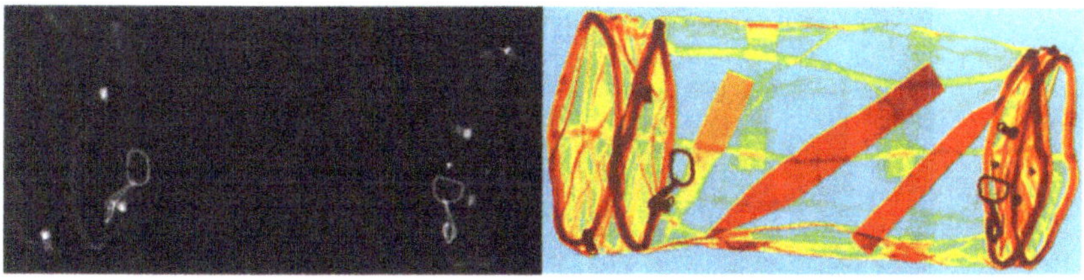

(a) (b)

Figure 13.18. *(a) Original image and (b) histogram equalized + jet colormap.*

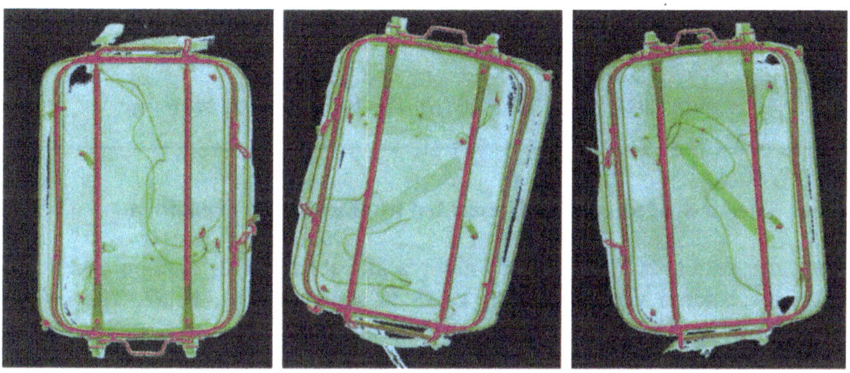

(a) (b) (c)

Figure 13.19. *Colored version of Fig. 13.13 generated by the algebraic transform of Eq. (13.6) [Abi06].*

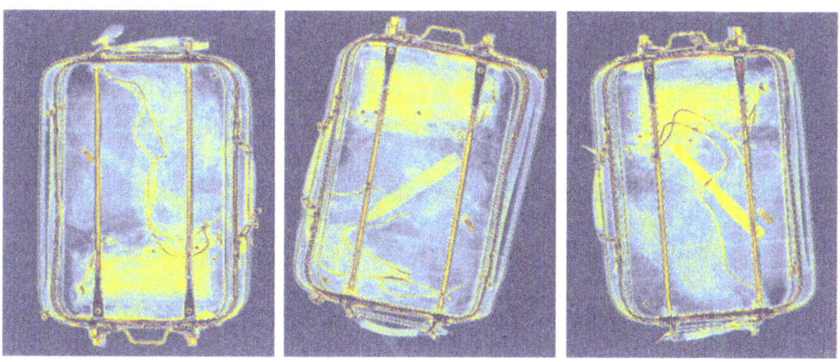

(a) (b) (c)

Figure 13.20. *Colored version of Fig. 13.13 generated using the sine transform [Abi06].*

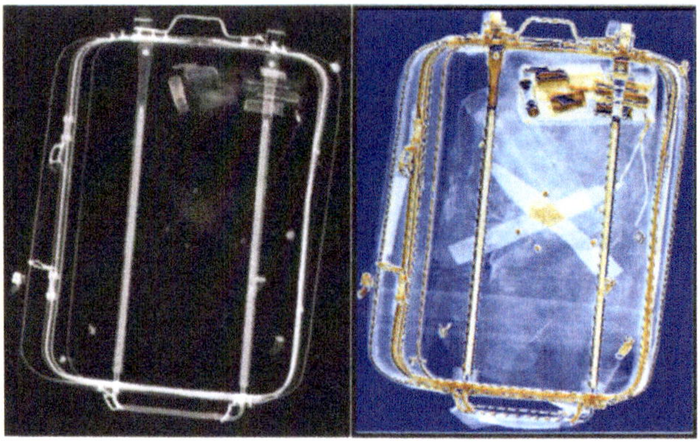

(a) (b)

Figure 13.21. *(a) Original and (b) sine-based color-coded image.*

13.6.2 Color-Coded Images Generated by *HSI*-Based Transforms

Using the nonlinear "Springtime" color scale described in Section 13.5.1, we obtain the colored version of Fig. 13.13 as shown in Fig. 13.22. The colored versions of Fig. 13.13 shown in Figs. 13.23 and 13.24 were produced using sets "CS1" and "CS2." A multilevel thresholding (*maximum entropy approach using ICM*) was performed to segment various objects in the scene into five classes. The red color was assigned to the class containing threat objects. The results of various selected color assignments are shown in Figs. 13.25 and 13.26. Figures 13.27 and 13.28 illustrate the results obtained from applying sets "NCS1" and "NCS2", respectively.

13.7 PERFORMANCE EVALUATION

A preliminary evaluation followed by a comparative study of the various pseudo-coloring methods designed and implemented were conducted in two steps. The preliminary in-lab and then online surveys were carried out to refine some general aspects as to the expected outcome and adjustment of initial results. A comprehensive airport testing was then conducted using actual airport screeners. The results of both studies are described in the following subsections.

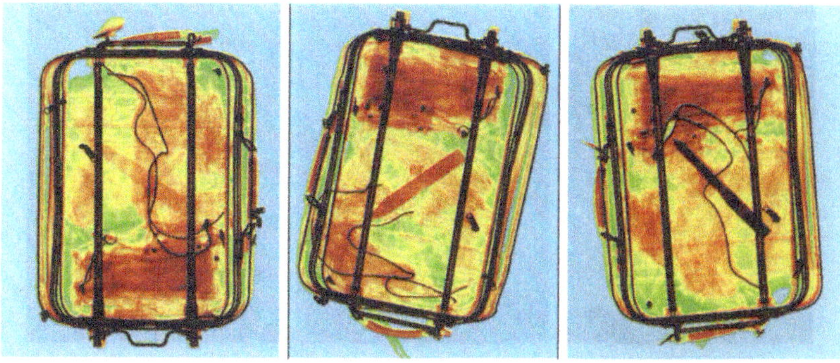

(a) (b) (c)

Figure 13.22. *Colored version of Fig. 13.13 generated by applying the "Springtime" color scale [Abi06].*

(a) (b) (c)

Figure 13.23. *Colored version of Fig. 13.13 generated by using set "CS1" with the constant saturation [Abi06].*

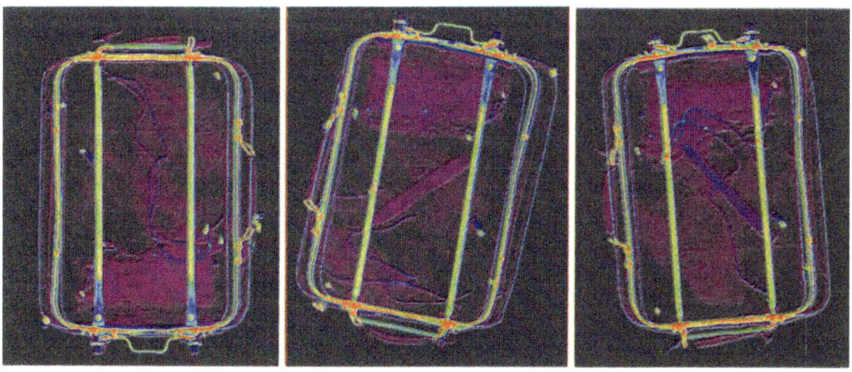

(a) (b) (c)

Figure 13.24. *Colored version of Fig. 13.13 generated by using set "CS2" with constant saturation [Abi06].*

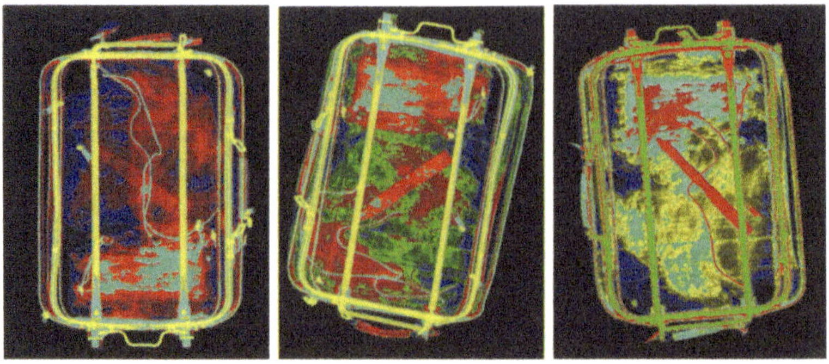

(a) (b) (c)

Figure 13.25. *Color-coded version of Fig. 13.13 obtained after segmentation, using a constant saturation and five different hues (complementary colors): (a) and (b) blue, green, red, cyan, and yellow, (c) blue, yellow, cyan, red, and green [Abi06].*

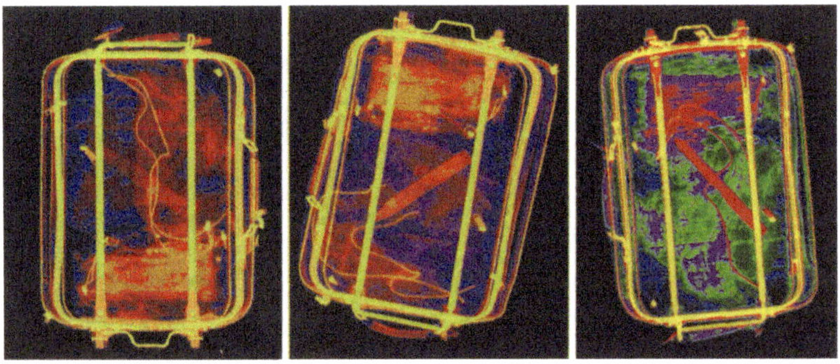

(a) (b) (c)

Figure 13.26. *Color-coded version of Fig. 13.13 obtained after segmentation, using a constant saturation and five different hues chosen along the color circle in clockwise direction from blue to yellow: (a) and (b) blue, magenta, red, orange, and yellow, (c) blue, green, magenta, red, and yellow [Abi06].*

(a) (b) (c)

Figure 13.27. *Colored version of Fig. 13.13 generated by using set "VS1" with variable saturation [Abi06].*

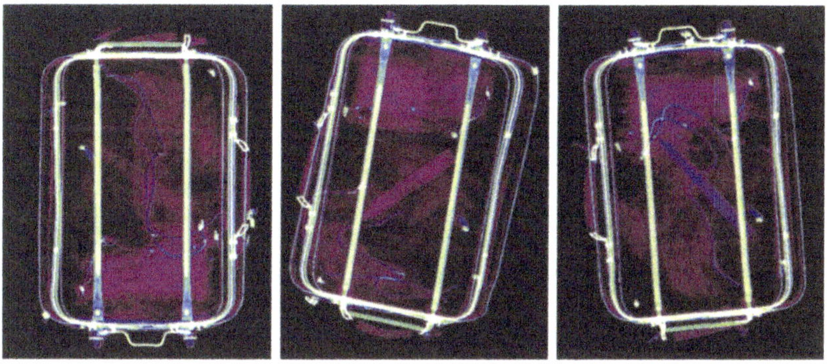

(a) (b) (c)

Figure 13.28. *Colored version of Fig. 13.13 generated by using set "VS2" with variable saturation [Abi06].*

13.7.1 Preliminary Online Survey

A preliminary online internet-based survey was conducted to compare people's responses to grayscale-enhanced images and their color-coded counterparts. Three color-coding methods and three grayscale methods were chosen for this evaluation. Ten test images with low-density threat objects were selected, and 132 people responded to the survey. The cosine, the *HSI* histogram-based, and the rainbow maps were used for color and the intensity-stretched grayscale, negative, and histogram-equalized images for grayscale enhancements. Each image was followed by three questions and at the end the overall preference among the six methods was also noted. A screen shot of the survey is shown in Fig. 13.29. Factors considered in this study were:

1. The ability to detect the threat
2. The visual appeal (how pleasant/helpful the method is)
3. The time used to identify the threat
4. The overall preference among the given methods

Each question is rated on a 1 - 10 scale with 10 being the highest rating. For the overall preference the choices were among all six different display schemes. The responses were noted and plotted as shown in Fig. 13.30 for item (1) and Fig. 13.31 for item (4). The results showed that color-coding was significantly more effective than grayscale images in allowing people to detect threat objects in x-ray scans. Eighty-six percent (86%) of the total 132 responses rated color as their preference. Among the different coloring schemes, the *HSV* scheme developed based on the results of the study on human perception of color was ranked highest by the greatest number of people.

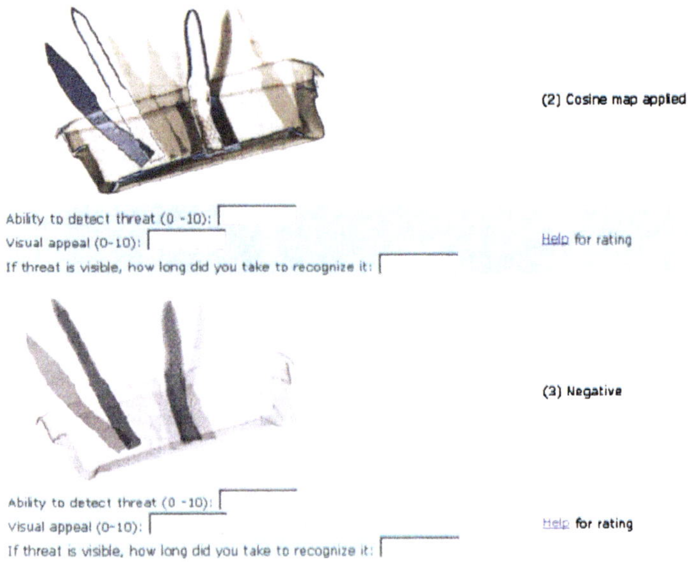

Figure 13.29. *Screen shot from the online survey showing the images followed by the questions.*

However, the other colormaps were ranked very close to the *HSV* map. The cosine colormap results were impressive. This cosine colormap produced very continuous and smooth results when compared with the other maps. Once it was established, through multiple evaluation trials with students and the staff population of the IRIS lab and with the online survey of a random population, that color-coded data is not only more appealing and boredom-proof but also more effective in detecting low-density threat objects in luggage scenes, a more formal performance study on actual airport screeners was designed and conducted.

13.7.2 Formal Airport Evaluation

Given the fact that airport screeners are the end users of any selected luggage coloring scheme, a natural step in this process is that the validation of the various color-coding approaches includes the responses of a representative section of the screeners' population. A fully automatic, portable, and interactive computer test was designed. A snapshot of one screen of this application is shown in Fig. 13.32. A set of 45 x-ray scans containing various low-density threat items in different configurations and levels of clutter were selected.

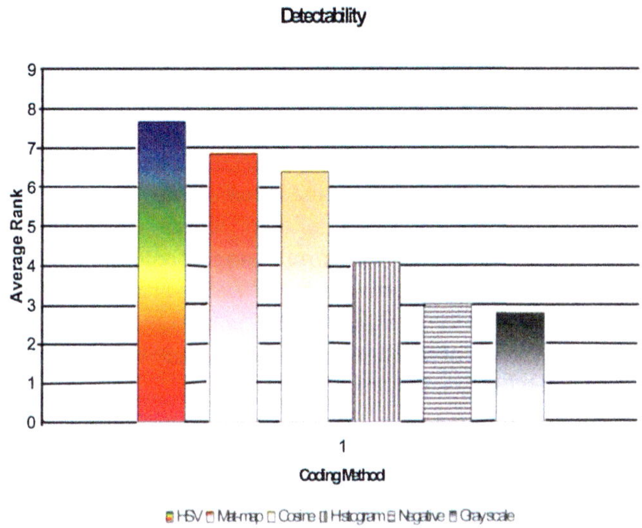

Figure 13.30. *Results of preliminary online survey on threat detectability change with the use of color in x-ray luggage scenes.*

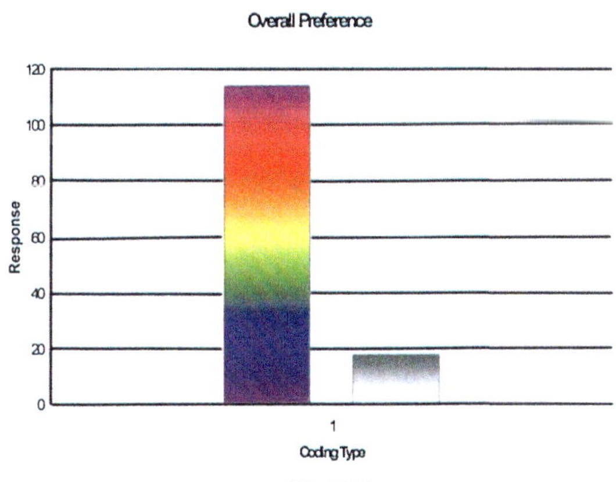

Figure 13.31. *Color vs. gray-level preferences by survey responders.*

Eight pseudocoloring approaches as described in Sections 13.4 and 13.5 were chosen according to preliminary evaluation of various transforms. The selected approaches were separately applied to the luggage scans containing low-density threats. All images were shown to screeners in random order using a random number generator, with the originals shown first in random order also. This ensures that no systematic influence is gained by colored images over the noncolored images in the detection of threats.

The screeners were asked not only to affirm seeing a threat but also to point and click on the threat to ensure they saw the actual threat. The screeners were also asked to rank the images in terms of their visual clarity and ease of interpretation, which is an important fact in relieving boredom and keeping the screener's level of concentration relatively high. The evaluation sessions were conducted at McGhee Tyson Airport in Knoxville, Tennessee and involved a total of 40 Transportation Security Administration (TSA) luggage screeners. Five types of information were collected for each image shown:

1. Did the screener see a threat object in the image?
2. If so, how many (1 or 2) threat items were seen?
3. Could the screener correctly click on the location of at threat object?
4. If two threat items were indicated, could the screener correctly click on the location of the second threat object?
5. A rating (from 1 to 10, with 10 being best) of how helpful the screener believed the displayed image was in detecting the threat object.

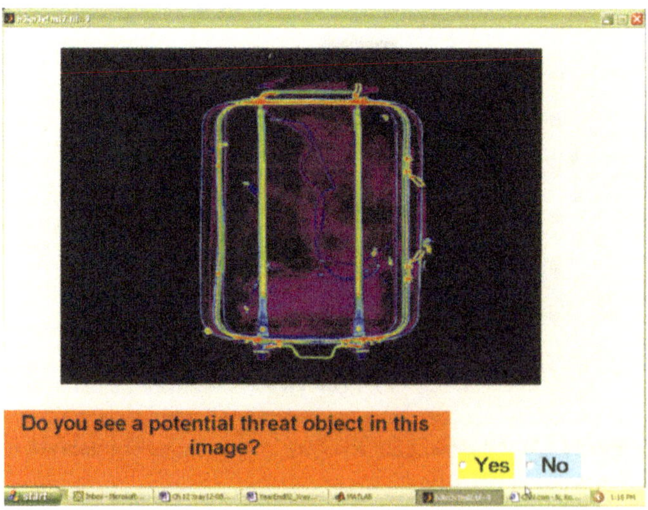

Figure 13.32. *Screen shot of the graphical user interface used in the airport evaluation study [Abi06].*

After all enhanced images had been shown as single windows, a montage image was shown for each original. In this montage the original image is shown side by side (for comparison purposes) with each of its colored versions and the screener is asked to rate each of the nine images on a 1-to-10 scale with 10 being the best in terms of ease of threat detection in the image.

An example of one montage window is shown in Fig. 13.33. The ability of the screener to correctly click on the threat item location within the luggage was determined through use of a binary mask image. When the screener clicks on a specific (x, y) location in the image being evaluated, the program checks the same location in the corresponding mask image. If this pixel location has a value of 1, the answer is recorded as being correct. Once all screeners completed the evaluation, the composite set of data obtained was evaluated in Excel to determine what trends might be apparent. Figure 13.34 presents a graphic showing the percent of all screeners who were able to correctly click on the threat location for the original image and for each of the color enhancement methods.

This graphic indicates that a low percentage (31%) of screeners were able to correctly locate the threat object in the original grayscale luggage scan. On the other hand, the color-enhanced images faired much better, ranging from 56.5 to 69.5% recognition. The image with enhancement method "Heqstrmap4" had the highest recognition rate. This enhancement consists of histogram equalization, followed by image stretching, followed by application of an in-house-developed colormap called "map4."

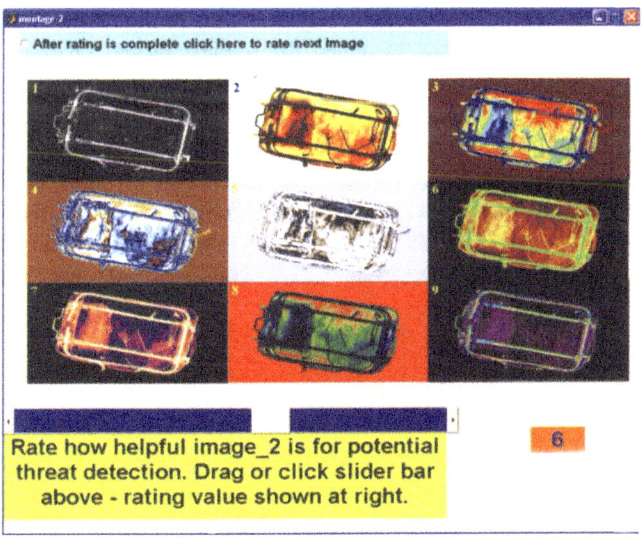

Figure 13.33. *Montage of original and all color-coded images for comparative rating [Abi06].*

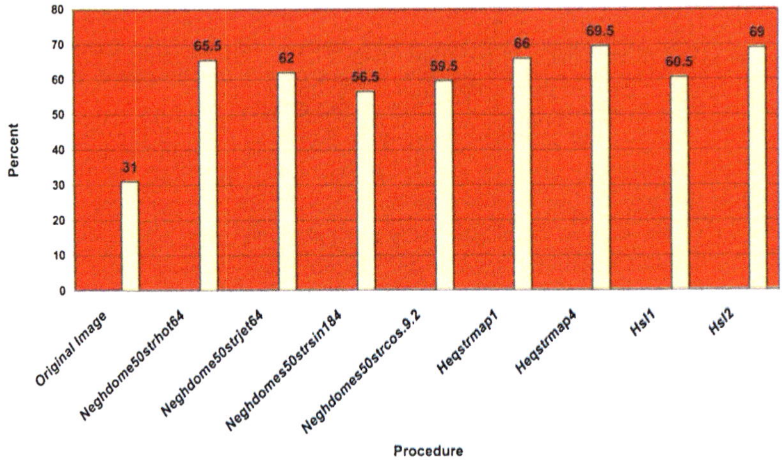

Figure 13.34. *Percentage of screeners able to correctly identify threat objects on each type of image.*

The other important evaluation criterion collected was the rating of each type of image color coding. Again, the original image scored lowest (1.64) while the enhanced images were all significantly higher, being in the range of 2.56 to 5.24. Color scale "Springtime" again provided the highest rating. Overall, the results obtained from the screener evaluations indicated a clear preference for color-enhanced x-ray scans over original, raw grayscale luggage scans as supplied from the scanning equipment. Four approaches, "Warm," "Springtime," "CS1," and "CS2," received higher average ratings than others. Of the four approaches, all except "Warm" were designed based on the *HSI* color model, which confirms earlier remarks that the *HSI* color model is more suitable to human interpretation and therefore more effective in revealing low-density threats concealed in x-ray luggage scans in this case.

13.8 CONCLUSION

A number of novel color transforms were introduced, applied to luggage scenes, and tested by screeners in an airport environment. Proper colormapping schemes have been designed based on perceptive and cognitive features without which it is impossible to produce an effective visualization. The expectation that pseudocoloring techniques can provide additional enhancement of x-ray luggage scans, better data visualization, increased screeners alertness, and longer attention/retention was demonstrated by experimental results and evaluations by actual airport screeners.

It was shown through visual interpretation, and more importantly through testing on TSA airport screeners, that newly developed colormapping techniques are very valuable tools in increasing the rate of low-density threat detection in x-ray luggage scans. A significant increase of up to 97%, as compared with results from the original data, in the rate of threat detection was obtained when color-coded data was used by screeners. Feedback from screeners also rated the color-processed data, on the average, as 219% more helpful in detecting a threat than the raw data. Not only did the testing show that color-processed data is more effective than grayscale data in detecting threats and keeping the screener's attention, but we were able to also rank the set of colormapping procedures as to which is most effective and most appealing to screeners. In comparing the *RGB*-based approaches with the *HSI*-based approaches, this latter color space proved superior, which was expected given the many known advantages of the *HSI* space in human-based applications [Wei97]. Experimental results show that the *HSI*-based methods produce results consistent with the human assessment. Future efforts would include more testing with the introduction of images containing no threat into the set of images evaluated to study performance in terms of the rate of false positives.

13.9 REFERENCES

[Abi04] B. Abidi, M. Mitckes, J. Liang, M. Abidi. Improving the detection of low density weapons in x-ray luggage scans using image enhancement and novel scene decluttering techniques. *J. Electronic Imaging* **13** (2004), pp. 523-538.

[Abi06] B. Abidi, Y. Zheng, A.V. Gribok, M.A. Abidi. Improving weapon detection in single energy x-ray images through pseudocoloring. *IEEE Transactions on Systems, Man, & Cybernetics, Part C: Aplications and Reviews* **36** (2006), pp. 784-796.

[But02] V. Butler, R.W. Poole, Jr. Rethinking checked-baggage screening. Policy Study No. 297, Reason Public Policy Institute, July 2002.

[Cla89] F.J. Clarke, J.K. Leonard. Proposal for a standardized continuous pseudocolor spectrum with optimal visual contrast and resolution. *Proc. 3rd Int. Conference on Image Processing and its Application*, pp. 687-691, 1989.

[Cze99] R.N. Czerwinski, D.L. Jones, W.D. O'Brien, Jr.. Detection of lines and boundaries in speckle images - Application to medical ultrasound. *IEEE Transactions on Medical Imaging* **18**, (1999), pp. 126-136.

[Dai96] J. Dai, S. Zhou. Computer Aided Pseudocolor coding of gray image: Complementary color coding technique. *Proc. SPIE* **2898**, pp. 186-191, 1996.

[Gon02] R.C. Gonzalez, R.E. Woods. *Digital Image Processing*. 2nd ed., Prentice-Hall, Upper Saddle River, New Jersey, 2002.

[Hea96] C.G. Healey. Choosing effective colours for data visualization. *Proc. IEEE Visualization*, pp. 263-270, 1996.

[Lev92] H. Levkowitz, G. Herman. Color scales for image data. *IEEE Computer Graphics and Applications* **12** (1992), pp. 72-80.

[Mac99] L.W. MacDonald. Using color effectively in computer graphics. *IEEE Computer Graphics and Applications* **19** (1999), pp. 20-35.

[May85] W.T. Mayo, P.V. Shankar, L.A. Ferrari. Color-coding medical ultrasonic images with frequency information. *Proc. SPIE* **575**, pp. 255-261, 1985.

[Mur84] G.M. Murch. Physiological principles for the effective use of volor. *IEEE Computer Graphics and Applications* **4** (1984), pp. 49-54.

[Oue04] N. Ouerhani, R. Wartburg, H. Hugli, R. Muri. Empirical validation of the saliency-based model of visual attention, *Electronics Letters on Computer Vision and Image Analysis* **3** (2004), pp.13-24.

[Shi02] X.Q. Shi, P. Sallstrom, U. Welander. A color coding method for radiographic images. *Image and Vision Computing* **20** (2002), pp. 761-767.

[Tay86] J.M. Taylor, G.M. Murch. The effective use of color in visual displays: Text and graphics applications. *Color Research and Applications* **11** (1986), Supplement pp. S3-10.

[War88] C. Ware. Color sequences for univariate maps: Theory, experiments and principles. *IEEE Computer Graphics and Applications* **8** (1988), pp. 41-49.

[Wei97] G.Q. Wei, K. Arbter, G. Hirzinger. Automatic tracking of laproscopic instruments by color coding. *Lecture Notes in Computer Science* **1205** (1997).

[Wri97] P. Wright, D. Mosser-Wooley, B. Wooley. Techniques & tools for using color in computer interface design. *ACM CrossRoads* (Spring 1997).

INDEX

Printed and bound by CPI Group (UK) Ltd, Croydon, CR0 4YY

16/09/2026

14963464-0001